ANALYTICAL THEORY OF FORCED OCEANIC WAVES

Analytical Theory of Forced Oceanic Waves

by
Wolfgang Fennel
and
Hans Ulrich Lass

with 88 figures and 3 tables

Akademie-Verlag Berlin 1989

Verfasser:

Dr. sc. nat. Wolfgang Fennel

Dr. sc. nat. Hans Ulrich Lass

Akademie der Wissenschaften der DDR
Institut für Meereskunde
DDR-2530 Rostock-Warnemünde

ISBN 3-05-500421-3

Erschienen im Akademie-Verlag Berlin, DDR-1086 Berlin, Leipziger Straße 3–4

Lizenznummer: 202 · 100/442/88
Printed in the German Democratic Republic
Gesamtherstellung: VEB Druckerei „Thomas Müntzer", 5820 Bad Langensalza
Lektor: Dipl.-Phys. Helga Müller
Hersteller: Alexander Herz
LSV 1465
Bestellnummer 763 762 3 (9082)

07500

Contents

Dedication

The scope of this book seems rather narrow. However, as outlined by the authors in their preface, developments and solutions of the analytical theory of forced oceanic waves are of great value in validating the numerical results for the realistic ocean circulation models to come.

Since several years both authors made vital contributions to a better understanding of ocean currents through a great number of papers on ocean dynamic field studies in the Baltic Sea and the Atlantic Ocean. On board our research vessels "Professor Albrecht Penck" and "A. v. Humboldt" with enthusiasm they expose themselves to the hard and risky experimental ocean research on small vessels.

I want to express my greatest appreciation to my collaborators for their devoted and reliable scientific work at sea and in our laboratory. In expressing my sincere congratulations to them for succeeding in summarizing and integrating many established experimental facts and theoretical approaches in their treatise "Analytical Theory of Forced Ocanic Waves", I hope their book will stimulate broad discussion on ongoing and future activities in the field of ocean dynamics through concerted action.

The circulation of the world ocean is the major problem which needs to be resolved in order to understand and to predict climate change and to assess its impact on the whole biosphere and on human activities.

With the envisaged World Ocean Circulation Experiment (WOCE) in the forthcoming decade, physical oceanography and marine scientific research will undergo a rather rapid development and will effectively contribute to the success of global scientific research programmes, such as the World Climate Research Programme or the Geo-Biosphere Programme and, of course, the Global Ocean Flux Programme.

The Marine Science Committee of the German Democratic Republic therefore highly acknowledges that the Akademie-Verlag Berlin gave priority to the technical completion of this volume as contribution to the international endeavours to support marine scientific research and education and training for the forthcoming World Ocean Circulation Experiment (WOCE).

Klaus Voigt

Preface

In the past 20 or 30 years a number of papers concerned with the linear forcing of the ocean by windstress have been published. It therefore seems to be appropriate to attempt a coherent account of the developments in this subject. In this book we treat the linear forcing problem on the basis of a Green's function technique. Since the corresponding mathematical problems are initial and boundary value problems it appears to be the most natural way to use Green's functions.

Because there exist many types of models of wind forcing we confine ourselves to a few examples. We have selected mostly switched on forces with certain simple spatial structures. However, the application of the techniques outlined in this book to other cases is straightforward for both deterministic and stochastic forcing. This is due to the key role of the Green's functions. If the associated Green's function is known then the forcing problem is formally solved, apart from the estimation of certain convolution integrals of Green's function and forcing functions.

We should explain why we are concerned with the analytical elaboration of the linear theory when large computers could be used to solve the hydrodynamic equations for rather realistic ocean models.

Our general point of view is that analytical theory and numerical modelling are complementary components of theoretical oceanography. Analytical results establish more conclusively the intricate interaction between the underlying physical processes than a series of numerical simulations. However, analytical theory is limited to certain idealizations which yield tractable sets of equations and so the numerical methods provide a fascinating tool to proceed further. Analytical solutions of the idealized models are of great value in validating the numerical results.

Moreover, many features of observations in the real oceans can qualitatively be understood in terms of linear theory. Even for processes affected by non-linearities many of linear features persist into the non-linear range, especially at large scales. Furthermore, when dealing with perturbation methods it is the linear theory which yields the first order approach. Thus it seems to be appropriate to write a book devoted to the further elaboration of analytical theory.

The book was written in two years starting in summer 1984 and ending in summer 1986. Although it is a two-authors book it is our wish to express that the major part of the manuscript is due to W. Fennel.

The original publications listed in the bibliography is representative but in no sense complete. Many of the listed books and review articles have extensive bibliographies of their own, which can be used to branch out further into the literature.

We wish to thank Renate Geißler, who typed the final manuscript, and Anke Kayser and Bernd Kayser, who drafted most of the figures. Moreover, we thank Dr. Torsten Seifert, who prepared several figures and performed the numerical estimation of the Lagrangian trajectories in Section 6.6.

In particular we thank Dr. Harry D. Dooley (ICES, Copenhagen) who read the manuscript and corrected the English.

Wolfgang Fennel
Hans Ulrich Lass

Rostock-Warnemünde, August 1986

CHAPTER 1

Introduction

1.1. INTRODUCTIONAL REMARKS

One of the fascinating subjects in geophysical fluid dynamics is the intimate interaction of the atmosphere and the ocean which strongly influences the global distribution and the variations of the climate on our planet.

The ocean responds to the atmosphere through the stress exerted by the surface wind, through the net flux of heat at the ocean surface, i.e., the net solar and infrared radiation and fluxes of sensible and latent heat, and through the balance of fresh water, i.e., the net evaporation and precipitation. The atmosphere responds to the ocean almost entirely through changes in the sea surface temperature. Because of the diversity of processes involved and because of the variety of the associated time and space scales in both the atmosphere and the ocean it is necessary to isolate certain processes in order to formulate tractable problems.

Hence, the research on this complex geophysical system has been split traditionally into two separate branches of geophysical fluid dynamics: atmospheric sciences and oceanography.

Physical oceanography consists of two parts

(1) observational oceanography which deals with observations and interpretation of oceanographic data in terms of physical laws, and
(2) theoretical oceanography which applies the laws of physics to the ocean, regarding it as a fluid on the rotating earth, acted upon by forces, and solves the corresponding mathematical problems.

Deep insight into the dynamics of the ocean requires a close interaction of both parts. Great physicists such as Newton, Bernoulli, Laplace, Stokes, Kelvin, Poincaré, and Taylor, to list a few, studied theoretically oceanic waves. However, for a long time there were no adaquate observations which could be accounted for by the theoretical investigations or which could test theories.

Experimental oceanography is commonly regarded as starting with the famous "H. M. S. Challanger" expedition which was conducted in the 1870's. Major advances in experimental techniques developed only slowly from this time: Indeed it took almost one century before technical progress in instrument design and data processing and in new experimental strategies, such as deep sea mooring arrays with self recording instruments and multi-ship programmes could significantly increase the resolution of oceanographic measurements.

Since the early 1970's the feedback loop between observations and theory has become tighter and has accelerated the progress in understanding the physics of the ocean. This book is devoted to one aspect of this understanding, namely the theoretical investigation of the response of the ocean to external forcing. Among the possible external forces, such as tide-generating forces, seismic disturbances, atmospheric pressure disturbances, and wind stress, we consider in particular the forcing through wind pattern. When discussing the response of the ocean to wind stress it is convenient and reasonable to regard the atmosphere as given, i.e., not being affected by the ocean, and this is the basic approach followed below. The investigations are almost exclusively concerned with the description of oceanic motions excited by various examples of wind patterns, where the wind has been considered as a body force evenly distributed within the upper mixed layer of the ocean.

From the mathematical point of view the forcing problem is governed by initial value and inhomogeneous boundary value problems. The appropriate mathematical tool to solve such problems is the Green's function technique which allows a systematic, unified approach.

The fascinating features of that concept consist in the possibility to express the response to external forcing completely by the associated Green's functions and the external forces in terms of source representations. Thus, if once the Green's function is known we have the formal solution of the forcing problem. The Green's function itselfes describes the response of a considered model to a very short pulse, i.e., Dirac's delta function, which excites all possible modes, and which depends only on the properties of the model.

1.2. BASIC EQUATIONS

The dynamical variables required to describe the oceanic motion are the velocity $\boldsymbol{u}$, the pressure p, the density ϱ, and certain further thermodynamic variables such as the temperature T and the salinity s. These observables generally depend on the time t and on the position vector $\boldsymbol{r}$. All positions will be referred to a right-handed orthogonal coordinate system which is uniformly rotating with the Earth's angular velocity $\boldsymbol{\Omega}$.

The oceanic motions are governed by the fundamental principles of fluid continuum theory and of thermodynamics, i.e., by the conservation laws for mass and momentum, an equation of state and the laws of thermodynamics.

The conservation of mass is expressed by the continuity equation

$$\frac{\partial}{\partial t}\varrho + \nabla \cdot \varrho \boldsymbol{u} = 0 .$$

With the aid of $\nabla \cdot \varrho \boldsymbol{u} = \varrho \nabla \boldsymbol{u} + \boldsymbol{u} \cdot \nabla \varrho$ this can be written as

$$\frac{\mathrm{d}}{\mathrm{d}t}\varrho + \varrho \nabla \cdot \boldsymbol{u} = 0 \tag{1.2.1}$$

where the Eulerian derivative

$$\frac{d}{dt} = \frac{\partial}{\partial t} + \boldsymbol{u} \cdot \nabla \tag{1.2.2}$$

was used. The conservation of momentum is expressed by Euler's equation in an uniformly rotating coordinate frame

$$\varrho \frac{\partial}{\partial t}\boldsymbol{u} + \varrho\boldsymbol{u} \cdot \nabla\boldsymbol{u} + 2\varrho\boldsymbol{\Omega} \times \boldsymbol{u} = \varrho\boldsymbol{g} - \nabla p + \boldsymbol{F} \tag{1.2.3}$$

where $\boldsymbol{g} = (0, 0, -g)$ denotes the gravitational acceleration ($g = 9{,}81\ \mathrm{m/s^2}$) and $\boldsymbol{F}$ represents the sum of external forces per unit volume acting on the fluid.

Sea water is a complex solution of may salts which can be quantified by means of the salinity s. The salinity s (in parts per thousand) is loosely defined as the mass in grams of all dissolved solids in one kilogram of sea water. The equation of state for sea water, i.e.

$$\varrho = \varrho(p, T, s) \tag{1.2.4}$$

has been found by experiment. Polynomial approximations that can be used to calculate ϱ to an accuracy better than five parts per million are given by the Joint Panel on Oceanographic Tables and Standards (UNESCO 1981). The formulas are also reported in the appendix of the book of GILL (1982).

With the aid of (1.2.4) the density derivative in (1.2.1) can be expanded as

$$\frac{d}{dt}\varrho = \left(\frac{\partial\varrho}{\partial p}\right)_{T,s}\frac{dp}{dt} + \left(\frac{\partial\varrho}{\partial s}\right)_{T,p}\frac{ds}{dt} + \left(\frac{\partial\varrho}{\partial T}\right)_{p,s}\frac{dT}{dt} \tag{1.2.5}$$

where for each partial derivative the other two variables of the set p, s, T are kept constant. To close the system of equations (1.2.1), (1.2.3), and (1.2.5) we require the conservation equations for T and s, e.g., conservation of internal energy and conservation of salt.

However, the first term in (1.2.5) is inversely proportional to the square of the sound speed. If we neglect high frequency acoustic waves with typical periods $T_s = 1$ to 10^{-5} seconds we may ignore the first term.

Moreover, if we neglect processes such as heat and salt diffusion, heating due to compression and cooling due to expansion, all heat transfer through the sea surface (solar heating, evaporative cooling, sensible heat flux to the atmosphere and long wave radiation into space), and all sinks and sources of salt due to precipitation, evaporation, ice melting and formation, we can drop the second and third term in (1.2.5). Then (1.2.5) reduces to the adiabatic equation

$$\frac{d\varrho}{dt} = \frac{\partial\varrho}{\partial t} + \boldsymbol{u} \cdot \nabla\varrho = 0\,. \tag{1.2.6}$$

With (1.2.6) the continuity equation (1.2.1) reduces to the incompressibility condition

$$\nabla \cdot \boldsymbol{u} = 0\,. \tag{1.2.7}$$

The set (1.2.6), (1.2.7) and (1.2.3) is referred to as the adiabatic equations. In this book all considerations below are based on this set.

The assumptions

$$\frac{dT}{dt} = 0 \quad \text{and} \quad \frac{ds}{dt} = 0$$

imply

$$\frac{\partial T}{\partial t} + \nabla \cdot T\boldsymbol{u} = 0 \quad \text{and} \quad \frac{\partial s}{\partial t} + \nabla \cdot s\boldsymbol{u} = 0\,,$$

respectively. Thus local changes of temperature and salinity are controlled entirely by advection. In reality substantial density variations may be caused by seasonal heating or cooling. Moreover, storm events often produces changes in the stratification by turbulent mixing. Clearly, one must be careful to exclude such processes from a comparison with theoretical results based on density conservation.

1.3. BOUSSINESQ EQUATIONS

Consider a Cartesian coordinate system with the position vector $\boldsymbol{r} = (x, y, z)$ and the velocity vector $\boldsymbol{u} = (u, v, w)$ where the origin $z = 0$ lies on the undisturbed sea surface and the z-axis points vertically upward, i.e., in the direction of $-\boldsymbol{g}$. The dominant terms by far in the equation of motion (1.2.3) are the gravitational acceleration $\boldsymbol{g}$ and the vertical pressure gradient, which approximately balances it. Hence, it is convenient to introduce the hydrostatic equilibrium state described by $\boldsymbol{u} = 0$, $\boldsymbol{F} = 0$, and

$$\frac{\partial}{\partial z} p_0(z) = -\varrho_0 g\,. \tag{1.3.1}$$

Thus for any given density distribution $\varrho_0(z)$, (1.3.1) can be integrated to find p_0. To describe motions which represent departures from the hydrostatic case we define the perturbations pressure p' and the perturbation density ϱ' by

$$p = p_0 + p'\,, \qquad \varrho = \varrho_0 + \varrho' \tag{1.3.2}$$

in which case (1.2.3) becomes

$$\frac{d}{dt}\boldsymbol{u} + 2\boldsymbol{\Omega} \times \boldsymbol{u} = -\frac{1}{\varrho}\nabla p' + \frac{\varrho'}{\varrho}\boldsymbol{g} + \frac{1}{\varrho}\boldsymbol{F}\,. \tag{1.3.3}$$

Thus only the density departure from the hydrostatic density distribution appears in the gravity term. The fluid is acted upon by a reduced gravitational acceleration.

From (1.3.2) we can write $\varrho = \varrho_0(1 + \varrho'/\varrho_0)$. Since $\varrho'/\varrho_0 \ll 1$ uniformly throughout the ocean (ϱ'/ϱ_0 typically 10^{-3}) we may put $\varrho \approx \varrho_0$ and get

$$\frac{d}{dt}\boldsymbol{u} + 2\boldsymbol{\Omega} \times \boldsymbol{u} = -\frac{1}{\varrho_0}\nabla p' + \frac{\varrho'}{\varrho_0}\boldsymbol{g} + \frac{1}{\varrho_0}\boldsymbol{F}\,. \tag{1.3.4}$$

The equation (1.2.6) takes the form

$$\frac{d}{dt}\varrho' + \left(\frac{\partial \varrho_0}{\partial z}\right) w = 0\,. \tag{1.3.5}$$

These equations are called the Boussinesq equations.
Introducing the buoyancy by

$$\boldsymbol{b} = \frac{\varrho'}{\varrho_0}\boldsymbol{g}$$

$$\boldsymbol{b} = \left(0, 0, -\frac{\varrho'}{\varrho_0} g\right)' = (0, 0, b) \text{ with } b = \frac{\varrho_0 - \varrho}{\varrho_0} g,$$

and the Brunt Väisälä Frequency (BVF) by

$$N^2(z) = -\frac{g}{\varrho_0}\frac{\partial}{\partial z}\varrho_0 \tag{1.3.6}$$

we may write

$$\frac{d}{dt}\boldsymbol{u} + 2\boldsymbol{\Omega} \times \boldsymbol{u} = -\frac{1}{\varrho_0}\nabla p' + \boldsymbol{b} + \frac{1}{\varrho_0}\boldsymbol{F} \tag{1.3.7}$$

$$\frac{d}{dt}b + N^2 w = 0 . \tag{1.3.8}$$

That is, we have together with (1.2.7) a set of five equations for the five variables u, v, w, p, and b.

1.4. THE f-PLANE AND THE β-PLANE APPROXIMATION

The so called "traditional" approximations of the Coriolis term appearing in (1.3.7), see e.g. ECKART (1960) and LEBLOND and MYSAK (1978), are discussed in this section. Using the Cartesian coordinate system plotted in Fig. 1.4.1 we can write

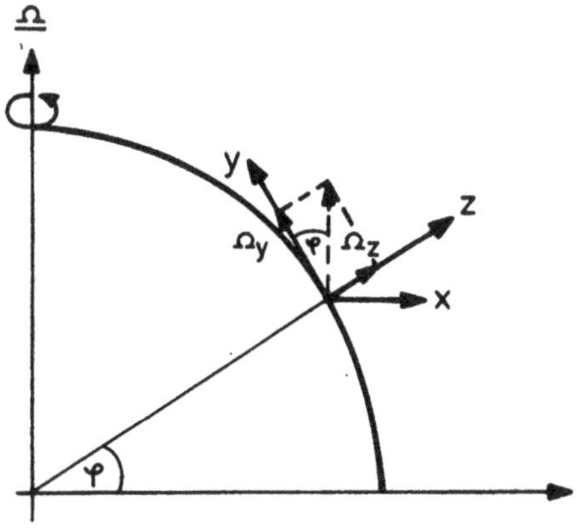

Fig. 1.4.1
The projection of $\boldsymbol{\Omega}$ onto the Cartesian coordinate frame fixed at the sea surface

$$2\boldsymbol{\Omega} \times \boldsymbol{u} = 2(w\Omega_y - v\Omega_z, u\Omega_z, -u\Omega_y) \tag{1.4.1}$$

with $\Omega_y = \Omega \cos \varphi$ and $\Omega_z = \Omega \sin \varphi$ where φ is the latitude. The magnitude of Ω is $\Omega = 2\pi/(1 \text{ day})$. Taking the smallness of vertical velocities compared with horizontal ones into account we may omit the w-term in the x-component of (1.4.1). The second part of the traditional approximation is to delete the vertical component of the Coriolis acceleration. The accuracy of this approximation is not always clear. However, these approximations simplify the equations to such an extent that they can easily be solved. For example the important method of separation of variables discussed below is based on those simplifications. Formally these approximations amounts to the following approximate form of the Earth's rotation vector

$$\boldsymbol{\Omega} = (0, 0, \Omega \sin \varphi)\,. \tag{1.4.2}$$

That is, the horizontal components of $\boldsymbol{\Omega}$ in the Cartesian coordinate system fixed at the undisturbed sea surface have been neglected. We introduce the Coriolis parameter or inertial frequency f

$$f = 2\Omega \sin \varphi\,. \tag{1.4.3}$$

At midlatitudes ($\varphi = \pm 45°$), $f = \pm 10^{-4}\ \mathrm{s}^{-1}$. In this book we consider positive f, i.e., motion on the northern hemisphere. However, the corresponding expression for the southern hemisphere can easily be obtained by replacing f with $-f$.

With the notation (1.4.3) the Coriolis acceleration takes the form

$$2\boldsymbol{\Omega} \times \boldsymbol{u} = (-fv, fu, 0)\,. \tag{1.4.4}$$

If the horizontal scales of the oceanic motion considered exceeds a typical scale, say 2000 km, the latitudinal variation of f must be considered. A mathematical convenient approximation, the so called β-plane approximation, was first introduced by ROSSBY (1939). This approximation is expressed by

$$f = f_0 + \beta y \tag{1.4.5}$$

where β gives the linear variation in the inertial frequency with latitude. The typical magnitude of β is $\beta = 2 \cdot 10^{-11}\ \mathrm{m}^{-1}\ \mathrm{s}^{-1}$. For a thorough justification of this approximation we refer to the book of PEDLOSKY (1979). This approximation consists in the introduction of a Cartesian metric centered at some reference latitude that locally approximates the spherical metric in this chosen neighbourhood of the Earth's surface. In this metric the effects of sphericity are retained by approximating f with a linear expansion concerning the latitudinal coordinate y. The symbol β has been traditionally used to denote $\partial f/\partial y$.

If the β effect can be neglected the resulting approximation of constant f is called the f-plane approximation.

The complete set of the Boussinesq equations (1.3.7), (1.3.8), (1.2.7) with (1.4.4) to the above mentioned approximations takes the form

$$\frac{\mathrm{d}u}{\mathrm{d}t} - fv + p_x = X \tag{1.4.6a}$$

$$\frac{\mathrm{d}v}{\mathrm{d}t} + fu + p_y = Y \tag{1.4.6b}$$

$$\frac{\mathrm{d}w}{\mathrm{d}t} + p_z - b = 0 \tag{1.4.6c}$$

$$\frac{\mathrm{d}b}{\mathrm{d}t} + N^2 w = 0 \tag{1.4.6d}$$

$$u_x + v_y + w_z = 0 \tag{1.4.6e}$$

where the pressure p means henceforth the perturbation pressure p' divided by ϱ_0. The external forces are expressed as

$$\frac{1}{\varrho_0}\boldsymbol{F} = (X, Y, 0)$$

and the subscripts x, y and z were used to denote partial derivatives.

1.5. TURBULENCE, DISSIPATION, AND LINEARIZATION OF THE BASIC EQUATIONS

Taking the relation (1.2.2) into account, the appearance of the total time derivatives in the set (1.4.6) indicates that these equations are non-linear. One phenomenon intimately connected with the non-linearity is turbulence, which is responsible for mixing and dissipation in the ocean. Since the oceanic processes cover a wide range of space- and time scales one cannot hope to find general solutions uniformly valid for a wide range of scales. It appears to be necessary to derive some approximations in order to get a tractable set of equations.

Let us assume that the observables may be decomposed into so-called mean parts and fluctuations. That is for example

$$u = \bar{u} + u' \tag{1.5.1}$$

where the overbar denotes the mean and the dash the pulsation. Let us, moreover, postulate the validity of the so-called Reynolds rules

$$\overline{cu} = c\bar{u} \tag{1.5.2a}$$

$$\overline{u + v} = \bar{u} + \bar{v} \tag{1.5.2b}$$

$$\overline{\bar{u}v} = \bar{u}\bar{v} \tag{1.5.2c}$$

$$\bar{\bar{u}} = \bar{u} \tag{1.5.2d}$$

$$\overline{u'} = 0 \tag{1.5.2e}$$

$$\overline{\bar{u}u'} = 0 \tag{1.5.2f}$$

$$\overline{\bar{u}\bar{u}} = \bar{u}\bar{u}\,. \tag{1.5.2g}$$

From (1.5.1) and (1.5.2c, f, g) it follows

$$\overline{uu} = \bar{u}\bar{u} + \overline{u'u'}\,. \tag{1.5.3}$$

Applying these relations to the equation (1.4.6a) and using (1.2.2) we obtain

$$\frac{\partial \bar{u}}{\partial t} + \bar{u}\frac{\partial \bar{u}}{\partial x} + \bar{v}\frac{\partial \bar{u}}{\partial y} + \bar{w}\frac{\partial \bar{u}}{\partial z} - f\bar{v} = -\frac{\partial}{\partial x}\bar{p} + \bar{X} - \frac{\partial}{\partial x}\overline{u'u'} - \frac{\partial}{\partial y}\overline{v'u'} - \frac{\partial}{\partial z}\overline{w'u'}\,. \tag{1.5.4}$$

Here the condition of incompressibility, which must hold separately for the mean and for the fluctuating velocity, was used.

However, caution must be exercised when applying the relations (1.5.2c) to (1.5.3). These conditions may be valid if the average process is thought to be an ensemble mean over a large number of realizations. Such a procedure is applicable to laboratory conditions, in which a series of identical experiments can be performed under a controlled environment. Oceanographic data, however, consist of time- or space series acquired under conditions beyond the control of the observer. Thus every reasonable averaging process must be over time or space, hence the relations (1.5.2c) to (1.5.3) are generally not valid. It appears that those conditions are approximately met if there exists a well pronounced gap in the frequency spectrum of the observables. Then the variations containing frequencies higher than the gap frequencies are defined to be fluctuations, i.e. u', and variations fluctuating on much longer time scales are defined as the mean values $\bar{u}$. For some explicit discussions of this problems see e.g. LASS and FENNEL (1980).

Such gaps exist for different physical processes in the ocean. For example, the typical time scale of small turbulence is of the order of few seconds. Small scale turbulence may be generated in the surface mixed layer by breaking surface waves and, moreover, induced by breaking internal waves in the entire water column. Those processes have been reviewed by PHILLIPS (1977), TURNER (1981), and MUNK (1981).

The most energetic process next to the small scale turbulence are the internal, particularly inertial waves, waves whose time scales are at least two orders of magnitude larger than those of the small scale turbulence. Another spectral gap exists in midlatitude oceans between inertial waves with a typical time scale of one day and planetary waves with time scales of some months.

Let us assume that the averaging procedure was chosen in such a way that the cut-off frequency is properly located within the frequency gap between inertial waves and the high frequency small scale turbulence within the surface mixed layer and in the interior of the ocean. Then the equation (1.5.4) applies and we may derive similar equations for the remaining dynamical fields.

The next difficulty arises from the correlations $\overline{u'u'}$, $\overline{v'u'}$, and $\overline{w'u'}$ which occur in (1.5.4) and which are called Reynolds stresses. In order to find a closed set of equations there exist two possibilities: The thorough way is to derive further equations for the Reynolds stresses. However, this procedure leads to equations containing correlations of third order, e.g. of the type $\overline{u'u'u'}$, $\overline{u'u'v'}$, and so on, and results ultimately in an infinite hierarchy of coupled equations with increasing order of correlations, similar to the hierarchies known in many particle physics, see e.g. MONIN and YAGLOM (1965). The investigation of these hierarchies, which must be terminated at some stage by means of some more or less well justified closure assumptions, is a branch of turbulence theory which is beyond the scope of this book. The second way is to choose an ad hoc ansatz which relates the Reynolds stresses in a simple manner to the mean current.

By means of such an approach some quantitative aspects of mixing and dissipation can be accounted for in an effective manner. One of the most common ways is to assume the Reynolds stresses to depend on the spatial derivatives of the mean flow. Among other possibilities a particularly simple model of this type is

$$\left.\begin{aligned}
&-\overline{u'u'} = 2A_h \frac{\partial \bar{u}}{\partial x}, \qquad -\overline{v'v'} = 2A_h \frac{\partial \bar{v}}{\partial y}, \qquad -\overline{w'w'} = 2A_v \frac{\partial \bar{w}}{\partial z},\\
&-\overline{u'v'} = -\overline{v'u'} = A_h\left(\frac{\partial \bar{v}}{\partial x} + \frac{\partial \bar{u}}{\partial y}\right), \qquad -\overline{u'w'} = -\overline{w'u'} = A_v \frac{\partial \bar{u}}{\partial z} + A_h \frac{\partial \bar{w}}{\partial x},\\
&-\overline{v'w'} = -\overline{w'v'} = A_v \frac{\partial \bar{v}}{\partial z} + A_h \frac{\partial \bar{w}}{\partial y}.
\end{aligned}\right\} \tag{1.5.5}$$

The coefficients A_h and A_v are called the horizontal and vertical turbulent viscosity coefficients, respectively. Since the properties of turbulence are ignored, A_h and A_v can be determined empirically only.

Estimates of turbulent viscosity coefficients cover a range of some orders of magnitude, $A_v \approx 10^{-1}$ to 10^3 cm²/s, $A_h \approx 10^5$ to 10^8 cm²/s.

This indicates that A_h and A_v cannot be determined in a way that is independent of the particular flow configuration and, therefore, A_h and A_v are not universal constants.

Recently OLBERS et al. (1985) could show that the intensity of the turbulent exchange of momentum and scalar properties appears to be low in the interior of the ocean.

In view of these difficulties and uncertainties we may also use a more simple ad hoc approach. This takes into account friction and dissipation associated with small scale turbulence caused by breaking internal waves in the interior of the ocean in a qualitative way. That is, we replace the time derivatives in the basic equations by $\partial/\partial t + r$, where the parameter r summarizes the dissipation processes governed by the Reynolds stresses. This kind of dissipation is called Rayleigh friction, with respect to the momentum equations (1.4.6a, b, c) and Newtonian cooling with respect to the buoyancy equation (1.4.6d). Clearly, the numerical value of r must be adjusted to the flow configuration considered, and hence it may vary from case to case. Despite the crudeness of such ad hoc assumptions, their use may give at least a qualitatively sensible picture of those aspects of forced oceanic motion which demand some dissipation processes to be present.

By means of such approximations of the Reynolds stresses we get a closed set of non-linear equations. In the frame of this book we wish to discuss the linear responses of the ocean to external forces. A linearization of the equations is obtained by neglecting all products of the observables, i.e. $\boldsymbol{u}, p$, and b. Generally, the validity of the linearization will depend on the time, length and velocity scales associated with the flow configuration under consideration. As an example we note that for inertial and subinertial wave motions the neglect of non-linear terms, such as $\boldsymbol{u} \cdot \nabla b$ and $\boldsymbol{u} \cdot \nabla \boldsymbol{u}$, is valid provided the Rossby number $R_0 = U/fL$ is small, i.e. $R_0 \ll 1$, where L and U are typical length and velocity scales. For a thorough discussion we refer to the book of PEDLOSKY (1979).

1.6. WIND STRESS AND BODY FORCE

One of the most important external forces acting upon the ocean is that imposed by the wind stress.

Although the details of the physics of the generation of current by winds, i.e. the momentum transfer from the atmospheric boundary layer near the ocean surface into the upper layers of the sea, are not yet well understood, there is no doubt that the momentum of the wind is transferred to the wind waves at the sea surface. A review of those processes was given by PHILLIPS (1977). Surface waves lose momentum by wave breaking, sometimes called "white capping", which injects small scale turbulence into the very surface layer. This turbulence penetrates downward, losing its kinetic energy by dissipation and by converting it into potential energy if the turbulence has to work against a stable stratification. A clear manifestation of this process is the well mixed surface layer generally found in all oceans and larger lakes. This draining of the energy from the wind wave induced turbulence confines it to a certain layer below the sea surface having a thickness of H_{mix}.

The Reynolds stresses $\overline{u'w'}$ and $\overline{v'w'}$ associated with small scale turbulence within the surface mixed layer are assumed to equal the wind stress components at the sea surface, implying

$$-\overline{u'w'} = \frac{\tau^{(x)}}{\varrho_0} = u_*^2; \qquad -\overline{v'w'} = \frac{\tau^{(y)}}{\varrho_0} = v_*^2 \quad \text{at} \quad z = 0. \tag{1.6.1}$$

Here the wind stress τ acting on the sea surface is defined by

$$\tau = (\tau^{(x)}, \tau^{(y)}) \tag{1.6.2}$$

where

$$\frac{\tau^{(x,y)}}{\varrho_0} = \frac{c_{10}\varrho_a}{\varrho_0} W_{10}^{(x,y)} W_{10} \tag{1.6.3}$$

with c_{10} being the drag coefficient, typically of the order of 10^{-3}, $W_{10}^{(x,y)}$ are the horizontal components of the wind velocity at ten meters above the sea surface, W_{10} is the magnitude of the wind velocity at this height and ϱ_a is the density of the air.

Since, as a rule, the horizontal scales of the wind pattern are much larger than H_{mix} we may neglect the horizontal derivatives in (1.5.4) and in the corresponding equations within the upper mixed layer. Thus the momentum balance becomes independent of the stratification and is given by

$$\frac{\partial \bar{u}}{\partial t} - f\bar{v} = -\frac{\partial}{\partial z}\overline{u'w'} \tag{1.6.4}$$

$$\frac{\partial \bar{v}}{\partial t} + f\bar{u} = -\frac{\partial}{\partial z}\overline{v'w'}. \tag{1.6.5}$$

Here the external forces which occur in (1.4.8) were neglected. Solving these equations requires additional information on how the Reynolds stresses vary with depth. On the other hand the net transport in the surface layer above $z = -H_{mix}$ is independent of the vertical distribution of the Reynolds stresses. To demonstrate this we introduce the transports

$$U_E = \int_{-H_{mix}}^{0} \bar{u}(z, t)\,dz, \qquad V_E = \int_{-H_{mix}}^{0} \bar{v}(z, t)\,dz. \tag{1.6.6}$$

Then with (1.6.1), it follows

$$\frac{\partial U_E}{\partial t} - fV_E = u_*^2, \tag{1.6.7}$$

$$\frac{\partial V_E}{\partial t} + fU_E = v_*^2. \tag{1.6.8}$$

The stationary part of the solution of (1.6.7) and (1.6.8) is

$$(U_E, V_E) = \frac{1}{f}(v_*^2, -u_*^2). \tag{1.6.9}$$

This is the well-known Ekman transport in the surface layer which is normal to the wind stress directed towards the right in the northern hemisphere. In his famous paper EKMAN

(1905) assumed a turbulent viscosity ansatz for the Reynolds stresses, and so he found as a superstructure the so-called Ekman spiral. However, the vertical integration of his solution yields, of course, (1.6.9).

As outlined by CHARNEY (1955) the horizontal variations of the windstress are important for the generation of currents in the interior of the ocean. Assume $w = 0$ at the sea surface and integrate the equation of continuity (1.4.6e) from $z = -H_{mix}$ to the surface. Then we get

$$\bar{w}(-H_{mix}) = \frac{\partial}{\partial x} U_E + \frac{\partial}{\partial y} V_E \qquad (1.6.10)$$

and with (1.6.9) we obtain

$$\bar{w}(-H_{mix}) = \frac{\partial}{\partial x}\left(\frac{v_*^2}{f}\right) - \frac{\partial}{\partial y}\left(\frac{u_*^2}{f}\right) = \frac{1}{\varrho_0} \operatorname{curl}_z \left(\frac{1}{f}\tau\right). \qquad (1.6.11)$$

This implies that the vertical component of the curl, denoted by curl_z, of the wind stress generates vertical motions at the base of the mixed layer which in turn disturb the stratification and hence induce horizontal pressure gradients. These pressure gradients generate currents in the layers below H_{mix}. This concept is generally known as Ekman pumping. From the mathematical point of view the wind stress enters the problem by means of a boundary condition.

We may also account for the action of the wind stress by means of a body force which is more or less evenly distributed over the upper layer. Since the distribution of the Reynolds stresses in the uppermost layer is practically independent of the currents within that layer, it is in fact reasonable to consider the vertical gradient of the Reynolds stresses as an external body force. The variation of the surface wave induced stresses in depth is not well known. Let us assume it decreases linearly from the surface to $z = -H_{mix}$ and vanishes below, i.e.,

$$-\overline{u'w'} = \frac{u_*^2}{H_{mix}} \theta(z + H_{mix})(z + H_{mix}), \qquad -\overline{v'w'} = \frac{v_*^2}{H_{mix}} \theta(z + H_{mix})(z + H_{mix}) \qquad (1.6.12)$$

where θ is the step function ($\theta(x) = 1$ for $x > 0$, $\theta(x) = 0$ for $x < 0$, and $\theta(0) = 1/2$).

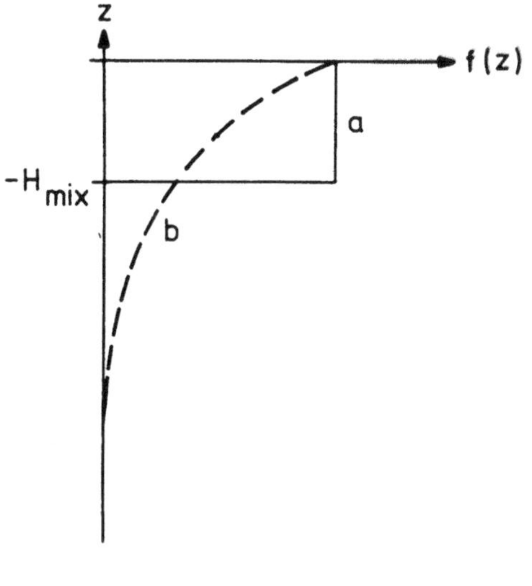

Fig. 1.6.1
Sketch of the vertical shape of the step-like (a) and an exponential (b) body force

An other possible choice is, for example, an exponentially decreasing stress such as

$$-(\overline{u'w'}, \overline{v'w'}) = (u_*^2, v_*^2)\, e^{z/H_{\rm mix}} \tag{1.6.13}$$

where $H_{\rm mix}$ is considered to be the e-folding mixing depth (see Fig. 1.6.1).

Assuming the vertical gradient of the Reynolds stresses in (1.5.4) to be an additional body force associated to the surface winds we obtain from (1.6.12)

$$X_{\rm wind} = -\frac{\partial}{\partial z}\overline{u'w'} = \frac{u_*^2}{H_{\rm mix}}\,\theta(z + H_{\rm mix}) \tag{1.6.14a}$$

$$Y_{\rm wind} = -\frac{\partial}{\partial z}\overline{v'w'} = \frac{v_*^2}{H_{\rm mix}}\,\theta(z + H_{\rm mix})\,. \tag{1.6.14b}$$

Using the ansatz (1.6.13) we find

$$(X_{\rm wind}, Y_{\rm wind}) = (u_*^2, v_*^2)\,\frac{e^{z/H_{\rm mix}}}{H_{\rm mix}}\,. \tag{1.6.15}$$

Throughout this book we will choose the body force according to (1.6.14). Henceforth we will drop the subscript "wind" in (1.6.14) and consider the body force imposed by the wind to be included in the sum of body forces X and Y.

This approach is similar to that one of Charney (1955) and Lighthill (1969) and it is equivalent to the concept of generation of ocean currents by Ekman pumping.

1.7. BOUNDARY CONDITIONS

The linearized version of the Boussinesq equations (1.4.6) may according to the discussion of the two foregoing sections be written as

$$u_t + ru - fv + p_x = X \tag{1.7.1a}$$

$$v_t + rv + fu + p_y = Y \tag{1.7.1b}$$

$$w_t + rw - b + p_z = 0 \tag{1.7.1c}$$

$$b_t + rb + N^2 w = 0 \tag{1.7.1d}$$

$$u_x + v_y + w_z = 0 \tag{1.7.1e}$$

where X and Y represent external body forces and r refers to Rayleigh friction and Newtonian cooling with the same relaxation parameter r, just for simplicity.

In order to find solutions of this set we have to consider boundary conditions which the solutions have to fulfil.

Although we deal to a certain extent with aspects of turbulent friction accounted for by the terms ru, rv, rw, and rb in (1.7.1) we assume the associated small scale turbulence processes, i.e. those induced by breaking internal waves, to have only minor effects on the vertical current structure. This amounts to neglecting processes such as the formation of bottom boundary layers etc., and so we may consider the fluid as a non-viscous one with regard to the boundary conditions.

First, we consider the conditions which express that the fluid can not penetrate through rigid boundaries such as the bottom of the ocean. The bottom may be expressed analytically by

$$A(x, y, z) = z - H(x, y) = 0\,. \tag{1.7.2}$$

Let us differentiate this by t. Then, with (1.2.2), it follows

$$\frac{\mathrm{d}A}{\mathrm{d}t} = \boldsymbol{u} \cdot \nabla A = -uH_x - vH_y + w = 0 \tag{1.7.3}$$

implying that the velocity normal to a flat bottom with $H_x = H_y = 0$ vanishes. In the case of a box shaped basin with a flat bottom and straight side walls (1.7.3) yields

$$w = 0 \quad \text{at} \quad z = -H; \qquad u = 0 \quad \text{at} \quad x = 0 \quad \text{and} \quad x = L_1; \qquad v = 0 \quad \text{at}$$

$$y = 0 \quad \text{and} \quad y = L_2 \tag{1.7.4}$$

where L_1 is the length and L_2 is the width of the basin.

For an ocean with a depth $H(x, y)$ being a function of x and y from (1.7.3) it follows

$$w(x, y, H(x, y)) = uH_x + vH_y\,. \tag{1.7.5}$$

Thus horizontal motions of water particles over an inclined bottom induce vertical velocities. In such case rather different motions may occur than in a flat bottomed ocean.

Next let us consider the air-sea interface which can be assumed to be a boundary between two fluids which cannot mix. This interface is analytically described by

$$A(x, y, z, t) = z - \eta(x, y, t) = 0 \tag{1.7.6}$$

where η denotes the variation of the sea level. Clearly, at the interface the pressure must be steady, i.e.,

$$p(x, y, \eta, t) = p_A \tag{1.7.7}$$

where p_A denotes the air pressure at sea level. Taking the derivative of (1.7.6) at both sides of the interface we have

$$\frac{\partial A}{\partial t} + \boldsymbol{u} \cdot \nabla A = \frac{\partial A}{\partial t} + \boldsymbol{u}_A \cdot \nabla A \tag{1.7.8}$$

or

$$(\boldsymbol{u} - \boldsymbol{u}_A) \cdot \nabla A = 0\,. \tag{1.7.9}$$

This condition is of importance when dealing with surface gravity waves. However, for processes with larger scales (1.7.9) becomes unimportant since then the wind component normal to the sea surface is negligible.

Next we differentiate (1.7.7)

$$\frac{\mathrm{d}p}{\mathrm{d}t} = \frac{\mathrm{d}p_A}{\mathrm{d}t} \quad \text{at} \quad z = \eta(x, y, t)\,. \tag{1.7.10}$$

Taking the linearized form of (1.7.10) and neglecting the vertical component of the wind at the sea surface we obtain with (1.3.1) the following surface boundary condition

$$\frac{\partial p}{\partial t} + w\frac{\partial p_0}{\partial z} = \frac{\partial p}{\partial t} - wg\varrho_0(0) = \frac{\partial P_A}{\partial t} \quad \text{at} \quad z = 0\,. \tag{1.7.11}$$

Obviously the generation of oceanic motions by air pressure enters the forcing problem by means of the inhomogeneous boundary condition (1.7.11) at the sea surface. The mathematical treatment of such a kind of forcing will be outlined in Section 2.4.1.

If the forcing through air pressure is ignored the right-hand side of (1.7.11) vanishes and then the sea surface may considered to be a free surface. Thus from (1.7.6) it follows

$$\frac{dA}{dt} = w - \frac{\partial \eta}{\partial t} - u\frac{\partial \eta}{\partial x} - v\frac{\partial \eta}{\partial y} \approx w - \frac{\partial \eta}{\partial t} = 0 \quad \text{at} \quad z = 0 \,. \tag{1.7.12}$$

The rightmost part of (1.7.12) is the linearized version of the kinematic boundary condition of the free sea surface.

In the water column large density jumps may occur within vertical scales of less than one meter, which may be idealized by density steps. We describe the surface of such a density step analytically by

$$A = z - \zeta(x, y, t) = 0 \,. \tag{1.7.13}$$

Using (1.7.7) and (1.7.8) we obtain

$$p_{\mathrm{I}} = p_{\mathrm{II}} \quad \text{and} \quad (\boldsymbol{u}_{\mathrm{I}} - \boldsymbol{u}_{\mathrm{II}}) \cdot \nabla A = 0 \quad \text{at} \quad z = \zeta(x, y, t) \tag{1.7.14}$$

where the subscripts I and II denote the layers above and below the interface.

Linearization of (1.7.14) yields

$$p_{\mathrm{I}} = p_{\mathrm{II}} \quad \text{and} \quad w_{\mathrm{I}} = w_{\mathrm{II}} \quad \text{at} \quad z = \bar{\zeta}(x, y, t) \tag{1.7.15}$$

where $\bar{\zeta}$ denotes the mean value of the interface. Thus at internal interfaces both the pressure and the vertical velocity are required to be steady.

1.8. HYDROSTATIC APPROXIMATION AND SEPARATION OF THE BASIC EQUATIONS

For dynamic processes with time scales $\tau \gg N^{-1}$, as a rule, the vertical component of the acceleration w_t and the friction associated with the vertical component of the velocity rw can be neglected in the vertical momentum balance (1.7.1 d). This treatment, known as the hydrostatic approximation, holds true for all dynamic features studied within this book except for internal waves, which will be considered in Chapter 4.

The linearized version of the Boussinesq equations (1.7.1) in the hydrostatic approximation is

$$u_t + ru - fv + p_x = X \tag{1.8.1 a}$$

$$v_t + rv + fu + p_y = Y \tag{1.8.1 b}$$

$$p_z - b = 0 \tag{1.8.1 c}$$

$$b_t + rb + N^2 w = 0 \tag{1.8.1 d}$$

$$u_x + v_y + w_z = 0 \,. \tag{1.8.1 e}$$

Eliminating the buoyancy b in (1.8.1 d) using (1.8.1 c) we may rewrite this set as

$$u_t + ru - fv + p_x = X \tag{1.8.2 a}$$

$$v_t + rv + fu + p_y = Y \tag{1.8.2 b}$$

$$u_x + v_y - \frac{\partial}{\partial z} N^{-2}(z) \frac{\partial}{\partial z}\left(\frac{\partial}{\partial t} + r\right) p = 0 \tag{1.8.2c}$$

$$w = -N^{-2}(z)\left(\frac{\partial}{\partial t} + r\right) p_z . \tag{1.8.2d}$$

In the following we consider ocean basins with a flat bottom and with shores approximated by straight vertical walls. Thus the boundary conditions (1.7.4), (1.7.11), and (1.7.15) apply. This has the very important consequence, that in the set (1.8.2) the horizontal and vertical variability are separable. At this stage we mention that the possibility of separation depends also on the way the Reynolds stresses have been approximated. The special ad hoc ansatz, $\overline{-u'w'} = ru$ etc., established in Section 1.5., ensures the separation of variables of the set (1.8.2). However, this is not the only possible approach. Let us apply the concept of turbulent viscosity for the Reynolds stresses as in (1.5.5). Neglecting the horizontal turbulent viscosity and assuming the ansatz $(-\overline{w'b'})_z = (A_v b)_{zz}$ the Boussinesq equations can be written as

$$u_t - fv + p_x = (A_v u_z)_z \tag{1.8.3a}$$

$$v_t + fu + p_y = (A_v v_z)_z \tag{1.8.3b}$$

$$p_z - b = 0 \tag{1.8.3c}$$

$$b_t + N^2(z)\, w = (A_v b)_{zz} \tag{1.8.3d}$$

$$u_x + v_y + w_z = 0 . \tag{1.8.3e}$$

The viscous fluid described by (1.8.3) is subject to the boundary conditions $A_v u_z = u_*^2$, $A_v v_z = v_*^2$, and $b = 0$ at the surface $z = 0$, and $A_v u_z = 0$, $A_v v_z = 0$, and $b = 0$ at the bottom $z = -H$. It will be apparent below that the set (1.8.3) is separable then and only then if the vertical austausch coefficient A_v is chosen as

$$A_v = \frac{a}{N^2(z)} . \tag{1.8.4}$$

The ansatz (1.8.4) was first utilized by FJELDSTAD (1933). A comprehensive discussion of the properties of the set (1.8.3) together with (1.8.4) has been given by McCREARY (1980). The approaches (1.8.2) and (1.8.3), (1.8.4) are very similar and describe nearly the same features especially in the transient regime. The most important differences may arise for the stationary solutions and in the near surface Ekman layer.

However, there is no observational evidence that the intensity of the small scale turbulence increases in the deep ocean as required by (1.8.4) due to the general decrease of $N(z)$ in the deeper layers of the ocean. In other words at present, each closure of the Reynolds stresses by the corresponding mean values of the observables is basically an ad hoc approach. That is the very reason why we confine ourselves in this book to the most simple and mathematically convenient approach (1.8.2) with the appropriate boundary conditions.

In order to separate the z-dependence let us consider the homogeneous versions of (1.8.2a, b, c). We assume solutions of the form

$$u(x, y, z, t) = u(x, y, t)\, F(z) \tag{1.8.5a}$$

$$v(x, y, z, t) = v(x, y, t)\, F(z) \tag{1.8.5b}$$

$$p(x, y, z, t) = p(x, y, t)\, F(z) . \tag{1.8.5c}$$

Substituting this into (1.8.2a, b, c) leads to the following differential equation for $F(z)$

$$\frac{d}{dz}\frac{1}{N^2(z)}\frac{d}{dz}F(z) + \lambda^2 F(z) = 0 \tag{1.8.6}$$

where λ^2 is a separation constant. Putting in (1.7.11) $p_A = 0$, and replacing $\partial/\partial t$ by $\partial/\partial t + r$, and inserting w according to (1.8.2d) we obtain the condition

$$F'(0) + \frac{N^2(0)}{g}F(0) = 0\,. \tag{1.8.7a}$$

Moreover, from (1.7.4) and (1.8.2d) we get

$$F'(-H) = 0\,. \tag{1.8.7b}$$

The differential equation (1.8.6) forms together with the boundary conditions (1.8.7) a Sturm Liouville eigenvalue problem for $F(z)$. The solution of such an eigenvalue problem, which will be discussed in detail in the following sections, consists of a complete set of orthogonal eigenfunctions $F_n(z)$ and their associated eigenvalues λ_n. Thus the separation of the vertical dependence of the dynamical fields amounts to expansion of those functions in terms of vertical eigenfunctions $F_n(z)$, i.e., by

$$(u, v, p) = \sum_{n=0}^{\infty}(u_n, v_n, p_n)\,F_n(z)\,. \tag{1.8.8}$$

Multiplying (1.8.2a, b, c) by $F_n(z)$, using the relation (1.8.6) in (1.8.2c) and integrating the resulting equations from the bottom to the surface we obtain

$$\left(\frac{\partial}{\partial t} + r\right)u_n - fv_n + p_{nx} = X_n \tag{1.8.9a}$$

$$\left(\frac{\partial}{\partial t} + r\right)v_n + fu_n + p_{ny} = Y_n \tag{1.8.9b}$$

$$u_{nx} + v_{ny} + \lambda_n^2\left(\frac{\partial}{\partial t} + r\right)p_n = 0 \tag{1.8.9c}$$

where

$$(X_n, Y_n) = \int_{-H}^{0} dz\,F_n(z)\,(X, Y) \tag{1.8.10}$$

are the projections of the volume force on the vertical eigenfunctions. Substituting (1.8.8) into (1.8.2d) and (1.8.2e) we get

$$w = -\sum_{n=0}^{\infty} N^{-2}(z)\,\frac{dF_n(z)}{dz}\left(\frac{\partial}{\partial t} + r\right)p_n(x, y, t) \tag{1.8.11a}$$

$$b = \sum_{n=0}^{\infty} p_n(x, y, t)\,\frac{dF_n(z)}{dz}\,. \tag{1.8.11b}$$

Thus the vertical velocity and the buoyancy can be expressed through derivation of the vertical eigenfunction. At this stage the forcing problem consists of two parts: First we have to solve the vertical eigenvalue problem (1.8.6), (1.8.7) for reasonably chosen $N^2(z)$ profiles, and, secondly, we have to determine the solutions for the horizontal problem, governed by (1.8.9) with the appropriate boundary conditions, and for any given forcing function.

1.9. THE VERTICAL EIGENVALUE PROBLEM

1.9.1. General relations

In the foregoing section it was shown that in the case of an ocean with a flat bottom the governing equations can be separated into Equations (1.8.9) describing the horizontal and Equation (1.8.6) describing the vertical variations of the dynamical fields, respectively.

The remaining part of this chapter is devoted to the evaluation of the vertical structure of the dynamic variables.

The governing equation is

$$\frac{\mathrm{d}}{\mathrm{d}z}\frac{1}{N^2(z)}\frac{\mathrm{d}}{\mathrm{d}z}F(z) + \lambda^2 F(z) = 0 \tag{1.9.1}$$

with the boundary conditions

$$\frac{\mathrm{d}F}{\mathrm{d}z} + \frac{N^2}{g}F = 0 \quad \text{at} \quad z = 0 \quad \text{and} \quad \frac{\mathrm{d}F}{\mathrm{d}z} = 0 \quad \text{at} \quad z = -H\,. \tag{1.9.2}$$

We define $Z(z)$ by

$$\frac{1}{N^2(z)}\frac{\mathrm{d}}{\mathrm{d}z}F(z) = Z(z)\,. \tag{1.9.3}$$

Then (1.9.1) can be transformed into

$$\frac{\mathrm{d}}{\mathrm{d}z}Z(z) + \lambda^2 F(z) = 0\,. \tag{1.9.4}$$

Differentiating (1.9.4) and using (1.9.3) supplies

$$\frac{\mathrm{d}^2 Z(z)}{\mathrm{d}z^2} + \lambda^2 N^2(z)\,Z(z) = 0\,. \tag{1.9.5}$$

This equation determines $Z(z)$ which describes the vertical variations of both the vertical velocity and the buoyancy, see (1.8.11). The boundary conditions for $Z(z)$ are readily obtained: From (1.7.4) we get

$$Z(-H) = 0 \tag{1.9.6}$$

and from (1.9.3) we obtain with (1.9.2)

$$Z(0) + F(0)/g = 0\,. \tag{1.9.7}$$

The differential equation (1.9.1) forms together with the boundary conditions (1.9.2) a Sturm Liouville eigenvalue problem. Such problems are extensively discussed in standard textbooks of mathematical physics, see e.g. COURANT and HILBERT (1968). Therefore, we present only a short summary of some general properties of eigenvalue problems which we use in the course of the book. An eigenvalue problem such as (1.9.1) with the boundary conditions (1.9.2) has solutions only for certain values of λ, the so-called eigenvalues $\lambda = \lambda_n$.

The associated $F_n(z)$, called the eigenfunctions, form as a rule a countable, complete set of orthogonal functions. In order to prove the orthogonality we consider two different eigenfunctions F_n and F_m belonging to two different eigenvalues λ_n and λ_m. Multiplying the corresponding equations of type (1.9.1) by F_m and F_n, integrating from the bottom $z = -H$ to the sea surface $z = 0$ and subtracting we find

$$\int_{-H}^{0} \mathrm{d}z \left(F_m \frac{\mathrm{d}}{\mathrm{d}z} \frac{1}{N^2} \frac{\mathrm{d}F_n}{\mathrm{d}z} - F_n \frac{\mathrm{d}}{\mathrm{d}z} \frac{1}{N^2} \frac{\mathrm{d}F_m}{\mathrm{d}z} \right) + (\lambda_n^2 - \lambda_m^2) \int_{-H}^{0} \mathrm{d}z\, F_n(z)\, F_m(z) = 0\,. \tag{1.9.8}$$

Integrating the first term of (1.9.8) by parts yields

$$[N^{-2}(z)\,(F_m F_n' - F_n F_m')]_{-H}^{0} + (\lambda_n^2 - \lambda_m^2) \int_{-H}^{0} \mathrm{d}z\, F_n(z)\, F_m(z) = 0\,. \tag{1.9.9}$$

The first term vanishes since the eigenfunctions F_n and F_m are subject to the boundary conditions (1.9.2). Thus

$$\int_{-H}^{0} \mathrm{d}z\, F_n(z)\, F_m(z) = 0\,, \qquad n \neq m\,. \tag{1.9.10}$$

This means that eigenfunctions of a Sturm Liouville problem belonging to different eigenvalues are orthogonal. The most important property of the eigenfunctions is that they form a complete set, i.e. an arbitrary function $f(z)$ can be expanded in an infinite series of the form

$$f(z) = \sum_{n=0}^{\infty} a_n F_n(z)\,. \tag{1.9.11}$$

The expansion coefficients a_n are determined by multiplying (1.6.9) by $F_m(z)$, integrating term by term, and using the orthogonality relation (1.9.10). It follows

$$a_m = \frac{\dfrac{1}{H} \displaystyle\int_{-H}^{0} \mathrm{d}z\, f(z)\, F_m(z)}{\dfrac{1}{H} \displaystyle\int_{-H}^{0} \mathrm{d}z\, (F_m(z))^2} \tag{1.9.12}$$

For convenience we have divided the integrals in (1.9.12) by the water depth H. Moreover, we normalize the eigenfunctions $F_n(z)$ according to

$$\frac{1}{H} \int_{-H}^{0} \mathrm{d}z\, F_n^2(z) = 1\,. \tag{1.9.13}$$

Then the eigenfunctions are dimensionless and the expansion coefficients a_n have the same dimension as the original function $f(z)$.

Inserting (1.9.12) into (1.9.11) and using the normalization (1.9.13) we obtain

$$f(z) = \int_{-H}^{0} dz' \, f(z') \left(\frac{1}{H} \sum_{n=0}^{\infty} F_n(z) \, F_m(z') \right). \tag{1.9.14}$$

Since $f(z)$ is an arbitrary function this implies

$$\frac{1}{H} \sum_{n=0}^{\infty} F_n(z) \, F_n(z') = \delta(z - z') \tag{1.9.15}$$

where $\delta(z - z')$ is Dirac's delta function. This relation expresses the completeness of the set of eigenfunctions.

Finally let us integrate the square of (1.9.11), i.e.,

$$\frac{1}{H} \int_{-H}^{0} dz \, f^2(z) = \sum_{n,m}^{\infty} a_n a_m \frac{1}{H} \int_{-H}^{0} dz \, F_n(z) \, F_m(z) \, .$$

Taking (1.9.10) into account the well-known Parseval's theorem follows

$$\frac{1}{H} \int_{-H}^{0} dz \, f^2(z) = \sum_{n=0}^{\infty} a_n^2 \, . \tag{1.9.16}$$

In practice the sums involved in (1.9.11) and (1.9.16) must be truncated at some finite value of n, say $n = M$, that is large enough to ensure the convergence of the solution. The theorem (1.9.16) appears to be useful to estimate the error which results from such a truncation.

If $N(z)$ appearing in (1.9.1) has been taken from measurements it is generally not possible to solve (1.9.1) analytically, because $N(z)$ can not be expressed by well known functions. Hence the vertical eigenvalue problem must be solved numerically. However to learn the most important properties of the eigenvalues and eigenfunctions it seems to be reasonable to approximate $N(z)$ by functions which are a good fit of the real stratification and allow analytical solutions of (1.9.1). We will accomplish this task in the following sections.

1.9.2. Constant Brunt Väisälä frequency

In this section we apply the general relations elaborated in the previous section for the rather simple stratification $N(z) = \text{const.}$

Then (1.9.1) becomes

$$\frac{d^2 F(z)}{dz^2} + \lambda^2 N^2 F(z) = 0 \, . \tag{1.9.17}$$

The general solution of (1.9.17) is

$$F(z) = A \sin(\lambda N z) + B \cos(\lambda N z) \tag{1.9.18}$$

where A and B are constants. Inserting (1.9.18) into the boundary conditions (1.9.2) supplies a homogeneous algebraic set for the coefficients A and B, namely,

$$\begin{pmatrix} \lambda N & \dfrac{N^2}{g} \\ \cos(\lambda NH) & \sin(\lambda NH) \end{pmatrix} \begin{pmatrix} A \\ B \end{pmatrix} = 0 \,. \qquad (1.9.19)$$

The existence of non-trivial solutions for A and B requires the determinant of (1.9.19) to vanish. This implies

$$\tan(\lambda NH) = \frac{N}{g\lambda} \,. \qquad (1.9.20)$$

This is an eigenvalue condition, which determines the eigenvalues λ_n. Let the first eigenvalue λ_0 be small and expand $\tan(\lambda_0 NH)$ in a power series. Then the approximated solution for λ_0 is

$$\lambda_0^2 = \frac{1}{gH} \quad \text{with} \quad \frac{N^2 H}{3g} \ll 1 \,. \qquad (1.9.21)$$

We call λ_0 the barotropic eigenvalue. It appears that even this simple treatment leads to a good approximation of λ_0 if parameters typically for oceans and marginal seas have been chosen.

Next we consider the solutions of (1.9.20) for large λ. Then we have approximately

$$\tan \lambda_n NH = 0 \qquad (1.9.22)$$

which implies

$$\lambda_n = \frac{n\pi}{NH} \,, \qquad n \geqq 1 \,. \qquad (1.9.23)$$

These λ_n are affected by the stratification and are denoted as baroclinic eigenvalues. If we adopt as a typically value of the BVF $N = 10^{-3}\ \text{s}^{-1}$ it appears that the barotropic eigenvalue λ_0 is one to two orders of magnitude smaller than the lowest baroclinic eigenvalue given by (1.9.23). We note that the inverse eigenvalue has the dimension of a velocity. We will show in Chapter 3, that certain classes of waves excited by initial disturbances propagate with phase speeds which equal the inverse vertical eigenvalues. This implies that the barotropic part of the oceanic response propagates much faster than the baroclinic one. In other words the response time of barotropic disturbances is much less than that one of baroclinic disturbances.

In order to estimate the normalized eigenfunctions we eliminate the A in (1.9.18) using (1.9.19) and take the eigenvalue equation (1.9.20) into account. This yields

$$F_n(z) = B_n \frac{\cos[\lambda_n N(z + H)]}{\cos(\lambda_n NH)} \,. \qquad (1.9.24)$$

Normalization of (1.9.24) according to (1.9.13) yields

$$F_n(z) = \sqrt{\frac{2}{1 + j_0(2\lambda_n NH)}} \cos[\lambda_n N(z + H)] \,, \qquad n = 0, 1, 2, \ldots \qquad (1.9.25)$$

where $j_0(x) = (\sin x)/x$ is a spherical Bessel function. Note that $j_0(2\lambda_0 NH) \approx 1$ and $j_0(2\lambda_n NH) = 0$ for $n \geqq 1$. Equation (1.9.25) expresses the n-th normalized vertical eigenfunction $F_n(z)$ for constant BVF.

Substituting (1.9.25) in (1.9.3) we get the $Z_n(z)$'s which describe the vertical structure of the vertical velocity

$$Z_n(z) = -\frac{\lambda_n}{N}\sqrt{\frac{2}{1 + j_0(2\lambda_n NH)}}\sin\left[\lambda_n N(z + H)\right]. \tag{1.9.26}$$

Expansion of $F_0(z)$ and using (1.9.21) yields approximately

$$F_0(z) = 1 - \frac{N^2 H}{g}\left(\frac{z}{H} + \frac{1}{2}\left(\frac{z}{H}\right)^2\right). \tag{1.9.27}$$

Thus the barotropic eigenfunction is nearly constant, apart from a minor variation due to the stratification. Thus the vertical structure of the barotropic part of the horizontal velocity components and the pressure variations are nearly constant over the whole water column.

Similarly, from (1.9.26) the vertical structure of the barotropic part of the vertical velocity follows to be

$$Z_0(z) = -\frac{1}{g}\left(1 + \frac{z}{H}\right). \tag{1.9.28}$$

Note that $Z_0(z)$ vanishes at the bottom $z = -H$ and increases linearly to the sea surface.

The vertical structure of the baroclinic modes, i.e. $n \geqq 1$, is given by (1.9.25). Using (1.9.23) we get

$$F_n(z) = (-1)^n \sqrt{2}\cos\left(n\pi\frac{z}{H}\right) \quad \text{for} \quad n \geqq 1, \tag{1.9.29}$$

and accordingly

$$Z_n(z) = -(-1)^n \sqrt{2}\left(\frac{n\pi}{HN^2}\right)\sin\left(n\pi\frac{z}{H}\right) \quad \text{for} \quad n \geqq 1. \tag{1.9.30}$$

From (1.9.29) and (1.9.30) it follows that the vertical structure of the baroclinic fields consists of a number of vertical modes with fixed positions of the nodes and antinodes whose number increase with mode number n.

Obviously, replacing the sea surface boundary condition (1.7.11) by $w = 0$ would filter out the barotropic mode and only the baroclinic motions described by (1.9.23), (1.9.29), and (1.9.30) would remain. The boundary condition $w = 0$ at $z = 0$ is known as the "rigid lid" condition. In those cases where the barotropic mode may be neglected the "rigid lid" condition is adventageous since it simplifies the evaluation of the vertical eigenvalue problem.

We mention that in the frame of the rigid lid approximation we can also assume that the barotropic eigenfunction degenerates to a constant, $F_0 = 1$, and the barotropic eigenvalue vanishes, $\lambda_0 = 0$, implying an infinite barotropic phase speed, $1/\lambda_0 \to \infty$.

Next we study a body force $f(z)$ uniformly distributed over the surface mixed layer from the surface to the bottom of the mixed layer at $z = -H_{mix}$ and zero elsewhere as discussed in Section 1.6, i.e.

$$f(z) = \frac{\theta(z + H_{mix})}{H_{mix}} \tag{1.9.31}$$

with $\theta(x)$ being the step function.

We determine the projection coefficients of the body force on the eigenfunctions by substituting (1.9.31) into (1.9.12)

$$\frac{1}{h_n} = \frac{1}{H}\int_{-H}^{0} dz\, \frac{\theta(z + H_{mix})}{H_{mix}} F_n(z) = \frac{1}{H}\int_{-H_{mix}}^{0} dz\, \frac{F_n(z)}{H_{mix}}. \qquad (1.9.32)$$

We shall denote the projection coefficient of forces of the type (1.9.31) by $1/h_n$ throughout the book. Using the eigenfunctions (1.9.25) from (1.9.32) we get

$$\frac{1}{h_n} = \left(\frac{2}{1 + j_0(2\lambda_n NH)}\right)^{1/2} \frac{\sin\left[\lambda_n N(H_{mix} - H)\right]}{\lambda_n N H H_{mix}}. \qquad (1.9.33)$$

With the approximated eigenvalues according to (1.9.21) and (1.9.23) we find for the barotropic and baroclinic projection coefficients, respectively,

$$\frac{1}{h_0} = \frac{1}{H} \quad \text{and} \quad \frac{1}{h_n} = \frac{\sqrt{2}}{H}(-1)^n\, j_0\left(n\pi \frac{H_{mix}}{H}\right). \qquad (1.9.34)$$

In shallow seas the ratio H_{mix}/H is of the order of 1/2. Hence the barotropic and the first few baroclinic projection coefficients dominate the wind induced motions.

In deep oceans $H_{mix}/H \ll 1$ can be assumed and it follows from (1.9.34) that the first baroclinic mode is more excited than the barotropic one and that several higher baroclinic modes must be taken into account.

Parseval's theorem (1.9.16) supplies a measure how good a finite number of vertical modes approximates the actual forcing function. The left hand side (1.9.16) becomes in the case of the forcing (1.9.31)

$$\frac{1}{H}\int_{-H}^{0} dz\, f^2(z) = \frac{1}{HH_{mix}} = \frac{1}{H^2}\left(1 + 2\sum_{n=1}^{\infty} j_0^2\left(n\pi \frac{H_{mix}}{H}\right)\right). \qquad (1.9.35)$$

In Fig. 1.9.1 the expression (1.9.35) has been plotted as a function of the mode number for a shallow sea with $H = 100$ m, $H_{mix} = 50$ m, label a, and for an ocean with the depth $H = 4000$ m, $H_{mix} = 75$ m, label c. In the shallow sea the barotropic and the first baroclinic mode represent 90% of the total sum, whereas in the ocean it needs the barotropic and some ten baroclinic modes to represent 90% of the sum (1.9.35). This can easily be realized because the vertical structure of the body force (1.9.31) behaves with increasing depth more and more like a delta function, and in the expansion of the delta function all modes contribute with the same weights, compare the completeness relation (1.9.15).

As mentioned in Section 1.6. the vertical structure of the body force is not well known. Hence it seems worth while to study the sensitivity of the model regarding the vertical structure of the volume force. We assume an exponentially decay of the volume force according to (1.6.15), i.e.,

$$f(z) = \frac{\exp(z/H_{mix})}{H_{mix}}. \qquad (1.9.36)$$

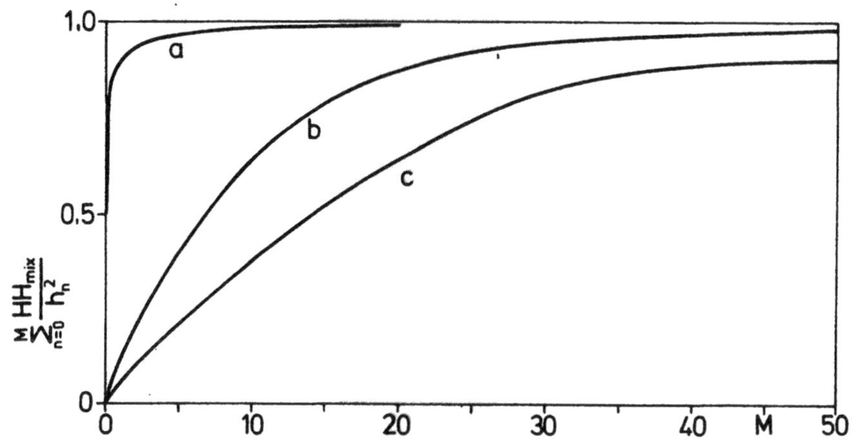

Fig. 1.9.1
Parseval's sum of the projection coefficients of body force versus the number of involved modes for a step-like force in a shallow sea, $H = 100$ m, and $H_{mix} = 50$ m (label a), and in an ocean, $H = 4$ km, $H_{mix} = 75$ m (label c). The label (b) indicates the curve resulting for the oceanic case, i.e., $H = 4$ km and $H_{mix} = 75$ m, if an exponentially decreasing body force is used.

Substituting (1.9.36) into (1.9.12) yields the projection coefficients

$$\frac{1}{h_n} = \sqrt{\frac{2}{1 + \mathrm{Sp}(2\lambda_n NH)}}\,\frac{1}{H}\,\frac{\cos(\lambda_n NH)}{1 + (\lambda_n NH_{mix})^2}\,. \tag{1.9.37}$$

With (1.9.21) and (1.9.23) we get explicitly

$$\frac{1}{h_0} = \frac{1}{H} \quad \text{and} \quad \frac{1}{h_n} = \frac{1}{h_0}\sqrt{2}\,\frac{\cos n\pi}{1 + (n\pi H_{mix}/H)^2}\,. \tag{1.9.38}$$

Parseval's theorem (1.9.16) yields in this case

$$\frac{1}{H}\int_{-H}^{0} \mathrm{d}z\, f^2(z) = \frac{1}{2HH_{mix}} = \left(\frac{1}{h_0}\right)^2 \left(1 + 2\sum_{n=1}^{\infty} \frac{1}{(1 + (n\pi H_{mix}/H)^2)^2}\right). \tag{1.9.39}$$

Table 1.9.1
Eigenvalues (λ_n), maximum phase velocities (λ_n^{-1}), and projection coefficients of the forcing (1.9.31) on the corresponding vertical eigenfunctions of the barotropic and the first four baroclinic modes for constant BVF. The numerical values of N, H, and H_{mix} are given in the text.

ocean				*shallow sea*		
n	λ_n s/m	λ_n^{-1} m/s	$\frac{100H_{mix}}{h_n}$	λ_n s/m	λ_n^{-1} m/s	$\frac{H_{mix}}{h_n}$
0	0.005	198	1.87	0.03	31.3	0.5
1	0.39	2.55	−2.65	1.57	0.68	−0.45
2	0.79	1.72	2.65	3.14	0.32	0
3	1.18	0.85	−2.64	4.71	0.21	−0.15
4	1.57	0.64	2.63	6.28	0.16	0

The expression (1.9.39) has been plotted in Fig. 1.9.1., label b, and depicts that a volume force of the form (1.9.36) excites significantly less modes than the volume force (1.9.31). Hence it would be of some importance to know the real vertical structure of the volume force. However, in the following we will confine ourselves to studies on forces of the type of Heaviside's step function. Table 1.9.1 displays the first five eigenvalues, inverse eigenvalues and projection coefficients for an ocean and a shallow sea. For the ocean it has been assumed $H = 4000$ m, $H_{mix} = 75$ m, and $N = 2 \cdot 10^{-3}\,s^{-1}$. For shallow sea the following parameters have been chosen $H = 100$ m, $H_{mix} = 50$ m, and $N = 2 \times 10^{-2}\,s^{-1}$. Clearly, the model of constant BVF is a rather crude approximation of the stratification of the real ocean. We consider some improved models of the BVF in the following sections.

1.9.3. Slowly varying stratification

Generally the stratification of the ocean varies more or less with depth. Let us assume $N(z)$ to be a slowly varying function of the vertical coordinate z. Then the Equation (1.9.5) may be solved by means of the WKB method, see e.g. Landau and Lifschitz (1967). In the WKB treatment the solution of (1.9.5) reads

$$Z(z) = \frac{A}{N^{1/2}} \sin\left(\lambda \int_0^z N\, dz'\right) + \frac{B}{N^{1/2}} \cos\left(\lambda \int_0^z N\, dz'\right). \tag{1.9.40}$$

Substituting (1.9.40) into (1.9.4) leads to

$$F(z) = \frac{-N^{1/2}}{\lambda} \left\{ A \cos\left(\lambda \int_0^z N\, dz'\right) - B \sin\left(\lambda \int_0^z N\, dz'\right) \right\}. \tag{1.9.41}$$

The expressions (1.9.40) and (1.9.41) are subject to the boundary conditions (1.9.6), (1.9.7) and (1.9.2), respectively. In order to estimate the coefficients A and B we get the algebraic set

$$\begin{pmatrix} \dfrac{N(0)}{g} & 1 \\ \sin\left(\lambda \int_0^{-H} N\, dz\right) & \cos\left(\lambda \int_0^{-H} N\, dz\right) \end{pmatrix} \begin{pmatrix} A \\ B \end{pmatrix} = 0\,. \tag{1.9.42}$$

From the requirement of vanishing determinant we obtain the following eigenvalue equation

$$\tan\left(\lambda \int_{-H}^{0} N(z)\, dz\right) = \frac{N(0)}{g\lambda}\,. \tag{1.9.43}$$

Estimating the coefficients A and B and normalizing the result according to (1.9.13) yields the following vertical eigenfunctions

$$F_n(z) = \left(\frac{2}{1 + j_0\left(2\lambda_n \int\limits_{-H}^{0} N\,\mathrm{d}z\right)}\right)^{1/2} \left(\frac{N(z)}{\frac{1}{H}\int\limits_{-H}^{0} N\,\mathrm{d}z}\right)^{1/2} \cos\left(\lambda_n \int\limits_{-H}^{z} N(z')\,\mathrm{d}z'\right) \quad (1.9.44)$$

and

$$Z_n(z) = -\left(\frac{2}{1 + j_0\left(2\lambda_n \int\limits_{-H}^{0} N\,\mathrm{d}z\right)}\right)^{1/2} \left(\frac{\lambda_n^2 H}{N(z)\int\limits_{-H}^{0} N\,\mathrm{d}z}\right)^{1/2} \sin\left(\lambda_n \int\limits_{-H}^{z} N(z')\,\mathrm{d}z'\right). \quad (1.9.45)$$

We note that for $N =$ const these results leads to those obtained in the preceeding section.

If we wish to analyze the barotropic mode in a manner analogously to the previous section we would assume the barotropic eigenvalue to be small, implying $\lambda_0 \int N\,\mathrm{d}z \ll 1$. This however contradicts the presumption of large eigenvalues required by the WKB treatment. Thus the WKB method fails for the barotropic mode. We may, however, use (1.9.21), (1.9.27), and (1.9.28) to obtain reasonable approximations.

For the baroclinic eigenvalues it follows from (1.9.43)

$$\lambda_n = \frac{n\pi}{\int\limits_{-H}^{0} N\,\mathrm{d}z} = \frac{n\pi}{H\bar{N}} \quad (1.9.46)$$

with $\bar{N}$ being the vertical mean value of the BVF.

Obviously, the eigenvalues given by (1.9.46) are identical with those of the constant BVF model provided the constant BVF has been replaced by the mean value of the BVF. Thus the baroclinic eigenvalues are functionals of the BVF profile. This implies the vertical baroclinic eigenvalues to be rather insensitive to the specific vertical pattern of the stratification.

For the baroclinic eigenfunctions we get from (1.9.44), (1.9.45) approximately

$$F_n(z) = \left(2\,\frac{N(z)}{\bar{N}}\right)^{1/2} \cos\left(n\pi\,\frac{\int\limits_{-H}^{z} N(z')\,\mathrm{d}z'}{\int\limits_{-H}^{0} N\,\mathrm{d}z}\right) \quad (1.9.47)$$

and

$$Z_n(z) = \frac{n\pi}{\bar{N}H}\left(\frac{2}{\bar{N}N(z)}\right)^{1/2} \sin\left[n\pi\,\frac{\int\limits_{-H}^{z} N(z')\,\mathrm{d}z'}{\int\limits_{-H}^{0} N\,\mathrm{d}z}\right], \quad (1.9.48)$$

respectively, for $n \geqq 1$.

Next we calculate the projection coefficients of the body force (1.9.31). We assume $N(z) = N_0 =$ constant within the upper layer, i.e., from $z = 0$ to $z = -H_{mix}$. Substituting (1.9.44) into (1.9.12) we obtain

$$\frac{1}{h_n} = \left(\frac{2}{1 + j_0(2\lambda_n H\bar{N})}\right)^{1/2} \left(\frac{1}{N_0\bar{N}}\right)^{1/2} \frac{1}{\lambda_n H H_{\mathrm{mix}}} \left[\sin\left(\lambda_n \int_{-H}^{0} N\,\mathrm{d}z\right) - \sin\left(\lambda_n \int_{-H}^{-H_{\mathrm{mix}}} N\,\mathrm{d}z\right)\right]. \qquad (1.9.49)$$

With (1.9.46) this yields

$$\frac{1}{h_n} = \frac{\sqrt{2}}{H}\left(\frac{N_0}{\bar{N}}\right)^{1/2} (-1)^n j_0\left(n\pi\,\frac{N_0 H_{\mathrm{mix}}}{\bar{N}H}\right). \qquad (1.9.50)$$

Moreover, we have approximately

$$\frac{1}{h_0} = \frac{1}{H}. \qquad (1.9.51)$$

In order to calculate an explicit example we adopt an exponential BVF profile as proposed by Garrett and Munk (1972) which has turned out to fit the stratification of midlatitudinal and tropical oceans quite well, i.e.,

$$N(z) = N_0 \mathrm{e}^{a(z + H_{\mathrm{mix}})} \quad \text{for} \quad z < -H_{\mathrm{mix}}. \qquad (1.9.52)$$

We choose $N_0 = 7.3 \cdot 10^{-3}\,\mathrm{s}^{-1}$, $a = 1/(1.3\,\mathrm{km})$, $H = 4\,\mathrm{km}$, and $H_{\mathrm{mix}} = 75\,\mathrm{m}$.

The corresponding eigenvalues, inverse eigenvalues and projection coefficients of the first 4 baroclinic modes are presented in Table 1.9.2.

Table 1.9.2
Eigenvalues (λ_n), maximum phase velocities (λ_n^{-1}), and projection coefficients of the forcing (1.9.31) on the corresponding vertical eigenfunctions of the first four baroclinic modes obtained by the WKB-method for a BVF profile (1.9.52). The numerical values of N_0, H, H_{mix}, and a are given in the text.

n	λ_n s/m	λ_n^{-1} m/s	$\frac{100 H_{\mathrm{mix}}}{h_n}$
1	0.35	2.88	−4.60
2	0.69	1.44	4.52
3	1.04	0.96	−4.39
4	1.39	0.72	4.22

The vertical eigenfunctions $F_n(z)$ and $Z_n(z)$ are depicted in Figs. 1.9.2 and 1.9.3, respectively. From the Fig. 1.9.2 we see that the variations of both the horizontal current and the pressure are more or less confined to the upper two thirds of the water column. This figure also demonstrates the approximate character of the WKB method because the vertical derivation of $F_n(z)$ do not exactly vanish at the bottom. On the other hand the eigenfunctions $Z_n(z)$ fit the bottom boundary condition as shown in Fig. 1.9.3.

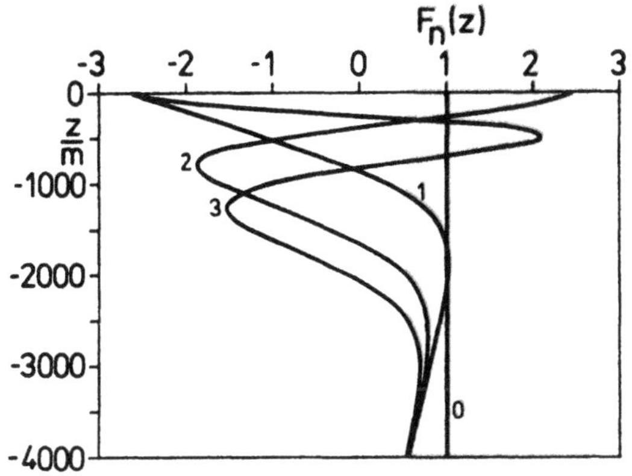

Fig. 1.9.2
The barotropic and the first three baroclinic vertical eigenfunctions $F_n(z)$ in the WKB treatment for a BVF $N(z) = \theta(z + H_{mix}) N_0 + \theta(-z - H_{mix}) N_0 \exp[a(z + H_{mix})]$ with $N_0 = 7{,}3 \cdot 10^{-3}\,s^{-1}$, $a^{-1} = 1{,}3$ km, $H = 4$ km, and $H_{mix} = 75$ m

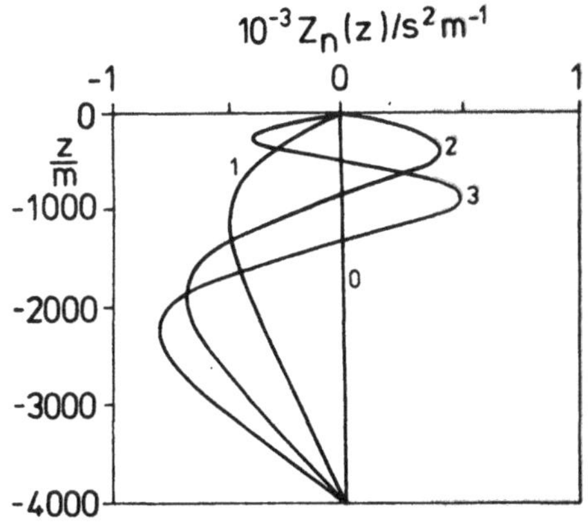

Fig. 1.9.3
The same as Fig. 1.9.2 for $Z_n(z)$

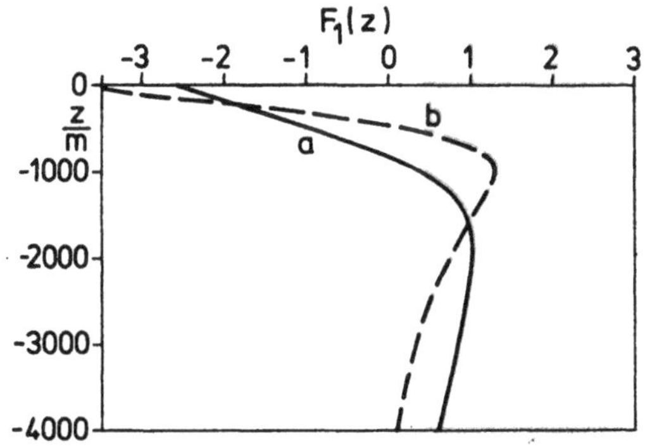

Fig. 1.9.4
(a) The first baroclinic eigenfunction $F_1(z)$ as in Fig. 1.9.2
(b) the first eigenfunction $F_1(z)$ which follows if $a^{-1} = 0{,}65$ km has been adopted.

The structure of the eigenfunctions $Z_n(z)$ indicates that the maximum vertical velocities occur more or less in the upper third of the water column.

Figure 1.9.4 demonstrates that the first baroclinic eigenfunction $F_1(z)$ is somewhat more confined to the sea surface if the e-folding distance of the BVF in (1.9.52) has been halved. In Fig. 1.9.5 we have plotted Parseval's sum as a function of the mode number. It appears that a stratification typical for midlatitudinal and tropical oceans requires less than the first 20 vertical modes to represent approximately 90% of the forcing function. This is significantly less than the necessary number of modes for the case of constant BVF as described in the previous section, compare Fig. 1.9.1, label c.

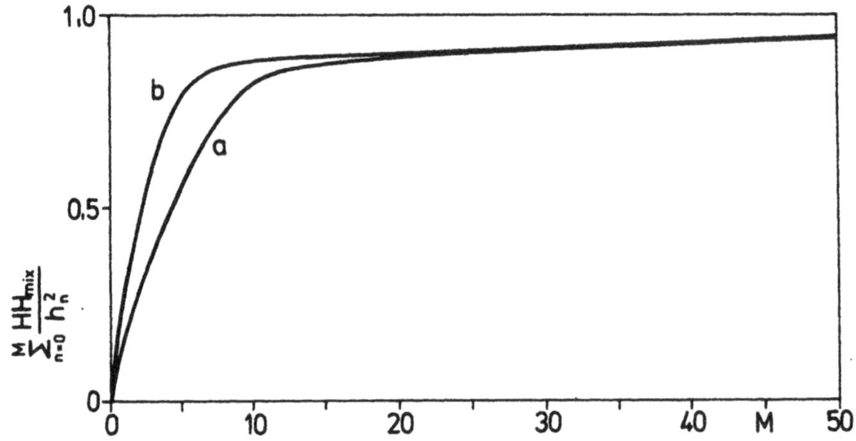

Fig. 1.9.5.
Parseval's sum of the projection coefficients of the step-like body force onto the WKB eigenfunctions as shown in Fig. 1.9.2.
Label (a) indicates the case $a^{-1} = 1.3$ km and (b) corresponds to $a^{-1} = 0.65$ km.

1.9.4. Constant BVF in the surface mixed layer and exponential BVF in the deep layer

The intensive wind induced turbulent mixing near the ocean surface implies a weak stratification within the mixed layer and strong vertical gradients of the density just below the mixed layer. Subsequently the vertical density gradient is observed to become gradually weaker with increasing depth down to the ocean bottom. It appears that the vertical BVF profile proposed by GARRETT and MUNK (1972), which may be written in the form

$$N(z) = \theta(z + H_{mix})\, N_1 + \theta(-z - H_{mix})\, N_2 e^{a(z + H_{mix})} \tag{1.9.53}$$

fits the observed BVF profile of large portions of the oceans and seas rather well. An example of a typical BVF profile for the Atlantic is shown in Fig. 1.9.6, and compared with an analytical model according to (1.9.53).

In this section we wish to discuss the vertical eigenvalue problem for the BVF model (1.9.53). We start with (1.9.5). The solution for the upper layer is readily obtained

$$Z(z) = A_1 \sin \lambda H_1 z + B_1 \cos \lambda N_1 z\,, \qquad -H_{mix} \leqq z \leqq 0\,. \tag{1.9.54}$$

Inserting this in (1.9.4) yields

$$F(z) = -\frac{N_1}{\lambda}(A_1 \cos \lambda N_1 z - B_1 \sin \lambda N_1 z)\,, \qquad -H_{mix} \leqq z \leqq 0\,. \tag{1.9.55}$$

The general solution of (1.9.5) within the lower layer is somewhat more complicated. We have explicitly

$$\frac{d^2 Z}{dz^2} + N_2^2 \lambda^2\, e^{2a(z + H_{mix})}\, Z(z) = 0\,. \tag{1.9.56}$$

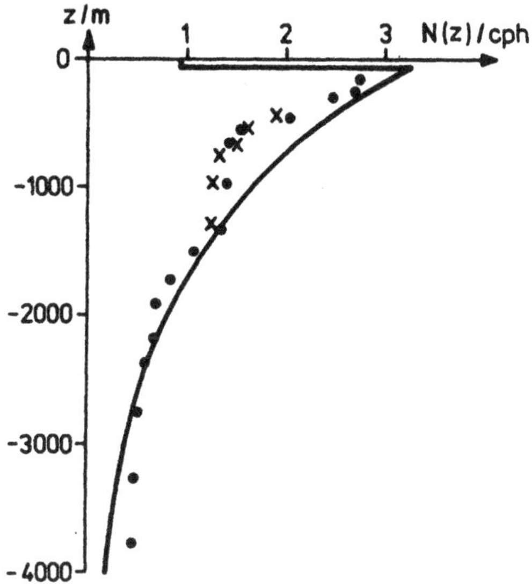

Fig. 1.9.6
The BVF model according to (1.9.53) (*line*), and measured values of "Meteor" stations from 1935 (*dots*) and of "Atlantis II" stations during GATE 1974 (*crosses*) (redrawn from FENNEL and LASS (1979))

With

$$x = \frac{\lambda}{a} N_2 \, e^{a(z+H_{mix})} \tag{1.9.57}$$

(1.9.56) transforms into

$$x^2 \frac{d^2Z}{dz^2} + x \frac{dZ}{dz} + x^2 Z(x) = 0 . \tag{1.9.58}$$

This is Bessel's differential equation of zeroth order which has the general solution

$$Z(z) = A\left[J_0\left(\frac{N_2\lambda}{a} e^{a(z+H_{mix})}\right) Y_0\left(\frac{N_2\lambda}{a} e^{-a(H-H_{mix})}\right) - J_0\left(\frac{N_2\lambda}{a} e^{-a(H-H_{mix})}\right) Y_0\left(\frac{N_2\lambda}{a} e^{a(z+H_{mix})}\right)\right] \tag{1.9.59a}$$

for $-H \leqq z \leqq -H_{mix}$. The coefficients of (1.9.59) have been chosen so that $Z(z)$ satisfies the boundary condition (1.9.6) at the ocean bottom. From (1.9.4) it follows

$$F(z) = -AN_2 \frac{e^{a(z+H_{mix})}}{\lambda} \left[J_1(x(z))\, Y_0(x(-H)) - J_0(x(-H))\, Y_1(x(z))\right] \tag{1.9.59b}$$

for $-H \leqq z \leqq -H_{mix}$. Here J_0, J_1 and Y_0, Y_1 denote the Bessel function and Neumann function of zeroth and first order, respectively. Moreover the relations $J_0'(x) = -J_1(x)$ and $Y_0'(x) = -Y_1(x)$ were used.

In order to estimate the unknown coefficients A_1, B_1 and A we have to apply the surface boundary condition (1.9.7) and (1.7.15) which requires the steadiness of $Z(z)$ and $F(z)$

at $z = -H_{mix}$. Then we get the following set of homogeneous algebraic equations

$$\begin{pmatrix} \frac{-N_1}{g\lambda} & 1 & 0 \\ -\sin \lambda N_1 H_{mix} & \cos \lambda N_1 H_{mix} & -\psi(-H_{mix}) \\ \cos \lambda N_1 H_{mix} & \sin \lambda N_1 H_{mix} & \frac{N_2}{N_1} \Psi(-H_{mix}) \end{pmatrix} \begin{pmatrix} A_1 \\ B_1 \\ A \end{pmatrix} = 0 . \tag{1.9.60}$$

Here ψ and Ψ were introduced to abbreviate the following expressions

$$\begin{aligned} \psi(z) &= J_0(x(z))\, Y_0(x(-H)) - J_0(x(-H))\, Y_0(x(z)) \\ \Psi(z) &= J_1(x(z))\, Y_0(x(-H)) - J_0(x(-H))\, Y_1(x(z)) . \end{aligned} \tag{1.9.61}$$

The system (1.9.60) has a nontrivial solution if its determinant vanishes. Hence, we get the following eigenvalue equation

$$\begin{aligned} &-\frac{N_1}{N_2} \Psi(-H_{mix}) \sin \lambda N_1 H_{mix} + \cos (\lambda N_1 H_{mix})\, \psi(-H_{mix}) \\ &+ \frac{N_1}{g} \left[\frac{N_2}{N_1} \Psi(-H_{mix}) \cos \lambda N_1 H_{mix} + \sin (\lambda N_1 H_{mix})\, \psi(-H_{mix}) \right] = 0 . \end{aligned} \tag{1.9.62}$$

From the previous sections the barotropic eigenvalue is known to be rather small. Hence, we expand (1.9.62) for small λ. Using the expansions of the Bessel functions for small arguments, e.g. ABRAMOWITZ and STEGUN (1984)

$$\begin{aligned} &J_0(x) = 1 - \frac{x^2}{4}; \qquad J_1(x) = \frac{x}{2} \\ &Y_0(x) = \frac{2}{\pi} \ln\left(\frac{x}{2}\right); \qquad Y_1(x) = -\frac{2}{\pi x} + \frac{2}{\pi} J_1(x) \left[\ln\left(\frac{x}{2}\right) + C \right] \end{aligned} \tag{1.9.63}$$

with $C = 0.5772\ldots$ being Euler's constant, we get approximately

$$\psi(-H_{mix}) = -\frac{2}{\pi} a(H - H_{mix}) . \tag{1.9.64}$$

Moreover, with the aid of $\lim_{x \to 0} (x \ln x) = 0$ we find

$$\Psi(-H_{mix}) = \frac{2}{\pi} \frac{a}{N_2 \lambda} . \tag{1.9.65}$$

Substituting (1.9.64) and (1.9.65) into (1.9.62) supplies

$$\frac{1}{g\lambda_0^2} = H \tag{1.9.66}$$

where orders of $(N_2 \lambda_0 H)^2/a$ have been neglected.

Thus the exact treatment of the vertical eigenvalue problem of a nearly realistic BVF reveals the result of Section 1.9.2, see in particular (1.9.21). Solving (1.9.60) in terms of A

yields the following expressions of the eigenfunctions

$$Z(z) = A\left[\theta(z+H_{\mathrm{mix}})\frac{\sin\lambda N_1 z + \dfrac{N_1}{g\lambda}\cos\lambda N_1 z}{-\sin\lambda N_1 H_{\mathrm{mix}} + \dfrac{N_1}{g\lambda}\cos\lambda N_1 H_{\mathrm{mix}}}\psi(-H_{\mathrm{mix}}) + \theta(-z-H_{\mathrm{mix}})\psi(z)\right] \tag{1.9.67 a}$$

$$F(z) = -A\left[\theta(z+H_{\mathrm{mix}})\frac{N_2}{\lambda}\frac{\cos\lambda N_1 z - \dfrac{N_1}{g}\sin\lambda N_1 z}{\cos\lambda N_1 H_{\mathrm{mix}} + \dfrac{N_1}{g\lambda}}\Psi(-H_{\mathrm{mix}})\right.$$

$$\left. + \theta(-z-H_{\mathrm{mix}})\frac{N_2}{\lambda}e^{a(z+H_{\mathrm{mix}})}\Psi(z)\right]. \tag{1.9.67 b}$$

From (1.9.67) we may derive the barotropic eigenfunctions by using the approximations (1.9.63) and the barotropic eigenvalue (1.9.66). It follows

$$Z_0(z) = -\frac{2aA}{\pi}(H+z) \tag{1.9.68 a}$$

$$F_0(z) = -\frac{2aA}{\pi\lambda_0^2}. \tag{1.9.68 b}$$

Obviously the expressions (1.9.68a, b) apart from a normalization constant equal the solutions for the barotropic eigenfunctions of the case of constant BVF profile as given by (1.9.27) and (1.9.28). Hence we can treat the barotropic and the baroclinic part of the vertical eigenvalue problem separately. This justifies a posteriori the use of the barotropic mode in the WKB treatment of the foregoing section.

The corresponding equations for the baroclinic part of the eigenvalue problem can be derived from (1.9.62) and (1.9.67) by assuming the eigenvalues to be large. This assumption is equivalent to the "rigid lid" assumption, which requires a vanishing vertical velocity component at the sea surface. Then we find the following baroclinic eigenvalue equation

$$N_1\cos(\lambda N_1 H_{\mathrm{mix}})\,\psi(-H_{\mathrm{mix}}) - N_2\sin(\lambda N_1 H_{\mathrm{mix}})\,\Psi(-H_{\mathrm{mix}}) = 0 \tag{1.9.69}$$

and the baroclinic eigenfunctions are

$$Z(z) = A\left[-\theta(z+H_{\mathrm{mix}})\frac{\sin\lambda N_1 z}{\sin\lambda N_1 H_{\mathrm{mix}}}\psi(-H_{\mathrm{mix}}) + \theta(-z-H_{\mathrm{mix}})\psi(z)\right] \tag{1.9.70 a}$$

$$F(z) = A\frac{N_2}{\lambda}\left[\theta(z+H_{\mathrm{mix}})\frac{\cos\lambda N_1 z}{\cos\lambda N_1 H_{\mathrm{mix}}}\Psi(-H_{\mathrm{mix}}) + \theta(-z-H_{\mathrm{mix}})e^{a(z+H_{\mathrm{mix}})}\Psi(z)\right]. \tag{1.9.70 b}$$

Next we determine the constant A by normalization of eigenfunctions. Since the normalization of the eigenfunctions (1.9.70) by direct integration as in (1.9.13) is rather difficult to obtain we employ another method.

Starting with (1.9.5) for the eigenvalues λ_n and λ_m we multiply the corresponding equations by Z_m and Z_n, substract the resulting equations and integrate from the bottom to the sea surface.
Then we get

$$\int_{-H}^{0} dz \left(Z_m \frac{d^2 Z_n}{dz^2} - Z_n \frac{d^2 Z_m}{dz^2} \right) + (\lambda_n^2 - \lambda_m^2) \int_{-H}^{0} N^2(z)\, Z_n Z_m \, dz = 0 . \tag{1.9.71}$$

Partial integration of (1.9.71) yields

$$\int_{-H}^{0} dz \frac{d}{dz} \left(Z_m \frac{dZ_n}{dz} - Z_n \frac{dZ_n}{dz} \right) + (\lambda_n^2 - \lambda_m^2) \int_{-H}^{0} dz\, N^2(z)\, Z_n Z_m = 0 . \tag{1.9.72}$$

Now we integrate the first term of (1.9.72) from the ocean bottom to $z = -H_{mix}$ and from $z = -H_{mix}$ to the sea surface and utilize the boundary conditions $Z(0) = Z(-H) = 0$. Then it follows

$$[Z_m(z)\, Z_n'(z) - Z_n(z)\, Z_m'(z)]_{z=-H_{mix}-\varepsilon} - [Z_m(z)\, Z_n'(z) - Z_n(z)\, Z_m'(z)]_{z=-H_{mix}+\varepsilon}$$
$$+ (\lambda_n^2 - \lambda_m^2) \int_{-H}^{0} N^2(z)\, Z_n(z)\, Z_m(z)\, dz = 0 \tag{1.9.73}$$

where the dash indicates z-differentiation.

Differentiating (1.9.73) by λ_n, taking the steadiness of Z_n and Z_n' at $z = -H_{mix}$ into account, and approaching λ_m to λ_n we get

$$Z_n(-H_{mix}) \frac{\partial}{\partial \lambda_n} (Z_n'(-H_{mix} - \varepsilon) - Z_n'(-H_{mix} + \varepsilon))$$
$$- Z_n'(-H_{mix}) \frac{\partial}{\partial \lambda_n} (Z_n(-H_{mix} - \varepsilon) - Z_n(-H_{mix} + \varepsilon)) + 2\lambda_n \int_{-H}^{0} Z_n^2(z)\, N^2(z)\, dz = 0 . \tag{1.9.74}$$

To proceed further it is convenient to put $Z(z)$ into the form

$$Z(z) = A \left[\theta(z + H_{mix}) \frac{\psi(-H_{mix})}{\Phi(-H_{mix})} \Phi(z) + \theta(-z - H_{mix})\, \psi(z) \right] \tag{1.9.75}$$

where, according to (1.9.70a), $\Phi(z) = \sin \lambda N_1 z$. We readily realize that the Wronskian of Φ and ψ at $z = -H_{mix}$ equals just the eigenvalue equation (1.9.69), i.e.

$$W_r(-H_{mix}) = \Phi(-H_{mix})\, \psi'(-H_{mix}) - \Phi'(-H_{mix})\, \psi(-H_{mix}) = 0 .$$

The dash indicates z-derivatives taken at $z = -H_{mix}$. In terms of (1.9.75) and (1.9.76) the Equation (1.9.74) reads

$$A_n^2 \frac{\psi_n(-H_{mix})}{\Phi_n(-H_{mix})} \left[\frac{d}{d\lambda_n} W_r(\lambda_n, z) \right]_{z=-H_{mix}} + 2\lambda_n \int_{-H}^{0} N^2(z)\, Z_n^2(z)\, dz = 0 . \tag{1.9.76}$$

Hence the normalization constant of the Z_n's is given by

$$A_n^2 = (2\lambda_n H)\left[\frac{\psi_n(-H_{mix})}{\Phi_n(-H_{mix})}\frac{d}{d\lambda_n}W_r(-H_{mix})\right]^{-1}. \tag{1.9.77}$$

Using the fact that the Wronskian of $J_1(x)$ and $Y_0(x)$ equals $2/\pi x$, and taking the identity $J_n'(x) = J_{n-1}(x) - (n/x)J_n(x)$ into account, which also holds true for $Y_n'(x)$, we find

$$\frac{d}{d\lambda_n}W_r(-H_{mix}) = \lambda_n \sin(\lambda_n N_1 H_{mix}) \times$$
$$\times\left\{\left[\frac{N_2^2 - N_1^2}{a} + \frac{N_1^2}{a}\frac{1+aH_{mix}}{\sin^2(\lambda_n N_1 H_{mix})} - \frac{N_1}{\lambda_n}\frac{1}{\tan(\lambda_n N_1 H_{mix})}\right]\psi(-H_{mix}) - \frac{4}{\pi^2}\frac{a}{\lambda_n^2}\frac{1}{\psi(-H_{mix})Y_0^2(-H_{mix})}\right\}. \tag{1.9.78}$$

The expression determines together with (1.9.77) the normalization constant $A = A_n$.

In order to normalize the F_n's we may express the $Z_n(z)$ in (1.9.76) by $F_n(z)$ using (1.9.3). After integration by parts and expressing $Z_n'(z)$ by (1.9.4) we get

$$B_n^2\frac{\psi_n(-H_{mix})}{\Phi_n(-H_{mix})}\frac{d}{d\lambda_n}W_r(-H_{mix}) + 2\lambda_n^3\int_{-H}^{0}F_n^2(z)\,dz = 0 \tag{1.9.79}$$

Here the letter B_n was used to denote the normalization constant of the $F_n(z)$'s

$$B_n^2 = \lambda_n^2 A_n^2$$
$$= \frac{2\lambda_n^2 H}{\left[\frac{N_2^2 - N_1^2}{a} + \frac{N_1^2}{a}\frac{1+aH_{mix}}{\sin^2(\lambda_n N_1 H_{mix})} - \frac{N_1}{\lambda_n}\frac{1}{\tan\lambda_n N_1 H_{mix}}\right]\psi^2(-H_{mix}) - \frac{4a}{\pi^2\lambda_n^2}Y_0^{-2}(-H_{mix})}. \tag{1.9.80}$$

Now we may calculate the projection coefficients of a body force as given by (1.9.31). The barotropic projection coefficient becomes approximately

$$\frac{1}{h_0} = \frac{1}{H}\int_{-H}^{0}F_0(z)\frac{\theta(z+H_{mix})\,dz}{H_{mix}} = \frac{1}{H}, \tag{1.9.81}$$

and for the baroclinic coefficients we get

$$\frac{1}{h_n} = \frac{1}{H}\int_{-H}^{0}F_n(z)\frac{\theta(z+H_{mix})\,dz}{H_{mix}} = -\frac{1}{H_{mix}H\lambda_n^2}\int_{H_{mix}}^{0}\frac{dZ_n(z)}{dz}\,dz = \frac{Z_n(-H_{mix})}{H_{mix}H\lambda_n^2}. \tag{1.9.82}$$

With the aid of (1.9.70) this amounts to

$$\frac{1}{h_n} = \frac{B_n}{HH_{mix}\lambda_n^2}\psi(-H_{mix}). \tag{1.9.83}$$

Finally we wish to derive asymptotic expressions of the eigenvalue equation (1.9.69) and the eigenfunctions (1.9.70) for large eigenvalues and to compare the results with those of the WKB-method applied for in Section 1.9.3. The asymptotic expressions of the Bessel functions are

$$J_n(x) = \left(\frac{2}{\pi x}\right)^{1/2} \cos\left(x - n\frac{\pi}{2} - \frac{\pi}{4}\right), \qquad Y_n(x) = \left(\frac{2}{\pi x}\right)^{1/2} \sin\left(x - n\frac{\pi}{2} - \frac{\pi}{4}\right).$$

Inserting these approximations into (1.9.69) and using standard trigonometric identities yields

$$(N_1 - N_2)(\lambda N_1 H_{\text{mix}}) - \frac{\lambda}{a}[N(-H_{\text{mix}}) - N(-H)]$$

$$- (N_1 + N_2) \sin\left\{\lambda_1 N H_{\text{mix}} + \frac{\lambda}{a}[N(-H_{\text{mix}}) - N(-H)]\right\} = 0 . \tag{1.9.84}$$

For a thin mixed layer $N_1 H_{\text{mix}} \ll (1/a)[N(-H_{\text{mix}}) - N(-H)]$ may hold and then from (1.9.84) it follows

$$\sin\left(\frac{\lambda}{a} N_2(1 - e^{-aH})\right) = 0 \tag{1.9.85}$$

which implies for the eigenvalues asymptotically

$$\lambda_n = \frac{n\pi a}{N_2(1 - e^{-aH})} \quad \text{for} \quad n \gg 1 . \tag{1.9.86}$$

To the same approximation the eigenfunctions (1.9.70) become

$$Z_n(z) = \frac{-A}{\sin\left(N_2 \frac{\lambda_n}{a} e^{-aH} - \frac{\pi}{4}\right)} \left(\frac{a}{2\lambda_n N_2 e^{az}}\right)^{1/2} \sin\left(\frac{\lambda_n}{a} N_2(e^{az} - e^{-aH})\right) \tag{1.9.87 a}$$

Table 1.9.3
Eigenvalues (λ_n), maximum phase velocities (λ_n^{-1}), projection coefficients of the forcing (1.9.31) on the corresponding vertical eigenfunctions, and values of the eigenfunctions in the surface mixed layer ($z = -10$ m) as well as in the thermocline ($z = -300$ m) of the first ten baroclinic modes for the BVF profile (1.9.53). The corresponding numerical values are given in the text.

n	λ_n s/m	λ_n^{-1} m/s	$\frac{100 H_{\text{mix}}}{h_n}$	$F_n(-10\text{ m})$	$F_n(-300\text{ m})$
1	0.36	2.78	4.64	2.47	2.10
2	0.76	1.32	−4.43	−2.37	−0.91
3	1.16	0.86	4.17	2.23	−0.54
4	1.55	0.65	−3.88	−2.02	1.71
5	1.95	0.51	−3.58	−1.92	2.23
6	2.35	0.43	3.29	1.78	−1.98
7	2.75	0.36	−3.02	−1.64	1.09
8	3.16	0.32	2.78	1.52	0.13
9	3.56	0.28	−2.56	−1.42	−1.29
10	3.97	0.25	2.37	1.32	2.07

$$F_n(z) = \frac{B_n}{\lambda_n \sin\left(\frac{\lambda n}{a} N_2 \, e^{-aH} - \frac{\pi}{4}\right)} \left(\frac{aN_2 \, e^{az}}{2\lambda_n}\right)^{1/2} \cos\left(\frac{\lambda_n}{a} N_2 (e^{az} - e^{-aH})\right). \tag{1.9.87b}$$

Thus, for an exponential BVF profile as well the eigenfunctions as the eigenvalues approach for large mode numbers the results of the WKB treatment. Moreover we note that the asymptotic value of the projection coefficients (1.9.83) tends to zero for large n. This indicates the convergence of the expansion of the body force in terms of the F_n's.

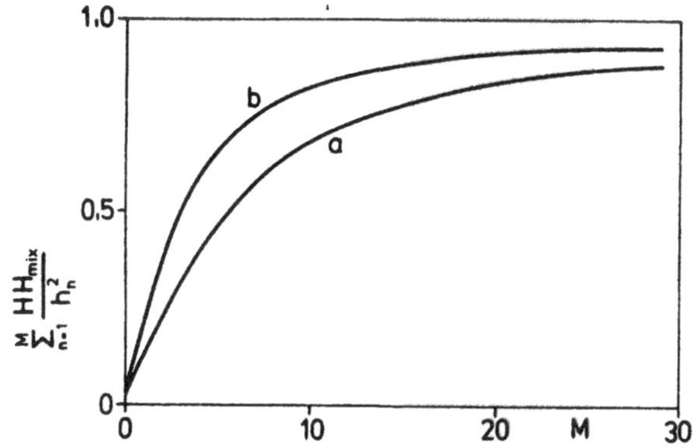

Fig. 1.9.7
Parseval's sum of the projection coefficients of the body force versus the mode number
(a) $a^{-1} = 1{,}3$ km, (b) $a^{-1} = 0{,}65$ km

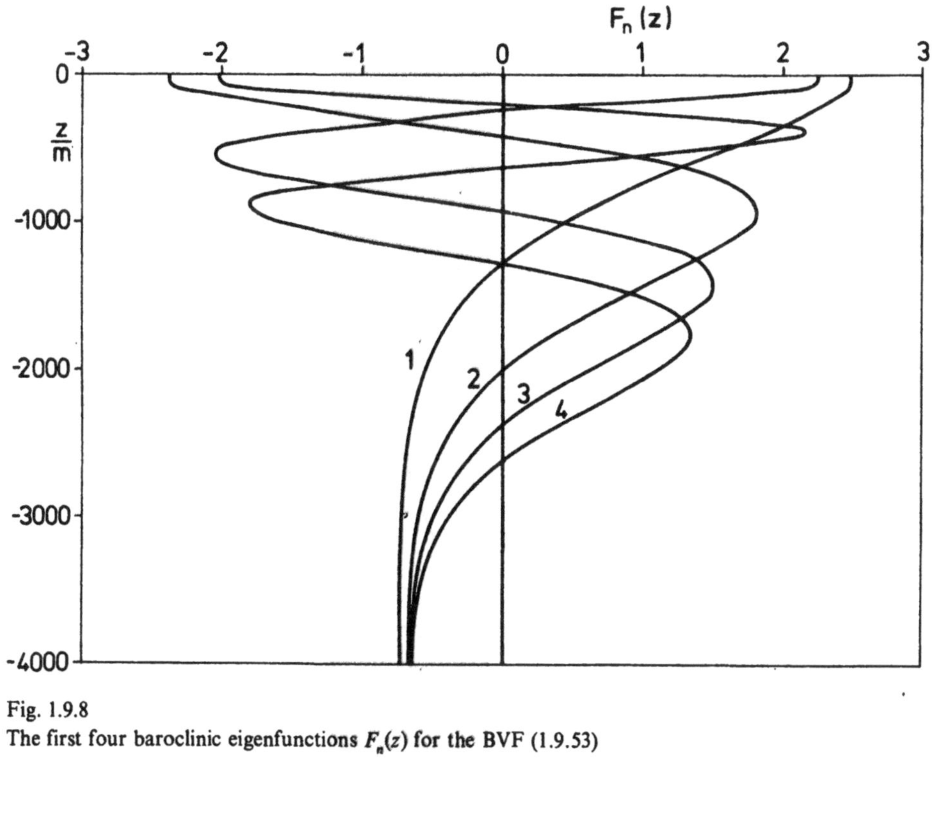

Fig. 1.9.8
The first four baroclinic eigenfunctions $F_n(z)$ for the BVF (1.9.53)

The eigenvalues, eigenfunctions and projection coefficients for a BVF profile typically for tropical oceans have been calculated, adopting the numerical values

$$N_1 = 1.8 \cdot 10^{-3}\,\mathrm{s}^{-1}, \qquad N_2 = 5.8 \cdot 10^{-3}\,\mathrm{s}^{-1}, \qquad H = 4000\,\mathrm{m}, \qquad H_{\mathrm{mix}} = 75\,\mathrm{m}$$

and $1/a = 1400$ m. The eigenvalues and projection coefficients are displayed in Table 1.9.3 for the first four baroclinic modes.

Both the projection coefficients and the eigenvalues given in Tables 1.9.2 and 1.9.3 are of the same order of magnitude and differ only by about 10% and less.

Figure 1.9.7, label a, depicts Parseval's sum as a function of mode number. Obviously some 10 vertical modes are nessecary to reproduce about 90% of the forcing signal in deep oceans. Stronger stratification (Fig. 1.9.7 label b) tends to reduce the number of vertical modes needed to approximate the vertical profile of (1.9.51) within a given accuracy.

The first four baroclinic eigenfunctions $F_n(z)$ and $Z_n(z)$ are shown in Figs. 1.9.8 and 1.9.9, respectively. These figures demonstrate that most of the variability of the horizontal current components is confined to the upper half of the ocean, whereas the vertical velocity may significantly be affected within the whole water column except near the bottom and the sea surface.

In Fig. 1.9.10 the first eigenfunction $F_1(z)$ is shown for two different values of the parameter a of the BVF (1.9.53); $a^{-1} = 1400$ m (label a), and $a^{-1} = 700$ m (label b). It appears that for a stronger stratification, i.e. label b, $F_1(z)$ is somewhat more confined

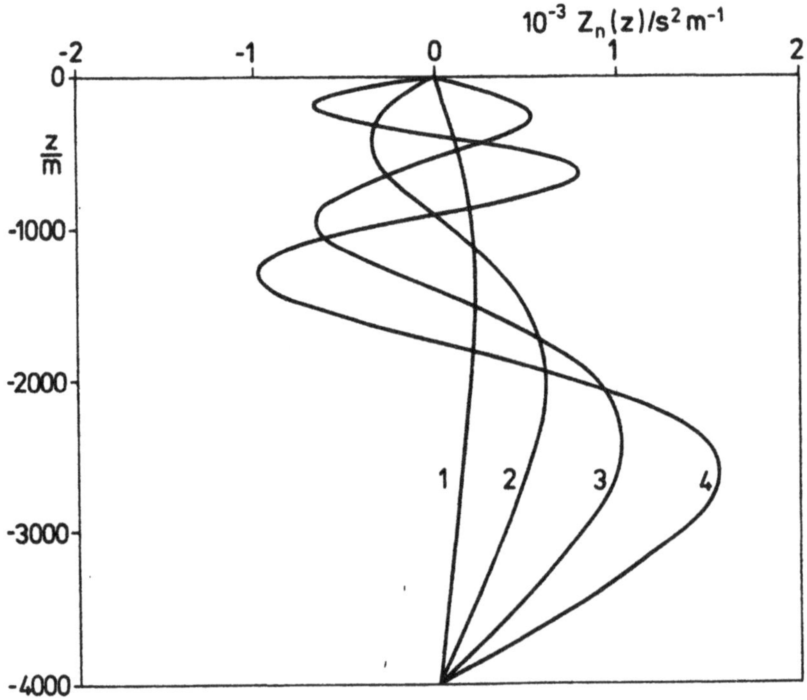

Fig. 1.9.9
The first four baroclinic eigenfunctions $Z_n(z)$ for the BVF (1.9.53)

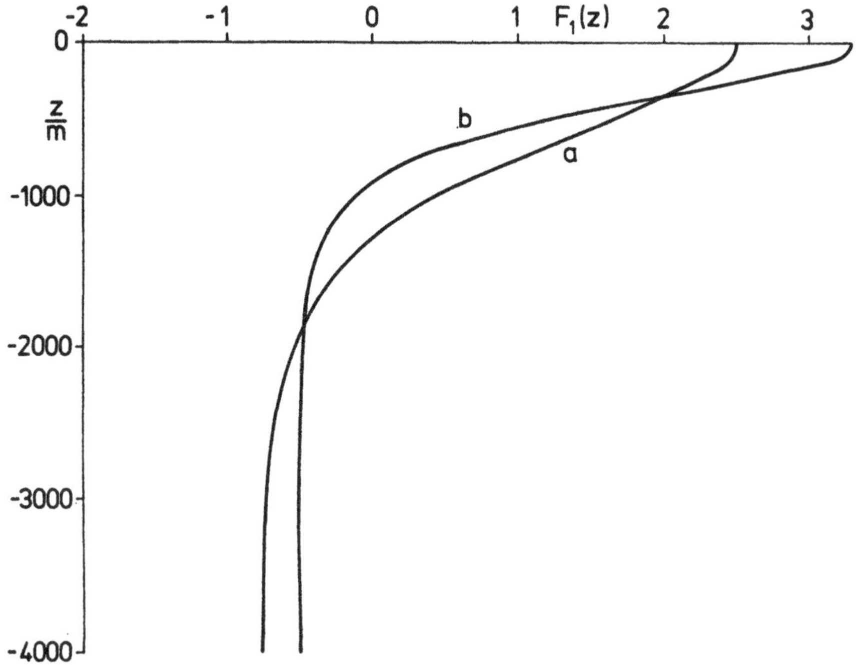

Fig. 1.9.10
The first baroclinic eigenfunction $F_1(z)$ for the BVF (1.9.53)
(a) $a^{-1} = 1.4$ km, (b) $a^{-1} = 0.65$ km

to the upper part of the ocean. However, even such a strong variation of the parameter a^{-1} has only a minor effect on the global shape of the eigenfunction. Thus the eigenfunctions as well as the eigenvalues are relatively insensitive to variations in the BVF.

CHAPTER 2

Green's function

2.1. GENERAL REMARKS

From the mathematical point of view the investigation of the dynamical response of the sea to external forcing corresponds to the treatment of initial value problems and inhomogeneous boundary value problems.

An elegant and powerful technique of solution of such differential equations is the concept of Green's function. Basically, the idea consists in the following: Let **L** be a linear differential operator describing a physical system and consider the inhomogeneous equation

$$\mathbf{L}u(x) = \varphi(x) . \tag{2.1.1}$$

Then the corresponding Green's function is defined by

$$\mathbf{L}G(x, x') = \delta(x - x') \tag{2.1.2}$$

with x' being a parameter and $\delta(x - x')$ is Dirac's delta function, i.e. the source $\varphi(x)$ was replaced by the delta function. Now the solution $u(x)$ of (2.1.1) can be expressed in terms of $G(x, x')$ as

$$u(x) = \int \mathrm{d}x' \, G(x, x') \, \varphi(x') . \tag{2.1.3}$$

To prove this let **L** act on $u(x)$

$$\mathbf{L}u(x) = \int \mathrm{d}x' \, \mathbf{L}G(x, x') \, \varphi(x') = \int \mathrm{d}x' \, \delta(x - x') \, \varphi(x') = \varphi(x) .$$

That is, (2.1.3) is the solution of (2.1.1). Thus, if once the Green's function of a problem is given its solution can be expressed in terms of a source representation (2.1.3).

The physical meaning of the Green's function becomes apparent if a resting physical system is excited by a pulse-like forcing, i.e., $\varphi(x) = \delta(x)$. Then the response of the system is, after (2.1.3), the Green's function itself. Hence, the Green's function is determined by the intrinsic properties of the system independent of the variety of possible forcing.

At a glance it may seem that the amount of labour would be the same whether we solve the problem (2.1.1) directly, or whether we calculate first the Green's function and use then a source representation of the kind of (2.1.3). However, the point is that the Green's function is independent of the structure of the source term. Thus, if G has been determined for a certain operator $\mathbf{L}$ we can discuss the solution for a variety of sources on the basis of (2.1.3). Moreover, the Green's function technique possesses various attractive features which will be discussed in this chapter. Although there exist several textbooks, e.g. Morse and Feshbach (1953), Wyld (1976), and Wladimirow (1972), which give introductions to Green's functions technique, it seems to be worthwhile to summarize those aspects and features of the Green's functions which are relevant for the investigations in this book. We mention that the Green's function are also known as transfer functions or propagators.

2.2. INHOMOGENEOUS INITIAL VALUE PROBLEMS

Let us consider the differential equation

$$\mathbf{D}_m u(t) = f(t) \tag{2.2.1 a}$$

where $\mathbf{D}_m$ is a linear differential operator with constant coefficients given by

$$\mathbf{D}_m = \sum_{n=0}^{m} a_n \frac{\mathrm{d}^n}{\mathrm{d}t^n} . \tag{2.2.1 b}$$

We assume that $f(t)$ starts to act at $t = 0$, i.e., $f(t) = 0$ for $t < 0$, and that $u(t)$ obeys $m - 1$ initial conditions

$$\begin{aligned} u(0) &= u_0 \\ u'(0) &= u_1 \\ u^{(2)}(0) &= u_2 \\ &\vdots \\ u^{(m-1)}(0) &= u_{m-1} . \end{aligned} \tag{2.2.1 c}$$

The general solution of this problem consists of the general solution of the homogeneous version of (2.2.1) plus a special solution of the inhomogeneous problem. We will show that the general solution can be expressed by means of the associated Green's function which is subject to

$$\mathbf{D}_m G(t) = \delta(t) . \tag{2.2.2}$$

With the ansatz

$$G(t) = \theta(t)\, g(t) \tag{2.2.3}$$

and taking the relations

$$\frac{\mathrm{d}}{\mathrm{d}t}\theta(t) = \delta(t) , \qquad \delta(t)\, g(t) = \delta(t)\, g(0)$$

into account we readily obtain an equivalent formulation of (2.2.3) in terms of the initial value problem

$$\mathbf{D}_m g(t) = 0 \tag{2.2.4a}$$

with

$$g(0) = g'(0) = g^{(2)}(0) = \dots = g^{(m-2)}(0) = 0\,, \qquad g^{(m-1)}(0) = 1\,. \tag{2.2.4b}$$

Thus, the Green's function $G(t)$ which corresponds to the operator $\mathbf{D}_m$ is just the solution of a homogeneous equation (2.2.4a), with the special initial conditions (2.2.4b), multiplied by $\theta(t)$.

We note three examples of Green's functions:

First example:

$$\mathbf{D}_1 G(t) = \left(\frac{\mathrm{d}}{\mathrm{d}t} + a\right) G(t) = \delta(t)\,. \tag{2.2.5a}$$

Inserting (2.2.3) yields $\delta(t)\, g(t) + \theta(t)\, \mathbf{D}_1 g(t) = \delta(t)$ or $\mathbf{D}_1 g(t) = 0$ and $g(0) = 1$. The solution is $g(t) = \mathrm{e}^{-at}$ and hence

$$G(t) = \theta(t)\, \mathrm{e}^{-at}\,. \tag{2.2.5b}$$

Second example (harmonic oscillator):

$$\mathbf{D}_2 G(t) = \left(\frac{\mathrm{d}^2}{\mathrm{d}t^2} + \omega_0^2\right) G(t) = \delta(t)\,. \tag{2.2.6a}$$

Inserting (2.2.3) yields $\delta'(t)\, g(0) + \delta(t)\, g'(0) + \theta(t)\, \mathbf{D}_2 g(t) = \delta(t)$, implying in accordance with (2.2.4b) $g(0) = 0$ and $g'(0) = 1$. The corresponding solution of $\mathbf{D}_2 g(t) = 0$ is

$$g(t) = \frac{\sin \omega_0 t}{\omega_0}, \text{ hence}$$

$$G(t) = \theta(t)\frac{\sin \omega_0 t}{\omega_0}\,. \tag{2.2.6b}$$

Third example (damped harmonic oscillator):

$$\mathbf{D}_2 G(t) = \left(\frac{\mathrm{d}^2}{\mathrm{d}t^2} + \beta\frac{\mathrm{d}}{\mathrm{d}t} + \omega_0^2\right) G(t) = \delta(t)\,. \tag{2.2.7a}$$

The corresponding homogeneous solution subject to $g(0) = 0$ and $g'(0) = 1$ is $g(t) = \mathrm{e}^{-\frac{\beta}{2}t}\dfrac{\sin bt}{b}$, where $b = (\omega_0^2 - \beta^2/4)^{1/2}$, and, hence,

$$G(t) = \theta(t)\, \mathrm{e}^{-\frac{\beta}{2}t}\frac{\sin bt}{b}\,. \tag{2.2.7b}$$

Next we consider a special solution of the inhomogeneous equation (2.2.1a). Suppose the associated Green's function is known, we may write

$$\hat{u}(t) = \int_0^\infty G(t - t')\, f(t')\, \mathrm{d}t'\,. \tag{2.2.8}$$

Let the operator $\mathbf{D}_m$ act on (2.2.8). Then it appears that u is in fact a solution of (2.1.1 a)

$$\mathbf{D}_m\hat{u} = \int_0^\infty \mathbf{D}_m G(t-t')\, f(t')\,\mathrm{d}t' = \int_0^\infty \delta(t-t')\, f(t')\,\mathrm{d}t' = f(t)\,.$$

Let us investigate how the general solution of the free equation, i.e. $\mathbf{D}_m u = 0$, can be expressed in terms of the Green's function and of the initial conditions (2.2.1c). We start with an example

$$u''(t) + \omega_0^2 u(t) = 0 \tag{2.2.9a}$$

$$u'(0) = u_1\,, \qquad u(0) = u_0\,. \tag{2.2.9b}$$

Let us continue $u(t)$ in the negative time domain, $t < 0$, by

$$\tilde{u}(t) = \theta(t)\, u(t)\,. \tag{2.2.10}$$

Then, it follows

$$\tilde{u}'(t) = \delta(t)\, u(0) + \theta(t)\, u'(t)$$

$$\tilde{u}''(t) = \delta'(t)\, u(0) + \delta(t)\, u'(0) + \theta(t)\, u''(t)$$

hence

$$\tilde{u}''(t) + \omega_0^2\tilde{u} = \delta(t)\, u_1 + \delta'(t)\, u_0\,. \tag{2.2.11}$$

Thus, the ansatz (2.2.10) transforms the homogeneous problem (2.2.9) in the inhomogeneous one (2.2.11). Applying (2.2.8) to (2.2.11) we have

$$\tilde{u}(t) = \int \mathrm{d}t' G(t-t')\,(\delta(t')\, u_1 + \delta'(t')\, u_0) \tag{2.2.12}$$

and, using $\delta'(t)\, G(t) = -\delta(t)\, G'(t)$,

$$\tilde{u}(t) = u_1 G(t) + u_0 G'(t)\,. \tag{2.2.13}$$

With (2.2.10), implying $\tilde{u}(t) = u(t)$ for $t > 0$, and with (2.2.6b) we obtain explicitly

$$u(t) = u_1\,\frac{\sin\omega_0 t}{\omega_0} + u_0\cos\omega_0 t\,. \tag{2.2.14}$$

For the example of a damped harmonic oscillator given by (2.2.6b)

$$u''(t) + \beta u'(t) + \omega_0^2 u(t) = 0 \tag{2.2.15a}$$

with

$$u'(0) = u_1 \quad \text{and} \quad u(0) = u_0\,, \tag{2.2.15b}$$

we obtain analogously

$$\tilde{u}''(t) + \beta\tilde{u}'(t) + \omega_0^2\tilde{u}(t) = u_0\delta'(t) + (u_1 + \beta u_0)\,\delta(t) \tag{2.2.16}$$

and hence

$$\tilde{u}(t) = (u_1 + \beta u_0)\, G(t) + u_0 G'(t)\,. \tag{2.2.17}$$

With $G(t)$ according to (2.2.7b) we arrive at

$$u(t) = \left(u_1 + \frac{\beta}{2}u_0\right) \mathrm{e}^{-\frac{\beta}{2}t}\,\frac{\sin bt}{b} + u_0\,\mathrm{e}^{-\frac{\beta}{2}t}\cos bt \tag{2.2.18}$$

where

$$b = \left(\omega_0^2 - \frac{\beta^2}{4}\right)^{1/2}$$

Thus, if a friction mechanism is present the response loses its memory of the initial condition after sufficiently long time.

The generalization of these examples is straightforward. With the initial conditions (2.2.1c) and with the ansatz (2.2.10) we may put $\mathbf{D}_m\tilde{u} = 0$ in the form

$$\mathbf{D}_m\tilde{u}(t) = \sum_{n=1}^{m} \sum_{j=0}^{n-1} a_n u_{n-j-1} \delta^{(j)}(t). \tag{2.2.19}$$

Treating the right side of (2.2.19) as source, we find from (2.2.8)

$$\tilde{u}(t) = \sum_{n=1}^{m} \sum_{j=0}^{n-1} a_n u_{n-j-1} G^{(j)}(t) \tag{2.2.20}$$

where the relation

$$\frac{\mathrm{d}^j}{\mathrm{d}t'^j} \delta(t')\, G(t - t') = \delta(t')\,(-1)^j \frac{\mathrm{d}^j}{\mathrm{d}t'^j} G(t - t') = \delta(t')\, G^{(j)}(t - t')$$

was used.

Substituting (2.2.3) and (2.2.10) in (2.2.20) and taking the initial conditions (2.2.4b) into account we obtain

$$u_{\text{hom}} = \sum_{n=1}^{m} \sum_{j=0}^{n-1} a_n u_{n-j-1} g^{(j)}(t). \tag{2.2.21}$$

With (2.2.8) and (2.2.21) we have expressed the complete solution of the problem (2.2.1) in terms of the Green's function $G(t)$, which is subject to (2.2.2), i.e.,

$$u(t) = \sum_{n=1}^{m} \sum_{j=0}^{n-1} a_n u_{n-j-1} g^{(j)}(t) + \int_0^t \mathrm{d}t'\, g(t - t')\, f(t'). \tag{2.2.22}$$

2.3. GREEN'S FUNCTION AND ONE-SIDED FOURIER TRANSFORMATION

As in the foregoing section we study inhomogeneous initial value problems, however, in this section we wish to employ the Fourier transformation technique. Such a treatment is of importance for two reasons: Firstly, the Fourier transforms of dynamical fields are the spectral functions which are of direct physical interest since they could be compared with Fourier analyzed measured time series in order to identify significant processes. Secondly, by means of Fourier transformation differential equations can be transformed into algebraic ones which are much easier to deal with. Thus more complicated problems can be solved in a systematic manner. As a rule the Fourier transforms can analytically be continued into parts of the complex frequency plane and thus the powerful tools of the theory of analytic functions can be applied in order to calculate the inverse Fourier transforms.

Usually, initial value problems are solved with the aid of Laplace transformation. We prefer, however, the one-sided Fourier transformation which is closely related to the Laplace transformation, but which has the formal advantage that the transformed quantities are already the spectral functions of the considered processes.

The Fourier transformation of $G(t)$ is defined by the pair of equations

$$G(\omega) = \int_{-\infty}^{+\infty} dt\, G(t)\, e^{i\omega t} \tag{2.3.1 a}$$

$$G(t) = \int_{-\infty}^{+\infty} \frac{d\omega}{2\pi} G(\omega)\, e^{-i\omega t} . \tag{2.3.1 b}$$

Because of the step function involved in $G(t)$, i.e. $G(t) = \theta(t)\, g(t)$, see (2.2.3) and (2.3.1 a), reads

$$G(\omega) = \int_{0}^{\infty} dt\, g(t)\, e^{i\omega t} . \tag{2.3.2}$$

Let us again start with some illustrative examples:

Firstly with (2.2.5b) we find the corresponding frequency spectrum

$$G(\omega) = \int_{0}^{\infty} dt\, e^{(i\omega - a)t} = \frac{i}{\omega + ia} . \tag{2.3.3}$$

This function can be analytical continued into the complex ω-plane where $G(\omega)$ is analytic, except in the lower half plane where $G(\omega)$ has a simple pole at $\omega = -ia$.

The Fourier inversion of $G(\omega)$ amounts to the calculation of

$$G(t) = \int_{-\infty}^{+\infty} \frac{d\omega}{2\pi} \frac{ie^{-i\omega t}}{\omega + ia} . \tag{2.3.4}$$

The integral is along the real ω-axis. In accordance with Jordan's Lemma we may close the contour by a large semicircle in the lower or in the upper half plane for $t > 0$ or $t < 0$, respectively. Closing the contour in the upper half plane, implying $t < 0$, the integral vanishes. Closing the contour in the lower half plane, i.e. $t > 0$, the residue of the pole $\omega = -ia$ contributes and we find

$$G(t) = 0 \quad \text{for} \quad t < 0 , \qquad G(t) = e^{-at} \quad \text{for} \quad t > 0 ,$$

i.e.

$$G(t) = \theta(t)\, e^{-at} . \tag{2.3.5}$$

From this example we see already, that causality and stability of the solution require the poles of $G(\omega)$ to lie in the lower half plane.

We note that in the limit $a \to 0$ (2.3.5) yields the step function. Thus we may employ (2.3.3) to find the Fourier transform of $\theta(t)$

$$\theta(\omega) = \int_{0}^{\infty} dt\, \theta(t)\, e^{i\omega t} = \lim_{\varepsilon \to 0} \int_{0}^{\infty} dt\, \theta(t)\, e^{-\varepsilon t + i\omega t} = \frac{i}{\omega + i\varepsilon} , \qquad \varepsilon \to +0 \tag{2.3.6}$$

where ε denotes a positive infinitesimal.

Another possibility to get the spectrum of the Green's function is to Fourier transform the differential equation (2.2.5a). Then it follows $(-i\omega + a)\,G(\omega) = 1$ which yields (2.3.3).

As a second example we consider the damped harmonic oscillator, see (2.2.7). To obtain the Fourier transform of the Green's function (2.2.7b) we have to calculate

$$G(\omega) = \int_0^\infty dt\, e^{-\left(\frac{\beta}{2}\right)t + i\omega t}\, \frac{\sin bt}{b}. \tag{2.3.7}$$

The transformation of the exponential is given by (2.3.3). Using (B 8), see appendix,

$$\int_{-\infty}^{+\infty} dt\, e^{i\omega t}\, \frac{\sin bt}{b} = \frac{\pi}{ib}(\delta(\omega + b) - \delta(\omega - b)) \tag{2.2.8}$$

we find with the aid of the convolution theorem

$$G(\omega) = \int_{-\infty}^{+\infty} \frac{d\omega'}{2\pi}\, \frac{i}{\omega - \omega' + i\beta/2}\, \frac{\pi}{ib}\, (\delta(\omega' + b) - \delta(\omega' - b)) \tag{2.3.9}$$

hence, with $b^2 = \omega_0^2 - (\beta/2)^2$

$$G(\omega) = \frac{-1}{\omega^2 + i\beta\omega - \omega_0^2}. \tag{2.3.10}$$

Fourier transformation of the differential equation (2.2.7a) yields the same result without further labour, i.e., $(-\omega^2 - i\omega\beta + \omega_0^2)\,G(\omega) = 1$. Again we observe that the poles of $G(\omega)$ are located in the lower half of the ω-plane.

As a third example we wish to study the case of undamped systems as, e.g., the harmonic oscillator (2.2.9a). First we calculate the Fourier transformation of the product $G(t) = \theta(t)\,g(t)$, which follows with (2.3.6) and with

$$g(\omega) = \int_{-\infty}^{+\infty} dt\, e^{i\omega t}\, g(t) \tag{2.3.11}$$

to be

$$G(\omega) = \int_{-\infty}^{+\infty} \frac{d\omega'}{2\pi}\, \frac{ig(\omega')}{\omega - \omega' + i\varepsilon}. \tag{2.3.12}$$

This expression is called spectral representation. Using Dirac's identity of singular kernels of Cauchy integrals

$$\frac{1}{x \pm i\varepsilon} = \frac{P}{x} \mp i\pi\delta(x) \tag{2.3.13}$$

where the letter P denotes Cauchy's principle value, we may rewrite (2.3.12) to

$$G(\omega) = P\int_{-\infty}^{+\infty} \frac{d\omega'}{2\pi}\, \frac{ig(\omega')}{\omega - \omega'} + \frac{1}{2}\, g(\omega'). \tag{2.3.14}$$

In the case of the harmonic oscillator we find after Fourier transformation of the differential equation (2.2.9a)

$$(-\omega^2 + \omega_0^2)\, G(\omega) = 1 \,. \tag{2.3.15}$$

Taking the identity $x\delta(x) = 0$ into account the solution of (2.3.15) is

$$G(\omega) = -\frac{1}{\omega^2 - \omega_0^2} + A(\omega)\, \delta(\omega^2 - \omega_0^2) \tag{2.3.16}$$

where $A(\omega)$ is an unspecified quantity which may depend on ω. If we wish to estimate the Fourier inversion of (2.3.16) we are faced with the problem that the poles of $G(\omega)$ are located on the real ω-axis, i.e., on the integration contour, and hence the integral does not exist in unique sense. In order to put (2.3.16) in a form consistent with the spectral representation (2.3.12) let us try

$$G(\omega) = \frac{-1}{(\omega + i\varepsilon)^2 - \omega_0^2} \,. \tag{2.3.17}$$

With (2.3.13) we find

$$G(\omega) = -\frac{P}{\omega^2 - \omega_0^2} - \frac{i\pi}{2\omega_0}\, (\delta(\omega - \omega_0) - \delta(\omega + \omega_0)) \,.$$

Using

$$\operatorname{sgn}(\omega)\, \delta(\omega^2 - \omega_0^2) = \frac{1}{2\omega_0}\, (\delta(\omega - \omega_0) - \delta(\omega + \omega_0))$$

it follows that $A(\omega) = i\pi \operatorname{sgn}(\omega)$, compare (2.3.16).

Thus (2.3.17) is the consistent solution of (2.3.15), which obeys the relation (2.3.12). Formally, (2.3.17) may be found by writing (2.3.15) in the form $G(\omega) = (-\omega^2 + \omega_0^2)^{-1}$ and replacing ω by $\omega + i\varepsilon$. This shifts the poles from the real axis into the lower half of the ω-plane and so the inverse Fourier transformation is well defined.

We add a more rigorous way to obtain (2.3.17) by solving the corresponding homogeneous problem $g''(t) + \omega_0^2 g(t) = 0$ with $g(0) = 0$ and $g'(0) = 1$. In the ω-domain the equations transform to $(-\omega^2 + \omega_0^2)\, g(\omega) = 0$ and the initial conditions become

$$g(0) = \int \frac{d\omega}{2\pi}\, g(\omega) = 0 \,, \qquad g'(0) = \int \frac{d\omega}{2\pi}\, (-i\omega\, g(\omega)) = 1 \,. \tag{2.3.18}$$

The solution is $g(\omega) = A\delta(\omega + \omega_0) + B\delta(\omega - \omega_0)$, and the conditions (2.3.18) yield $A + B = 0$ and $(A - B)\, i\omega_0/2\pi = 1$, implying $A = -B = -i\pi/\omega_0$.

Thus we get

$$g(\omega) = -\frac{i\pi}{\omega_0}\, [\delta(\omega + \omega_0) - \delta(\omega - \omega_0)] \,.$$

Inserting this into (2.3.12) yields (2.3.17).

The physical background is the requirement of causality which in turn requires that the singularities of the Fourier transforms have to be located in lower half of the complex frequency plane. For systems with dissipation this is automatically fulfilled. For systems without dissipation there exist singularities on the real ω-axis, which have to be shifted into the lower half-plane by replacing ω by $\omega + i\varepsilon$, as it is prescribed by the spectral representation (2.3.12).

Finally we summarize some properties of the general problem (2.2.2). The Fourier transformation of (2.2.2) yields

$$D_m(\omega)\, G(\omega) = 1 \tag{2.3.19}$$

where $D_m(\omega)$ is a polynominal of m-th order

$$D_m(\omega) = \sum_{n=0}^{m} a_n(-i\omega)^n . \tag{2.3.20}$$

Since any polynomial of m-th order has m complex or real roots $\omega = \sigma_n$ there exists a product representation of the form

$$D_m(\omega) = \prod_{n=1}^{m} (\omega - \sigma_n) . \tag{2.3.21}$$

We ignore the cumbersome algebraic problems which may occur when dealing with equations of higher orders. The solution (2.3.19) reads

$$G(\omega) = \frac{1}{D_m(\omega)} . \tag{2.3.22}$$

In case of real roots σ_n we have to replace ω by $\omega + i\varepsilon$. Let us expand $G(\omega)$ in terms of partial fractions, i.e.,

$$G(\omega) = \sum_{n=1}^{m} \frac{A_n}{\omega - \sigma_n} \tag{2.3.23 a}$$

with

$$A_n = \left. \frac{1}{\dfrac{\partial}{\partial\omega} D_m(\omega)} \right|_{\omega=\sigma_n} \tag{2.3.23 b}$$

There may exist higher order roots of $D_m(\omega) = 0$, e.g., $\omega = \sigma_j$ with $D'(\sigma_j) = 0$. This would imply the occurrence of a term

$$G \sim \frac{A_j}{(\omega - \sigma_j)^2}$$

in (2.3.23). In case of a non-vanishing real part of σ_j the associated peak would be more pronounced than those of the simple poles. In the time domain this corresponds to a response increasing linearly with time, since

$$\int \frac{d\omega}{2\pi} A_j e^{-i\omega t} \frac{\partial}{\partial\omega} \frac{-1}{\omega - \sigma_j} = A_j t \int \frac{d\omega}{2\pi i} \frac{e^{-i\omega t}}{\omega - \sigma_j}$$

$$= -A_j t\theta(t) \exp(-it \operatorname{Re}\sigma_j - t \operatorname{Im}\sigma_j) \tag{2.3.24}$$

where $\operatorname{Im}\sigma_j < 0$ was assumed.

As example we note the damped harmonic oscillator with $\omega_0 = \beta/2$. Then from (2.3.10) it follows $G(\omega) = -1/(\omega + i\omega_0)^2$ and in the time domain we have, according to (2.3.24), $G(t) = t\theta(t)\,e^{-t\beta/2}$.

In summary, the dynamic features of the system under investigation may readily be inferred from the spectral representation of the Green's function. The system is completely described by the location of the poles of the Green's function in the lower half of the complex ω-plane. Every location of a pole is associated with two characteristic time scales: The imaginary part of σ_n describes the damping rate, i.e., the time scale after which the system forgets its initial state. The real part of σ_n determines the intrinsic frequency of the system.

2.4. GREEN'S FUNCTIONS FOR BOUNDARY VALUE PROBLEMS

2.4.1. Formulation of the problem

Let us assume, that after some manipulations, such as separation of variables and Fourier transforms with respect to some of the variables, the problem is reduced to a boundary value problem for one of the dynamical observables, say e.g. $u(x)$. Here $u(x)$ stands for one of the fields u, v, w or p.

$$\mathbf{L}u(x) = \frac{d}{dx} r(x) \frac{d}{dx} u(x) - Q(x)\,u(x) = \varphi(x) \tag{2.4.1}$$

where $r(x)$, $Q(x)$ and $\varphi(x)$ are specified known functions in the interval $a \leqq x \leqq b$. Let the boundary conditions be given by

$$A_1 u(a) + B_1 u_x(a) = C_1 \tag{2.4.2}$$

$$A_2 u(b) + B_2 u_x(b) = C_2\,. \tag{2.4.3}$$

In various practical cases the remaining observables may be expressed in terms of $u(x)$ and the forcing functions and therefore, a great part of the forcing problem is solved if the solution of (2.4.1) has been determined.

Let $\mathbf{L}$ be a self-adjoint second order differential operator and u and v arbitrary twice differentiable functions. Then Green's theorem applies

$$\int_a^b dx[v\,\mathbf{L}u - u\,\mathbf{L}v] = [r(vu_x - uv_x)]_a^b\,. \tag{2.4.4}$$

We note that any second order linear differential operator can be put in self-adjoint form by multiplication by a suitable factor (see e.g. COURANT and HILBERT 1968).

In order to solve (2.4.1) for a general source $\varphi(x)$ we introduce the Green's function $G(x, x')$ which is the solution for a point source at x'

$$\mathbf{L}G(x, x') = \delta(x - x')\,, \tag{2.4.5}$$

and which fulfils appropriate boundary conditions. In order to elaborate the rules for choice of convenient boundary conditions of $G(x, x')$ let us suppose the solution can be found; we show how to do this below.

Apply Green's theorem (2.4.4) to u and $v = G$

$$\int_a^b dx\,[G(x, x')\,\mathrm{L}u(x) - u(x)\,\mathrm{L}G(x, x')] = [r(x)\,(G(x, x')\,u_x - u(x)\,G_x(x, x'))]_{x=a}^{b}$$
$$= \int_a^b dx\,G(x, x')\,\varphi(x) - u(x') \tag{2.4.6}$$

where (2.4.1), (2.4.5), and the integral properties of delta functions were used. Interchanging x and x', from (2.4.6) we get

$$u(x) = \int_a^b dx'\,G(x', x)\,\varphi(x')$$
$$- \left[r(x')\left(G(x', x)\,\frac{d}{dx'}\,u(x') - u(x')\,\frac{d}{dx'}\,G(x', x)\right)\right]_{x'=a}^{b}. \tag{2.4.7}$$

We can now consider several possibilities for the boundary conditions on $u(x)$ at $x = a$ and $x = b$.

1. Let in (2.4.2) and (2.4.3) be $A_1 = A_2 = 1$, $B_1 = B_2 = 0$ and C_1, C_2 given.

$$u(a) = C_1\,, \qquad u(b) = C_2 \tag{2.4.8a}$$

for this case choose as boundary conditions on $G(x, x')$

$$G(a, x) = G(b, x) = 0\,. \tag{2.4.8b}$$

Obviously, this eliminates the unknown quantities $du\,(x')/dx'$ at $x' = a, b$.

Then (2.4.7) reduces to

$$u(x) = \int_a^b dx'\,G(x', x)\,\varphi(x') + r(b)\,C_2 G_{x'}(b, x) - r(a)\,C_1 G_{x'}(a, x)\,. \tag{2.4.8c}$$

This is the solution of (2.4.1) and (2.4.8a) in terms of known quantities.

2. Consider now $A_1 = B_2 = 1$, $A_2 = B_1 = 0$ and C_1, C_2 given. For this case choose as boundary conditions on $G(x, x')$

$$G(a, x) = G_{x'}(b, x) = 0\,. \tag{2.4.9a}$$

Then the unknown quantities $u(b)$ and $u_{x'}(a)$ are eliminated from the formula (2.4.7) and we obtain the solution in terms of known quantities

$$u(x) = \int_a^b dx'\,G(x', x)\,\varphi(x') - r(a)\,C_1 G_{x'}(a, x) - r(b)\,G(b, x)\,C_2\,. \tag{2.4.9b}$$

3. Let A_1, A_2, B_1, B_2 and C_1, C_2 be given. Then choose

$$A_1 G(a, x) + B_1 G_{x'}(a, x) = 0 \tag{2.4.10a}$$

$$A_2 G(b, x) + B_2 G_{x'}(b, x) = 0\,. \tag{2.4.10b}$$

Through these relations the derivatives of the Green's function in (2.4.7) can be expressed in terms of G and we have for the quantity needed in (2.4.7)

$$r(b)\,G(b,x)\left[u_{x'}(b)+u(b)\,\frac{A_2}{B_2}\right]-r(a)\,G(a,x)\left[u_{x'}(a)+u(a)\,\frac{A_1}{B_1}\right]$$

$$=G(b,x)\,r(b)\,\frac{C_2}{B_2}-G(a,x)\,r(a)\,\frac{C_1}{B_1}$$

which is a known quantity. The solution of (2.4.1) is in this case

$$u(x)=\int_a^b \mathrm{d}x'\,G(x',x)\,\varphi(x')+G(a,x)\,r(a)\,\frac{C_1}{B_1}-G(b,x)\,r(b)\,\frac{C_2}{B_2}\,. \tag{2.4.11}$$

In each case we can choose the boundary conditions on G so as to eliminate unknown values of u and u_x at the boundaries. In this way we obtain a solution of the inhomogeneous boundary value problem (2.4.1), (2.4.2) and (2.4.3) once we find the Green's function $G(x, x')$.

Next we show that the Green's function is symmetric in its two arguments. Applying the Green's theorem (2.4.4) to $u = G(x, x')$ and $v = G(x, x'')$ it follows

$$G(x',x'')-G(x'',x')=[r(x)\,(G(x,x'')\,G_x(x,x')-G(x,x')\,G_x(x,x''))]_{x=a}^{b} \tag{2.4.12}$$

For the general homogeneous boundary conditions (2.4.10) the right-hand side of (2.4.12) vanishes and we find

$$G(x',x'')=G(x'',x')\,. \tag{2.4.13}$$

Finally we note that in the case of homogeneous boundary conditions on u

$$A_1u(a)+B_1u_x(a)=0 \tag{2.4.14a}$$

$$A_2u(b)+B_2u_x(b)=0 \tag{2.4.14b}$$

the second term on the right hand side of (2.4.7) vanishes and we have

$$u(x)=\int_a^b G(x,x')\,\varphi(x')\,\mathrm{d}x\,. \tag{2.4.15}$$

Generally there exists a formulation of the problem (2.4.1) to (2.4.3) which is equivalent to that with the homogeneous conditions (2.4.14a, b), however, with a somewhat modified source $\varphi(x)$. It is easy to check that the following modified sources $\tilde{\varphi}(x)$ lead to equivalent solutions of the three cases discusses above:

Case 1

$$\tilde{\varphi}(x')=\varphi(x')+r(b)\,C_2\,\frac{\mathrm{d}}{\mathrm{d}x'}\,\delta(x'-b)-r(a)\,C_1\,\frac{\mathrm{d}}{\mathrm{d}x'}\partial(x'-a) \tag{2.4.16a}$$

Case 2

$$\tilde{\varphi}(x')=\varphi(x')+r(a)\,C_1\,\frac{\mathrm{d}}{\mathrm{d}x'}\,\delta(x'-a)-r(b)\,C_2\delta(x'-b) \tag{2.4.16b}$$

Case 3

$$\tilde{\varphi}(x')=\varphi(x')+r(a)\,\frac{C_1}{B_2}\,\delta(x'-a)-r(b)\,\frac{C_2}{B_2}\,\delta(x'-b)\,. \tag{2.4.16c}$$

Then a modified equation follows

$$\mathbf{L}u(x) = \tilde{\varphi}(x)\,. \tag{2.4.17}$$

With the homogeneous boundary conditions (2.4.14a, b) the equation (2.4.17) has the solutions (2.4.8c), (2.4.9b) or (2.4.11), respectively.

2.4.2. Calculations of Green's functions

Let us return to the differential equation for $G(x, x')$, i.e. (2.4.5), which reads explicitly

$$\frac{\mathrm{d}}{\mathrm{d}x}\left[r(x)\frac{\mathrm{d}}{\mathrm{d}x}G(x, x')\right] - Q(x)\,G(x, x') = \delta(x - x')\,. \tag{2.4.18}$$

For $x \neq x'$ this reduces to a homogeneous equation and so we can use whatever techniques are available for homogeneous equations in solving for $G(x, x')$ in the two regions $x > x'$ and $x < x'$. At $x = x'$ we must match the two solutions so obtained. Using the step function $\theta(x)$ ($\theta(x) = 1$ for $x > 0$, $\theta(x) = 0$ for $x < 0$ and $\theta(0) = 1/2$) we may try the ansatz

$$G(x, x') = \theta(x - x')\,G^{>}(x, x') + \theta(x' - x)\,G^{<}(x, x')\,. \tag{2.4.19}$$

The functions $G^{>}$ and $G^{<}$ are solutions of the homogeneous differential equation in the two regions $x > x'$ and $x < x'$. The first derivative of $G(x, x')$ contains a term proportional to $\delta(x - x')$

$$G_x(x, x') = \delta(x - x')\,(G^{>}(x, x') - G^{<}(x, x')) + \theta(x - x')\,G_x^{>}(x, x') + \theta(x' - x)\,G_x^{<}(x, x')\,.$$

Since at the right-hand side of (2.4.18) no term proportional to $\mathrm{d}\delta(x - x')/\mathrm{d}x$ appears, we have the condition

$$G^{>}(x', x') = G^{<}(x', x')\,. \tag{2.4.20}$$

The second derivative gives

$$G_{xx}(x, x') = \delta(x - x')\,(G_x^{>}(x, x') - G_x^{<}(x, x')) + \theta(x - x')\,G_{xx}^{>}(x, x') + \theta(x' - x)\,G_{xx}^{<}(x, x')\,. \tag{2.4.21}$$

This implies a second condition, namely

$$r(x')\,(G_x^{>}(x', x') - G_x^{<}(x', x')) = 1\,. \tag{2.4.22}$$

The homogeneous solutions $G^{>}$ and $G^{<}$ of (2.4.18) in the two regions $x > x'$ and $x < x'$ plus the two conditions (2.4.20), (2.4.22), plus the two boundary conditions at $x = a$ and $x = b$ discussed previously determine the Green's function completely. The four conditions fix the two pairs of integration constants obtained on solving the differential equation in the two regions.

Let $u^{(1)}(x)$ and $u^{(2)}(x)$ be two linear independent solutions of the homogeneous differential equation and may $u^{(1)}$ fulfil the boundary condition at $x = a$ and $u^{(2)}$ that at $x = b$. Then the Green's function may be constructed in terms of

$$G^{>}(x, x') = \frac{u^{(2)}(x)\, u^{(1)}(x')}{r(x')\, W(x')} \tag{2.4.23}$$

and

$$G^{<}(x, x') = \frac{u^{(2)}(x')\, u^{(1)}(x)}{r(x')\, W(x')} \tag{2.4.24}$$

where $W(x')$ is the Wronskian of $u^{(1)}$ and $u^{(2)}$

$$W(x) = u^{(1)} u_x^{(2)} - u_x^{(1)} u^{(2)}\,. \tag{2.4.25}$$

It appears that $r(x)\, W(x)$ is a constant. To prove this we write the differential equations for $u^{(1)}$ and $u^{(2)}$ explicitly

$$r u_{xx}^{(1)} + r_x u_x^{(1)} - Q u^{(1)} = 0$$
$$r u_{xx}^{(2)} + r_x u_x^{(2)} - Q u^{(2)} = 0\,.$$

Multiply the former by $u^{(2)}$ and the latter by $u^{(1)}$ and substract to obtain

$$\left[\frac{\mathrm{d}}{\mathrm{d}x} + \frac{r_x}{r}\right] W(x) = 0\,.$$

This has the solution

$$W(x) = W(x_0) \exp\left(-\int_{x_0}^{x} \frac{r_x}{r}\, \mathrm{d}x\right) = W(x_0)\, \frac{r(x_0)}{r(x)}\,,$$

where x_0 is an arbitrary point within the interval $a \leqq x \leqq b$. Thus, it follows

$$W(x)\, r(x) = W(x_0)\, r(x_0) = \text{const} \tag{2.4.26}$$

and, hence,

$$G^{>}(x, x') = \frac{u^{(2)}(x)\, u^{(1)}(x')}{W(x_0)\, r(x_0)} = G^{<}(x', x)\,. \tag{2.4.27}$$

Let us consider the following example

$$u_{xx} + \alpha^2 u = \varphi(x) \tag{2.4.28}$$

with the homogeneous boundary conditions $u(0) = u(L) = 0$. The Green's function is subject to the differential equation

$$G_{xx}(x, x') + \alpha^2 G(x, x') = \delta(x - x') \tag{2.4.29}$$

and to the boundary conditions, which have been chosen $G(0, x') = G(L, x') = 0$ in accordance with (2.4.8b). In the region $x > x'$ the homogeneous solution is $u^{(2)}(x) = \sin \alpha(x - L)$ and in the region $x < x'$ we have $u^{(1)}(x) = \sin \alpha x$. From (2.4.25) and (2.4.26) we get

$$\mathrm{W}_r(\alpha) = \alpha \cos \alpha x \sin \alpha(L - x) + \alpha \cos \alpha(L - x) \sin \alpha x = \alpha \sin \alpha L \tag{2.4.30}$$

and so we find explicitly

$$G(x, x') = \theta(x - x') \frac{\sin \alpha(x - L) \sin \alpha x'}{\alpha \sin \alpha L} + \theta(x' - x) \frac{\sin \alpha(x' - L) \sin \alpha x}{a \sin \alpha L}. \tag{2.4.31}$$

Using the formula (2.4.10) with $C_1 = C_2 = 0$ we find the solution of (2.4.28), i.e.,

$$u(x) = \int_0^L dx'\, G(x, x')\, \varphi(x') = \frac{\sin \alpha(L - x)}{\alpha \sin \alpha L} \int_0^x \varphi(x') \sin \alpha x'\, dx'$$

$$+ \frac{\sin \alpha x}{\alpha \sin \alpha L} \int_x^L \sin \alpha(L - x')\, \varphi(x')\, dx'. \tag{2.4.32}$$

2.4.3. Expansion of Green's functions in eigenfunctions

Next we wish to study an alternative method of estimation of Green's functions in terms of sums of eigenfunctions of the corresponding Sturm Liouville eigenvalue problem.

Putting in (2.4.1) $Q(x) = q(x) + \lambda\varrho(x)$ we have

$$\mathbf{L}' u_n(x) = \frac{d}{dx}\left[r(x) \frac{d}{dx} u_n(x)\right] - q(x)\, u_n(x) = -\lambda_n \varrho(x)\, u_n(x). \tag{2.4.33}$$

As mentioned in Section 1.9.1. a Sturm Liouville eigenvalue problem like (2.4.33) with the boundary conditions of the general type

$$A_1 u_n(a) + B_1 \frac{d}{dx} u_n(a) = 0 \tag{2.4.34a}$$

$$A_2 u_n(b) + B_2 \frac{d}{dx} u_n(b) = 0 \tag{2.4.34b}$$

has a complete set of orthonormal eigenfunctions $u_n(x)$ with

$$\int_a^b dx\, \varrho(x)\, u_n(x)\, u_m(x) = \delta_{nm} \tag{2.4.35}$$

where δ_{nm} is the Kronecker delta. To each $u_n(x)$ corresponds an eigenvalue λ_n of the operator $\mathbf{L}'$

$$\lambda_n = -\int_a^b u_n(x)\, \mathbf{L}' u_n(x)\, dx. \tag{2.4.36}$$

The completeness may be expressed in terms of the delta function

$$\sum_n \varrho(x)\, u_n(x)\, u_n(x') = \delta(x - x'). \tag{2.4.37}$$

An arbitrary function $\varphi(x)$ can be expanded in an eigenfunction series which converges absolutely and uniformly

$$\varphi(x) = \sum_n a_n u_n(x) \tag{2.4.38}$$

where

$$a_n = \int_a^b \mathrm{d}x\, \varrho(x)\, u_n(x)\, \varphi(x)\,. \tag{2.4.39}$$

Now let us return to our inhomogeneous boundary value problem (2.4.1) which in the notation of (2.4.33) is

$$\mathbf{L}'u + \lambda\varrho u = \varphi \tag{2.4.40}$$

with the homogeneous boundary conditions (2.4.14a, b). In the case of inhomogeneous boundary conditions we modify the source φ to obtain the equivalent problem with homogeneous conditions, as discussed at the end of Section 2.4.1. The corresponding equation of the Green's function is

$$\mathbf{L}'G(x, x') + \lambda\varrho(x)\, G(x, x') = \delta(x - x') \tag{2.4.41}$$

with the boundary conditions (2.4.10a, b).

Using (2.4.37) and expanding $G(x, x')$ in an eigenfunction series we find with (2.4.38), (2.4.39), and with

$$G_n(x') = \int \varrho(x)\, G(x, x')\, u_n(x)\, \mathrm{d}x$$

$$\begin{aligned}\sum_n (\mathbf{L}'G_n(x') + \lambda\varrho(x)\, G_n(x'))\, u_n(x) &= \sum_n (-\lambda_n + \lambda)\, \varrho(x)\, u_n(x)\, G_n(x') \\ &= \sum_n \varrho(x)\, u_n(x)\, u_n(x')\,.\end{aligned} \tag{2.4.42}$$

Multiply (2.4.42) by $u_m(x)$ and integrate over the interval a to b, and then use (2.4.35) to obtain

$$G_n(x') = \frac{u_n(x')}{\lambda - \lambda_n}$$

and, therefore,

$$G(x, x') = \sum_n \frac{u_n(x)\, u_n(x')}{\lambda - \lambda_n}\,. \tag{2.4.43}$$

It is easy to check that (2.4.43) satisfies the boundary conditions (2.4.10a, b) since the eigenfunctions $u_n(x)$ fulfil (2.4.34a, b). Obviously, the symmetry property (2.4.13) applies to (2.4.43). Since the representation (2.4.43) involves an infinite series, a closed expression of $G(x, x')$ may generally be more useful if available.

Let us consider again our illustrative example (2.4.28), (2.4.29). The corresponding Sturm Liouville problem reads

$$\frac{\mathrm{d}^2}{\mathrm{d}x^2} u_n(x) + \varkappa_n^2 u(x) = 0$$

with $u_n(0) = u_n(L) = 0$. The normalized eigenfunctions are

$$u_n(x) = (2/L)^{1/2} \sin \varkappa_n x\,,$$

and the eigenvalues are $\varkappa_n = n\pi/L$. Thus from (2.4.43) it follows

$$G(x, x') = \frac{2}{L} \sum_n \frac{\sin(\varkappa_n x)\, \sin(\varkappa_n x')}{\alpha^2 - \varkappa_n^2}\,. \tag{2.4.44}$$

The solution (2.4.32) of the inhomogeneous problem (2.4.28) may written as

$$u(x) = \sum_{n=1}^{\infty} \frac{\sin \varkappa_n x}{\alpha_2 - \varkappa_n^2} \varphi_n$$

where $\varphi_n = \frac{2}{L} \int_0^L \varphi(x') \sin \varkappa_n x' \, dx'$ determines the coefficients of an odd Fourier series of the source $\varphi(x)$.

Finally we mention that the eigenvalues $\alpha = \varkappa_n$ being the zeros of the Wronskian (2.4.30) are given by

$$W_r(\varkappa_n) = \varkappa_n \sin \varkappa_n L = \varkappa_n \sin n\pi = 0 .$$

Hence we could obtain the expression (2.4.44) also by partial fraction expansion of (2.4.31) with respect to the zeros of the Wronskian.

CHAPTER 3

Forcing of an unbounded ocean on the f-plane

3.1. INERTIAL OSCILLATIONS

Waves or oscillations having the frequency f usually are called inertial waves or oscillations. This motion was first observed explicitly by GUSTAFSON and OTTERSTEDT (1932) in the Baltic, although some earlier indications of their existence have been reported, e.g. NANSEN (1902).

Inertial waves are known to be a common feature in the ocean and in lakes. Most spectra of ocean currents show a predominant peak at the inertial frequency f. Typical velocities are few centimeter per second, but after a strong forcing event velocities up to 70 centimeter per second has been observed, see KRAUSS (1979).

First we consider a trivial example which usually has been discussed in textbooks in order to illustrate the phenomenon. Let an unbounded ocean at rest suddenly be forced by a spatially constant delta-pulse, e.g., $X = u_0\,\delta(t)$, $Y = 0$, with $u_0 = \text{const}$. Then the equations of motion reduce to

$$u_t = fv = u_0\delta(t) \tag{3.1.1}$$

$$v_t + fu = 0\,. \tag{3.1.2}$$

The solutions are readily obtained

$$u = \theta(t)\,u_0 \cos ft \tag{3.1.3}$$

$$v = -\theta(t)\,u_0 \sin ft\,. \tag{3.1.4}$$

The hodograph rotates in clockwise direction with the angular frequency f. This rotation is due to the Coriolis force which acts perpendicular to the direction of motion. The water particles move on circles governed by

$$(x - x_0)^2 + (y - y_0)^2 = \frac{u_0^2}{f^2}\,. \tag{3.1.5}$$

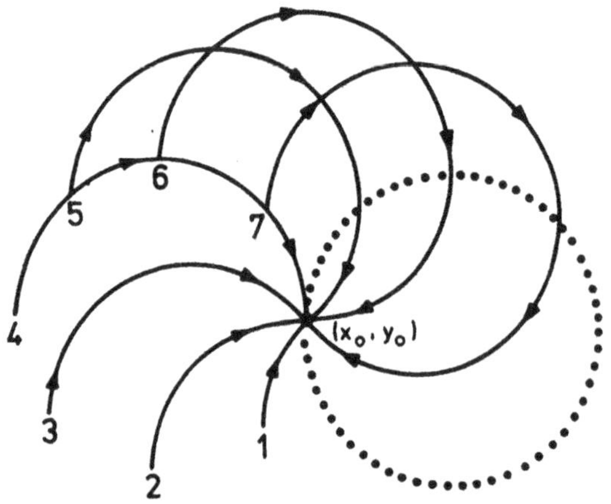

Fig. 3.1.1
Progressive vector diagram of a pure inertial oscillation
In this special case the progressive vector diagram coincides with the Lagrangian trajectory. Shown is the trajectory of a particle starting from (x_0, y_0) at $t = 0$ (dotted line). Moreover, the trajectories of some particles are drawn which start from different positions and which subsequently reach the point (x_0, y_0) where they produce a clockwise rotating Eulerian current.

The radius depends only on the strength of the pulse divided by the inertial frequency. In this trivial example the progressive vector diagram of the Eulerian current measured at any location (x_0, y_0) is identical with the Lagrangian trajectory. In Fig. 3.1.1 the trajectory of a water particle which starts from (x_0, y_0) at $t = 0$ is shown by a dotted line. Moreover, the trajectories of an ensemble of enumerated water particles have been depicted. The initial position of each particle has been chosen such that the particles subsequently pass the point (x_0, y_0) where they produce a clockwise rotating Eulerian current.

3.2. GOVERNING EQUATIONS

In the following sections we investigate the response of an unbounded ocean on a f-plane to external forces on the basis of the linearized Boussinesq equations in the hydrostatic approximation. This approximation neglects the time derivative in the momentum balance of the vertical current and is appropriate to describe the oceanic response in the inertial and subinertial frequency range. The hydrostatic approximation amounts to filtering out high frequency motion with frequencies close to the Brunt Väisälä Frequency (BVF) $N(z)$, usually being large compared to f. This can be seen explicitly in the Section 4.12., where the constraint of the hydrostatic approximation will be removed in order to investigate internal waves.

Thus our basic set is

$$u_t - fv + p_x = X \tag{3.2.1}$$

$$v_t + fu + p_y = Y \tag{3.2.2}$$

$$-b + p_z = 0 \tag{3.2.3}$$

$$b_t + N^2 w = 0 \tag{3.2.4}$$

$$u_x + v_y + w_z = 0\,. \tag{3.2.5}$$

Since we consider a laterally unbounded ocean the only boundary conditions are in the vertical axis

$$w = \frac{1}{g} p_t \qquad \text{at } z = 0 \tag{3.2.6}$$

$$w = 0 \qquad \text{at } z = -H \tag{3.2.7}$$

where the bottom has assumed to be flat. From (3.2.3) and (3.2.4) it follows

$$w = \frac{-1}{N^2} p_{tz} \,. \tag{3.2.8}$$

After differentiation of (3.2.1) and (3.2.2) with respect to x and to y we may combine the resulting equations to

$$\frac{\partial^2}{\partial t^2}(u_x + v_y) + f^2(u_x + v_y) = -\Delta p_t + \frac{\partial}{\partial t}\operatorname{div}(X, Y) + f \operatorname{curl}_z (X, Y) \tag{3.2.9}$$

where Δ is the horizontal Laplacian $\Delta = \frac{\partial^2}{\partial x^2} + \frac{\partial^2}{\partial y^2}$. With (3.2.5) and (3.2.8) we obtain from (3.2.9)

$$\left(\frac{\partial^2}{\partial t^2} + f^2\right) w_{zz} + N^2 \Delta w = -\frac{\partial}{\partial z}\left(\frac{\partial}{\partial t}\operatorname{div}(X, Y) + f \operatorname{curl}_z (X, Y)\right). \tag{3.2.10}$$

Thus the vertical motion will be forced by the vertical component of the curl of the wind, and by the temporal variations of the wind divergence. This implies that the vertical velocity will not be affected by a spatially constant wind.

Introducing the abbreviation

$$\mathbf{Z} = \frac{\partial}{\partial z} \frac{1}{N^2} \frac{\partial}{\partial z} \tag{3.2.11}$$

and utilizing (3.2.8) we may write

$$w_z = -\mathbf{Z} p_t \,. \tag{3.2.12}$$

With (3.2.12) we find the following equation for the pressure

$$\left(\frac{\partial^2}{\partial t^2} + f^2\right) \mathbf{Z} p_t + \Delta p_t = \frac{\partial}{\partial t}\operatorname{div}(X, Y) + f \operatorname{curl}_z (X, Y) \tag{3.2.13}$$

The boundary conditions on the pressure follows from (3.2.6) to (3.2.8)

$$p_z = -\frac{N^2}{g} p \qquad \text{for } z = 0 \tag{3.2.14}$$

$$p_z = 0 \qquad \text{for } z = -H \,. \tag{3.2.15}$$

The velocities u and v can be expressed in terms of p and of the external forces X and Y by means of the following differential equations, which are derived from (3.2.1) and (3.2.2)

$$u_{tt} + f^2 u = -f p_y - p_{xt} + X_t + f Y \tag{3.2.16}$$

$$v_{tt} + f^2 v = f p_x - p_{yt} + Y_t - f X \,. \tag{3.2.17}$$

Thus, if once p has been estimated we may use (3.2.16), (3.2.17), and (3.2.8) to calculate the velocities u, v, and w, respectively.

Next we estimate the Fourier transforms of (3.2.13), (3.2.16), and (3.2.17) according to

$$F(\omega, k, \varkappa, z) = \int_{-\infty}^{+\infty} dt\, dx\, dy\, e^{i\omega t - ikx - i\varkappa y}\, F(x, y, z, t) \tag{3.2.18}$$

$$F(x, y, z, t) = \int_{-\infty}^{+\infty} \frac{d\omega}{2\pi} \frac{dk}{2\pi} \frac{d\varkappa}{2\pi}\, e^{-i\omega t + ikx + i\varkappa y}\, F(\omega, k, \varkappa, z) \tag{3.2.19}$$

and we get

$$\mathbf{Z}p + \frac{k^2 + \varkappa^2}{\omega^2 - f^2} p = \frac{-1}{\omega^2 - f^2}\left[\left(ik + \frac{\varkappa f}{\omega}\right) X + \left(i\varkappa - \frac{kf}{\omega}\right) Y\right] \tag{3.2.20a}$$

$$u = \frac{1}{\omega^2 - f^2}\left[(i\varkappa f + \omega k)\, p + i\omega X - fY\right] \tag{3.2.20b}$$

$$v = \frac{1}{\omega^2 - f^2}\left[(\omega\varkappa - ikf)\, p + fX + i\omega Y\right]. \tag{3.2.20c}$$

Moreover, from (3.2.8) it follows

$$w = \frac{i\omega}{N^2} p_z\,. \tag{3.2.20d}$$

Later we introduce a Green's function, which is governed by

$$\left(\mathbf{Z} + \frac{k^2 + \varkappa^2}{\omega^2 - f^2}\right) K(z, z') = \delta(z - z') \tag{3.2.21}$$

with the boundary conditions

$$K_z(0, z') = -\frac{N_2}{g} K(0, z') \tag{3.2.22a}$$

$$K_z(-H, z') = 0\,. \tag{3.2.22b}$$

If the Green's function exists the solution describing the oceanic response can be expressed in terms of $K(z, z')$ and of the external forces X and Y, i.e.,

$$p = \frac{-1}{\omega^2 - f^2}\left[\left(ik + \frac{f\varkappa}{\omega}\right) K * X + \left(i\varkappa - \frac{fk}{\omega}\right) K * Y\right] \tag{3.2.23a}$$

$$u = \frac{1}{\omega^2 - f^2}(i\omega X - fY) + \frac{1}{(\omega^2 - f^2)^2} \times \left\{-\frac{i}{\omega}(\varkappa^2 f^2 + \omega^2 k^2)\, K * X + \left[\frac{k\varkappa}{\omega}(\omega^2 - f^2) + f^2(k^2 + \varkappa^2)\right] K * Y\right\} \tag{3.2.23b}$$

$$v = \frac{1}{\omega^2 - f^2}(fX + i\omega Y) - \frac{1}{(\omega^2 - f^2)^2} \times \left\{\left[(\omega^2 - f^2)\frac{ik\varkappa}{\omega} + f(k^2 + \varkappa^2)\right] K * X + \frac{i}{\omega}(\omega^2\varkappa^2 + f^2k^2)\, K * Y\right\}. \tag{3.2.23c}$$

Here the convolutions $K * X$ and $K * Y$ have been introduced which are defined by

$$K * X = \int_{-H}^{0} dz' \, K(z, z') \, X(z') \tag{3.2.24}$$

and

$$K * Y = \int_{-H}^{0} dz' \, K(z, z') \, Y(z') \, . \tag{3.2.25}$$

Thus if once the Green's function is known we may estimate the convolutions, and then the complete solution is given in the Fourier space in terms of the set (3.2.23).

The set (3.2.23) describes the solution for an ocean which was at rest before the external forces start to act. Otherwise the initial condition must be included in the manner outlined in Section 2.2. However, we will consider throughout the book an ocean which was initially at rest.

3.3. ESTIMATION OF THE GREEN'S FUNCTION

In the previous section the general solution of the forcing problem has been expressed in terms of the Green's function. Now we are left with the task to estimate the Green's function which is governed by (3.2.21) and (3.2.22a, b), i.e.,

$$\left(\frac{d}{dz}\frac{1}{N^2}\frac{d}{dz} + \lambda^2\right) K(z, z') = \delta(z - z')$$

$$K_z(0, z') = -\frac{N^2}{g} K(0, z')$$

$$K_z(-H, z') = 0 \, .$$

Here the notation $\lambda^2 = (k^2 + \varkappa^2)/(\omega^2 - f^2)$ has been introduced.

First we consider the case of constant BVF, i.e., $N^2 = \text{const}$. Then we deal with an ordinary second order differential equation with constant coefficients

$$\left(\frac{1}{N^2}\frac{d^2}{dz^2} + \lambda^2\right) K(z, z') = \delta(z - z') \, . \tag{3.3.1}$$

The two linear independent solutions of the homogeneous version of (3.3.1) are $\cos(\lambda N z) - [N^2/(\lambda g)] \sin(\lambda N z)$ and $\cos \lambda N(z + H)$. The former fulfil the boundary condition at the surface, and the latter that at the bottom. According to (2.4.26) the Wronskian is independent of z

$$W_r(z) = W_r(0) = N\left(\lambda \sin \lambda N H - \frac{N}{g} \cos \lambda N H\right) . \tag{3.3.2}$$

Consider the Wronskian to be a function of λ, and recognize that the zeros of $W_r(\lambda)$ are just the vertical eigenvalues of the vertical eigenvalue problem discussed in Subsection 1.9.2.

$$W_r(\lambda_n) = 0 \tag{3.3.3}$$

with the λ_n's being the enumerated roots of

$$\tan \lambda_n NH = \frac{N}{g\lambda_n}, \qquad n = 0, 1, 2, 3, \dots . \tag{3.3.4}$$

Now it is easy to construct the functions $K^>$ and $K^<$, compare Section 2.4, and we obtain

$$K^<(z, z') = \frac{N^2}{W_r}\left(\cos \lambda Nz - \frac{N}{g\lambda}\sin \lambda Nz\right)\cos \lambda N(z' + H) = K^>(z', z). \tag{3.3.5}$$

The Green's function is composed of $K^<$ and $K^>$, i.e.,

$$K(z, z') = \theta(z - z')\, K^<(z, z') + \theta(z' - z)\, K^>(z, z'). \tag{3.3.6}$$

Note that z varies along the negative z-axis, hence, $K^>$ and $K^<$ are interchanged compared to (2.4.19).

Since the equations (3.2.21), (3.2.22) correspond to those which govern p with the right-hand side replaced by the delta function, compare (3.2.14), (3.2.15), and (3.2.20 a), we may denote $K(z, z')$ as the Green's function of the pressure.

An alternative expression of the Green's function may be given in terms of the normalized vertical eigenfunctions

$$F_n(z) = \sqrt{2}\, \frac{\cos \lambda_n(z + H)\, N}{(1 + j_0(2\lambda_n NH))^{1/2}} \tag{3.3.7}$$

with $j_0(x) = \dfrac{\sin x}{x}$. Using the completeness of the $F_n(z)$'s

$$\delta(z - z') = \frac{1}{H}\sum_{n=0}^{\infty} F_n(z)\, F_n(z') \tag{3.3.8}$$

it follows

$$K(z, z') = \frac{1}{H}\sum_{n=0}^{\infty} \frac{F_n(z)\, F_n(z')}{\lambda^2 - \lambda_n^2} \tag{3.3.9}$$

as it was outlined in Subsection 2.4.3.

The expression (3.3.9) can also be obtained directly from (3.3.5) and (3.3.6) by means of a partial fraction expansion with respect to the zeros of the Wronskian. To this end we evaluate the Taylor series of $W_r(\lambda)$ near the λ_n's and take (3.3.3) into account, i.e.,

$$W_r(\pm\lambda) = W_r'(\pm\lambda_n)(\lambda \pm \lambda_n) + \dots \tag{3.3.10}$$

with

$$W_r'(\lambda_n) = \frac{\lambda_n N^2 H}{\cos \lambda_n NH}\,[1 + j_0(2\lambda_n NH)]. \tag{3.3.11}$$

The prime denotes the λ-derivative. Obviously, $W_r'(\lambda_n)$ is an odd function of λ_n. Moreover, we observe that the following combinations occur in the partial fraction expansion of $K^<$ and $K^>$

$$\frac{1}{W_r'(\lambda_n)(\lambda - \lambda_n)} + \frac{1}{W_r'(-\lambda_n)(\lambda + \lambda_n)} = \frac{2\lambda_n}{W_r'(\lambda_n)(\lambda^2 - \lambda_n^2)}.$$

Hence, we obtain

$$K^<(z, z') = \sum_{n=0}^{\infty} \frac{N^2 \cdot 2\lambda_n}{W_r'(\lambda_n)(\lambda^2 - \lambda_n^2)} \cos \lambda_n N(z' + H) \left(\cos \lambda_n N z - \frac{N^2}{g\lambda_n} \sin \lambda_n N z \right)$$

Using (3.3.4) we get

$$K^<(z, z') = \sum_{n=0}^{\infty} \frac{N^2 \cdot 2\lambda_n}{W_r'(\lambda_n)(\lambda^2 - \lambda_n^2)} \cos \lambda_n N(z' + H) \frac{\cos \lambda_n N(z + H)}{\cos(\lambda_n N H)} = K^>(z', z)$$

Inserting $W_r'(\lambda_n)$ according to (3.3.11) we arrive at

$$K(z, z') = \sum_{n=0}^{\infty} \frac{2 \cos \lambda_n N(z + H) \cos \lambda_n N(z' + H)}{H[1 + j_0(2\lambda_n N H)](\lambda^2 - \lambda_n^2)} \cdot \tag{3.3.12}$$

Notably, this procedure yields the exact normalization constant of the $F_n(z)$'s, compare (3.3.7).

For practical estimations we may approximately use $\lambda_0^2 = 1/(gH)$, $\lambda_n = n\pi/(NH)$, and $F_0 = 1$, $F_n(z) = \sqrt{2}\cos(n\pi z/H)$, $n = 1, 2, 3, \ldots$ as outlined in Subsection 1.9.2. However, caution must be exercised if the approximation of a constant barotropic mode is used, because then z derivatives of p, and consequently of K, are to be calculated. In these cases the full $F_0(z)$ or reasonable approximations including some orders of z must be taken into account, see (1.9.27).

Inserting $\lambda^2 = (k^2 + \varkappa^2)/(\omega^2 - f^2)$ in (3.3.9) it follows

$$K(z, z') = (\omega^2 - f^2) \frac{1}{H} \sum_{n=0}^{\infty} \frac{F_n(z)\, F_n(z')}{(\omega_n^2(\boldsymbol{k}) - \omega^2)\, \lambda_n^2} \tag{3.3.13}$$

with

$$\omega_n^2(\boldsymbol{k}) = f^2 + \frac{\boldsymbol{k}^2}{\lambda_n^2} \cdot \tag{3.3.14}$$

Here the notation $\boldsymbol{k}^2 = k^2 + \varkappa^2$ has been introduced.

The equation (3.3.14) defines the dispersion relation of free rotational waves in an unbounded ocean on the f-plane. We will follow GILL (1982), who denotes this type of wave "Poincaré waves" or "inertial waves" though they are sometimes referred to as Sverdrup waves. The term Poincaré waves is often confined to waves in the coastal ocean or in channels. In the latter case one of the wave numbers becomes discrete as we shall see below in Section 6.2.

3.4. PHYSICAL IMPLICATIONS OF THE ANALYTICAL PROPERTIES OF THE GREEN'S FUNCTION

In this section we discuss the analytical properties of the Green's function given by (3.3.5), (3.3.6) or, equivalently, by (3.3.13) in the complex ω-plane. First, it can be seen from (3.3.13) that the Green's function has poles at $\omega = \omega_n(\boldsymbol{k})$, see (3.3.14), i.e. at those frequencies that intersect the dispersion curves of the free Poincaré waves in the unbounded f-plane ocean. The corresponding frequencies are bounded by $\omega \geqq f$ because of $\boldsymbol{k}^2/\lambda_n^2 \geqq 0$.

Moreover, from (3.3.5) it follows that in this frequency range the Green's function shows an oscillatory behaviour in the vertical. However, in the subinertial range, i.e. $\omega < f$, the Green's function turns out to be a smooth function of ω. In fact, putting $\lambda = \mathrm{i}|\lambda|$, and using the identities $\sin \mathrm{i}x = \mathrm{i} \sinh x$ and $\cos \mathrm{i}x = \cosh x$ it follows

$$W_r(\mathrm{i}\,|\lambda|) = -N\left[\sinh(|\lambda|\,NH) + \frac{N^2}{g}\cosh(|\lambda|\,NH)\right]. \tag{3.4.1}$$

Hence, the Wronskian has no zeros. The Green's function appears to be localized around $z = z'$, as it can be seen from

$$K^<(z,z') = \frac{N^2}{W_r}\left[\cosh(|\lambda|\,Nz) - \frac{N^2}{g}\sinh(|\lambda|\,Nz)\right]\cosh(|\lambda|\,N(z'+H)) = K^>(z',z). \tag{3.4.2}$$

Summarily, the analytical properties of the Green's function exhibit vertically oscillating and localized response patterns in the inertial and subinertial frequency domain, respectively. This indicates the important part of Poincaré waves for the vertical propagation of an oceanic response to surface forcing. Only waves with frequencies exceeding the inertial can carry information downward. The subinertial portion of the response appears to be trapped in the upper part of the ocean.

Next we consider the complete response of the dynamical fields u, v, and p. Inserting (3.3.13) into the set (3.2.23) we obtain

$$p(\omega, \mathbf{k}, z) = \sum_{n=0}^{\infty} F_n(z)\left[\left(\mathrm{i}k + \frac{f\varkappa}{\omega}\right)\frac{F_n * X}{\lambda_n^2(\omega_n^2 - \omega^2)} + \left(\mathrm{i}\varkappa - \frac{kf}{\omega}\right)F_n * Y\right] \tag{3.4.3 a}$$

$$u(\omega, \mathbf{k}, z) = \frac{\mathrm{i}\omega X - fY}{\omega^2 - f^2} + \sum_{n=0}^{\infty} F_n(z)\left[\frac{\mathrm{i}(\varkappa^2 f^2 + \omega^2 k^2)}{\omega(\omega^2 - f^2)}F_n * X + \left(\frac{\mathrm{i}k\varkappa}{\omega} - \frac{fk^2}{\omega^2 - f^2}\right)F_n * Y\right]\frac{1}{\lambda_n^2(\omega_n^2 - \omega^2)} \tag{3.4.3 b}$$

$$v(\omega, \mathbf{k}, z) = \frac{fX + \mathrm{i}\omega Y}{\omega^2 - f^2} + \sum_{n=0}^{\infty} F_n(z)\left[\left(\frac{\mathrm{i}k\varkappa}{\omega} + \frac{fk^2}{\omega^2 - f^2}\right)F_n * X + \frac{\mathrm{i}(\varkappa^2\omega^2 + f^2k^2)}{\omega(\omega^2 - f^2)}F_n * Y\right]\frac{1}{\lambda_n^2(\omega_n^2 - \omega^2)} \tag{3.4.3 c}$$

where the abbreviations $F_n * X$ and $F_n * Y$ denote the convolutions according to (3.2.24) and (3.2.25). Evidently, these expressions are just the coefficients of the expansions of X and Y in terms of the vertical eigenfunctions $F_n(z)$, i.e.,

$$F_n * X = \frac{1}{H}\int_H \mathrm{d}z'\, F_n(z')\, X(z') = X_n$$

$$F_n * Y = \frac{1}{H}\int_H \mathrm{d}z\, F_n(z')\, Y(z) = Y_n\,.$$

From the set (3.4.3) it follows that the spectral respresentations of the pressure and of the horizontal velocities have in addition to the poles of the Green's function further poles at $\omega = 0$ and $\omega^2 = f^2$. The former correspond to steady motion while the latter are due to inertial oscillations. Furthermore there may exist poles of the Fourier transforms of the external forcing fields. Thus the set (3.4.3) shows a catalogue of possible response pattern. However, it depends on the special structure of assumed forcing which of the possibilities will be realized.

We note the lack of those poles in (3.4.3) which can be attributed to Kelvin waves. This implies that within an unbounded ocean on a f-plane information of response patterns cannot be communicated by means of Kelvin waves. Thus, the response of such an ocean differs essentially from those of, e.g., the coastal ocean or an ocean on the equatorial β-plane, where Kelvin waves are possible as described in the Chapters 5 and 8.

Introducing the normal mode expansions of u, v, p, X, and Y we may rewrite (3.4.3) to find the modal components of the spectral functions of u, v, and p explicitly

$$p_n(\omega, \boldsymbol{k}) = -\frac{(\mathrm{i}k\omega + f\varkappa)}{\lambda_n^2\omega(\omega_n^2 - \omega^2)} X_n(\omega, \boldsymbol{k}) - \frac{(\mathrm{i}\varkappa\omega - fk)}{\lambda_n^2\omega(\omega_n^2 - \omega^2)} Y_n(\omega, \boldsymbol{k}) \tag{3.4.4 a}$$

$$u_n(\omega, \boldsymbol{k}) = \frac{\mathrm{i}(\varkappa^2 - \omega^2\lambda_n^2)}{\lambda_n^2\omega(\omega_n^2 - \omega^2)} X_n(\omega, \boldsymbol{k}) + \frac{(f\omega\lambda_n^2 + \mathrm{i}k\varkappa)}{\lambda_n^2\omega(\omega_n^2 - \omega^2)} Y_n(\omega, \boldsymbol{k}) \tag{3.4.4 b}$$

$$v_n(\omega, \boldsymbol{k}) = \frac{(f\omega\lambda_n^2 + \mathrm{i}k\varkappa)}{\lambda_n^2\omega(\omega_n^2 - \omega^2)} X_n(\omega, \boldsymbol{k}) + \frac{\mathrm{i}(k^2 - \omega^2\lambda_n^2)}{\lambda_n^2\omega(\omega_n^2 - \omega^2)} Y_n(\omega, \boldsymbol{k}) . \tag{3.4.4 c}$$

This form turns out to be convenient for practical estimations.

This result can be obtained directly from the basic set (3.2.1), (3.2.2), and (3.2.5). According to (3.2.15) we may replace w_z by p_t, and obtain after Fourier transformation with respect to x, y, and t, and after expansion of all observables in terms of the vertical eigenfunctions the algebraic set

$$\begin{pmatrix} -\mathrm{i}\omega & -f & \mathrm{i}k \\ f & -\mathrm{i}\omega & \mathrm{i}\varkappa \\ \mathrm{i}k & \mathrm{i}\varkappa & -\mathrm{i}\omega\lambda_n^2 \end{pmatrix} \begin{pmatrix} u_n \\ v_n \\ p_n \end{pmatrix} = \begin{pmatrix} X_n \\ Y_n \\ 0 \end{pmatrix} .$$

The determinant of the matrix is

$$\mathrm{i}\omega\lambda_n^2(\omega^2 - \omega_n^2(\boldsymbol{k})) .$$

The solution of this algebraic set with the aid of Cramer's rule yields (3.4.4).

3.5. GREEN'S FUNCTION FOR VARIABLE BVF, WKB TREATMENT

In this and in the following section the modifications caused by variable BVF's are briefly studied. We confine this discussion to a rigid lid boundary condition, i.e.,

$$w = 0\,, \qquad p_z = 0 \quad \text{for} \quad z = 0\,. \tag{3.5.1}$$

It turns out to be convenient to discuss the Green's function associated with the vertical velocity instead of the pressure Green's function. In fact, if we rewrite (3.2.13)

$$\left(\frac{\mathrm{d}}{\mathrm{d}z}\frac{1}{N^2}\frac{\mathrm{d}}{\mathrm{d}z} + \lambda^2\right) p = \text{forcing} \tag{3.5.2}$$

the corresponding equation for w reads, see (3.2.10),

$$\left(\frac{\mathrm{d}^2}{\mathrm{d}z^2} + \lambda^2 N^2\right) w = \frac{\partial}{\partial z}\,(\text{forcing})\,. \tag{3.5.3}$$

Since we consider volume forces evenly distributed in the upper layer of thickness H_{mix} the forcing is proportional to $\theta(z + H_{\text{mix}})$. Hence, the right-hand side of (3.5.3) has with

$$\frac{\partial}{\partial z}\,\theta(z + H_{\text{mix}}) = \delta(z + H_{\text{mix}})$$

automatically the structure of a Green's function equation. If w is estimated we may use the set (4.3.9) which will be derived in Section 4.3. of the following chapter to determine the complete response to any volume forcing.

Let us consider

$$\left(\frac{\mathrm{d}^2}{\mathrm{d}z^2} + \lambda^2 N^2\right) \tilde{K}(z, z') = \delta(z - z') \tag{3.5.4a}$$

with

$$\tilde{K}(0, z') = \tilde{K}(-H, z') = 0 \tag{3.5.4b}$$

where $\tilde{K}$ is the Green's function associated with w.

First we estimate $\tilde{K}$ for a slowly varying $N(z)$. Then the WKB method applies and we have the two linearly independent solutions

$$\varphi(z) = \frac{1}{N^{1/2}}\sin\left[\lambda\int_z^0 N(z')\,\mathrm{d}z'\right], \qquad \psi(z) = \frac{1}{N^{1/2}}\sin\left[\lambda\int_{-H}^{z} N(z')\,\mathrm{d}z'\right] \tag{3.5.5}$$

with the Wronskian

$$W_r = \sin(\lambda \bar{N} H) \tag{3.5.6}$$

where

$$\bar{N} = \frac{1}{H}\int_{-H}^{0} N(z')\,\mathrm{d}z' \tag{3.5.7}$$

is the mean value of the BVF. Since the zeros of the Wronskian determine the eigenvalues λ_n we find in accordance with (1.9.46)

$$\lambda^2 = \frac{k^2}{\omega^2 - f^2} = \lambda_n^2 = \left(\frac{n\pi}{N\bar{H}}\right)^2 .$$

With the aid of (3.5.5) and (3.5.6) the Green's function is readily constructed

$$\tilde{K}^<(z, z') = \frac{\varphi(z)\,\psi(z')}{W_r} = \tilde{K}^>(z', z) . \tag{3.5.8}$$

3.6. GREEN'S FUNCTION FOR AN EXPONENTIAL BVF

Our next example is the model of an exponential BVF

$$N(z) = N_1\theta(z + H_{mix}) + N_2\,e^{a(z+H_{mix})}\,\theta(-z - H_{mix}) . \tag{3.6.1}$$

We wish to construct the Green's function subject to the equations (3.5.4a) and (3.5.4b). As outlined in the foregoing section we may put $z' = -H_{mix}$. Then we have the two linearly independent solutions, compare Subsection 1.9.4.,

$$\varphi(z) = \sin \lambda N_1 z \quad \text{for} \quad 0 \geqq z \geqq -H_{mix} \tag{3.6.2}$$

and

$$\psi(z) = J_0\left(\frac{\lambda N(z)}{a}\right) - \frac{J_0\left(\frac{\lambda N(-H)}{a}\right)}{Y_0\left(\frac{\lambda N(-H)}{a}\right)}\,Y_0\left(\frac{\lambda N(z)}{a}\right), \qquad -H_{mix} \geqq z \geqq -H \tag{3.6.3}$$

The Wronskian follows to be

$$\begin{aligned} W_r(-H_{mix}) = {} & \lambda N_1 \cos(\lambda N_1 H_{mix}) \left[J_0\left(\frac{\lambda N(-H_{mix})}{a}\right) - \frac{J_0\left(\frac{\lambda N(-H)}{a}\right)}{Y_0\left(\frac{\lambda N(-H)}{a}\right)}\,Y_0\left(\frac{\lambda N(-H_{mix})}{a}\right)\right] \\ & - \lambda N(-H_{mix}) \sin(\lambda N_1 H_{mix}) \left[J_1\left(\frac{\lambda N(-H_{mix})}{a}\right) - \frac{J_0\left(\frac{\lambda N(-H)}{a}\right)}{Y_0\left(\frac{\lambda N(-H)}{a}\right)}\,Y_1\left(\frac{\lambda N(-H_{mix})}{a}\right)\right]. \end{aligned} \tag{3.6.4}$$

We observe that the zeros of the Wronskian yield the eigenvalue condition (1.9.69), which has been discussed in Subsection 1.9.4. With the aid of (3.6.2), (3.6.3), and (3.6.4) we may construct the Green's function

$$\tilde{K}^<(z, -H_{mix}) = \frac{\varphi(z)\,\psi(-H_{mix})}{W_r} = \tilde{K}^>(-H_{mix}, z) . \tag{3.6.5}$$

Clearly, an equivalent expression of this Green's function can be obtained by means of the set of eigenfunctions $Z_n(z)$ given in Subsection 1.9.4. Then we find

$$K(z, -H_{\text{mix}}) = \sum_{n=1}^{\infty} \frac{Z_n(z)\, Z_n(-H_{\text{mix}})}{\lambda^2 - \lambda_n^2} . \tag{3.6.6}$$

3.7. LARGE SCALE FORCING

In remaining sections of this chapter we will apply the general expressions derived in the previous sections to examples of external forcing. In the following sections we consider some reasonably chosen wind forces, whereby we proceed from simple spatially constant winds to subsequently more complicated structures.

We start with a pulse-like large scale forcing

$$X = \frac{u_*^2}{H_{\text{mix}}}\, \delta(tf)\, \theta(z + H_{\text{mix}}) \tag{3.7.1 a}$$

$$Y = \frac{v_*^2}{H_{\text{mix}}}\, \delta(tf)\, \theta(z + H_{\text{mix}}) . \tag{3.7.1 b}$$

In order to keep the argument of the delta function dimensionless the factor f has been introduced. In the Fourier space we obtain

$$(X, Y) = \frac{1}{fH_{\text{mix}}}\, (u_*^2, v_*^2)\, (2\pi)^2\, \delta(\boldsymbol{k})\, \theta(z + H_{\text{mix}}) . \tag{3.7.2}$$

Inserting (3.7.2) into the set (3.4.3), and utilizing the identity $x\delta(x) = 0$, it follows

$$p = 0 \tag{3.7.3 a}$$

$$u = (2\pi)^2\, \delta(\boldsymbol{k})\, \frac{\theta(z + H_{\text{mix}})}{fH_{\text{mix}}(\omega^2 - f^2)}\, (\mathrm{i}\omega u_*^2 - f v_*^2) \tag{3.7.3 b}$$

$$v = (2\pi)^2\, \delta(\boldsymbol{k})\, \frac{\theta(z + H_{\text{mix}})}{fH_{\text{mix}}(\omega^2 - f^2)}\, (f u_*^2 + \mathrm{i}\omega v_*^2) . \tag{3.7.3 c}$$

Inverting the Fourier transforms, where the ω's occurring in the denominator have been replaced by $\omega + \mathrm{i}\varepsilon$ to ensure causality, compare Section 2.3., we arrive at

$$u(t) = \theta(t)\, \frac{\theta(z + H_{\text{mix}})}{H_{\text{mix}}} \left(\frac{u_*^2 \cos ft}{f} + \frac{v_*^2 \sin ft}{f} \right) \tag{3.7.4 a}$$

$$v(t) = \theta(t)\, \frac{\theta(z + H_{\text{mix}})}{H_{\text{mix}}} \left(\frac{-u_*^2 \sin ft}{f} + \frac{v_*^2 \cos ft}{f} \right) . \tag{3.7.4 b}$$

This solution describes a clockwise rotating inertial oscillation. Putting $u_0 = [u_*^2/(fH_{\text{mix}})]\, \theta(z + H_{\text{mix}})$ and $v_*^2 = 0$ the expressions (3.7.4) equal the solution obtained in Section 3.1., i.e., (3.1.3) and (3.1.4).

An important aspect of (3.7.4) is that the inertial oscillations are confined to the upper mixed layer, i.e., they do not penetrate into the deep ocean. Neither the pressure nor the vertical velocity respond to a spatially constant wind in an unbounded f-plane ocean.

Now we replace the delta function in (3.7.1) by the step function $\theta(t)$. That means the wind starts to act at $t = 0$ and then it remains constant. For this kind of forcing we obtain in the Fourier domain

$$(X, Y) = \frac{i}{\omega}(2\pi)^2\, \delta(\boldsymbol{k})\, \frac{\theta(z + H_{mix})}{H_{mix}}\, (u_*^2, v_*^2)\,. \tag{3.7.5}$$

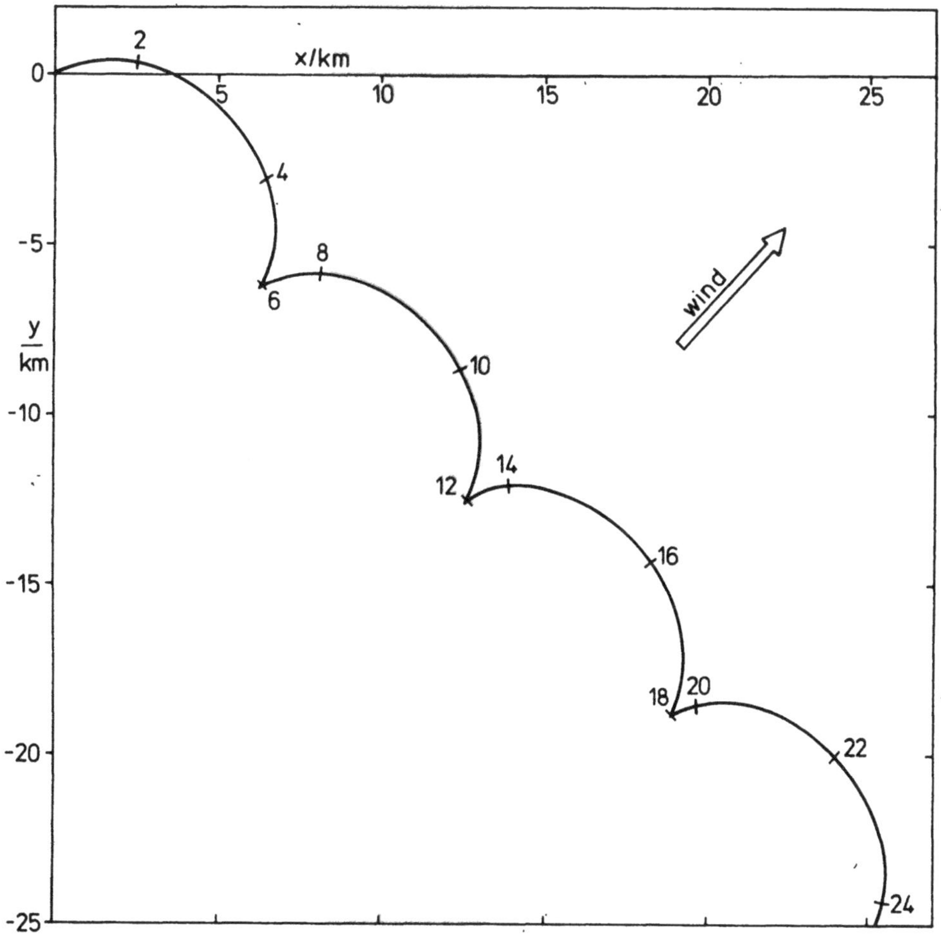

Fig. 3.7.1

The progressive vector diagram of the solution (3.7.7a, b)

Note that in this example the progressive vector diagram coincides with the Lagrangian trajectory of a water particle starting from $x = y = 0$ at $t = 0$. The numbers indicate multiples of ft.

With (3.7.5) we obtain from (3.4.3)

$$p = 0 \tag{3.7.6a}$$

$$u = -(2\pi)^2\,\delta(\boldsymbol{k})\,\frac{\theta(z + H_{\text{mix}})}{H_{\text{mix}}(\omega^2 - f^2)}\left(u_*^2 + \frac{\mathrm{i}fv_*^2}{\omega}\right) \tag{3.7.6b}$$

$$v = (2\pi)^2\,\delta(\boldsymbol{k})\,\frac{\theta(z + H_{\text{mix}})}{H_{\text{mix}}(\omega^2 - f^2)}\left(\frac{\mathrm{i}fu_*^2}{\omega} - v_*^2\right). \tag{3.7.6c}$$

Inverse Fourier transformation of (3.7.6b, c) yields

$$u(t) = \theta(t)\,\frac{\theta(z + H_{\text{mix}})}{fH_{\text{mix}}}\,(u_*^2 \sin ft + v_*^2(1 - \cos ft)) \tag{3.7.7a}$$

$$v(t) = \theta(t)\,\frac{\theta(z + H_{\text{mix}})}{fH_{\text{mix}}}\,(-u_*^2(1 - \cos ft) + v_*^2 \sin ft)\,. \tag{3.7.7b}$$

This solution consists of inertial oscillations and of a steady Ekman transport perpendicular to, and to the right, the wind (see Fig. 3.7.1)

$$(u_{\text{E}}, v_{\text{E}}) = \theta(t)\,\frac{\theta(z + H_{\text{mix}})}{fH_{\text{mix}}}\,(v_*^2, -u_*^2)\,. \tag{3.7.8}$$

Again the response is completely confined to the upper layer, and p and w are not affected. Evidently, this is due to the fact that p and w are forced through the divergence and the curl of the wind, which both vanish for a spatially constant wind, compare (3.2.10) and (3.2.13).

3.8. FORCING OVER A HALF PLANE, DIVERGENT NON-ROTATIONAL WIND

As further example we consider a wind blowing over a half plane. Such a forcing is no longer homogeneous concerning the horizontal variables x and y, hence vertical velocities may occur. The problems discussed in the following sections govern the adjustment of the response to the geostrophic equilibrium. The problem of geostrophic adjustment was originally formulated by Rossby (1939) to interpret the occurrence of geostrophically balanced motions in the ocean and atmosphere. Rossby introduced the fundamental length scale of the problem defined as the inverse product of Coriolis parameter and vertical eigenvalue and referred to as the radius of deformation.

In this and in the next section we consider fixed wind edges. In Section 3.11. the case of a moving wind edge will be investigated. Such models may be applied to study the responses in the vicinity of ice edges, which practically establish wind edges. Moreover, such models may be used to study aspects of the oceanic responses near the boundaries of wind systems such as the trade wind belts.

First we assume

$$X = \frac{u_*^2}{H_{\text{mix}}}\,\theta(z + M_{\text{mix}})\,\theta(t)\,\theta(x) \tag{3.8.1a}$$

$$Y = 0\,. \tag{3.8.1b}$$

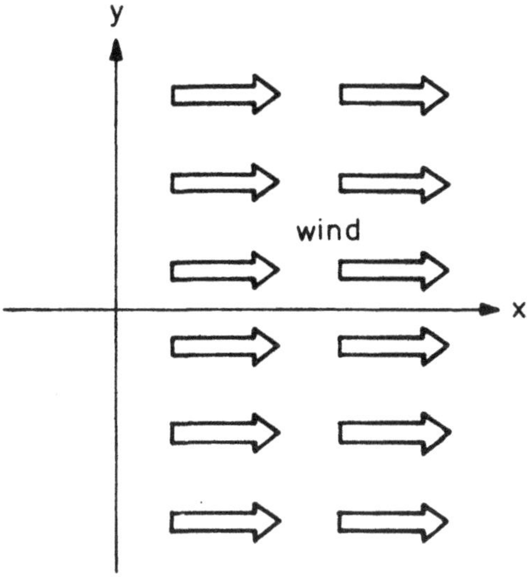

Fig. 3.8.1
Divergent non-rotational wind pattern according to (3.8.1)

That is, we consider a non-rotational divergent wind blowing over the half plane $x > 0$ parallel to the x-axis, and which has a discontinuity line along the y-axis, i.e., $x = 0$, see Fig. 3.8.1.

In order to apply for the technique outlined in Sections 3.3. and 3.4. we Fourier transform (3.8.1) with respect to x, y, and t, and expand X in terms of the vertical eigenfunctions $F_n(z)$. Then the projection coefficients are given by

$$X_n(\omega, k) = \frac{u_*^2}{h_n} \frac{\mathrm{i}}{(\omega + \mathrm{i}\varepsilon)} \frac{-\mathrm{i}}{(k - \mathrm{i}\varepsilon)} 2\pi\, \delta(\varkappa)\,. \tag{3.8.2}$$

Here the Fourier transforms of $\theta(t)$ and $\theta(x)$ are written explicitly including the positive infinitesimals ε, which give the prescription how to go around the poles $\omega = 0$ and $k = 0$ when estimating the inverse Fourier transformation. We drop the ε below, bearing in mind this prescription whenever dealing with the inverse Fourier transformation. Moreover, in (3.8.2) the modal forcing depth h_n has been introduced according to (1.9.32).

Inserting (3.8.2) into the set (3.4.4) it follows

$$p_n(\omega, \boldsymbol{k}) = -\frac{u_*^2}{h_n} \frac{2\pi\, \mathrm{i}\delta(\varkappa)}{\lambda_n^2 \omega(\omega_n^2 - \omega^2)} \tag{3.8.3 a}$$

$$u_n(\omega, \boldsymbol{k}) = -\frac{u_*^2 f}{h_n} \frac{2\pi\, \mathrm{i}\delta(\varkappa)}{k(\omega_n^2 - \omega^2)} \tag{3.8.3 b}$$

$$v_n(\omega, \boldsymbol{k}) = -\frac{u_*^2 f}{h_n} \frac{2\pi\delta(\varkappa)}{k\omega(\omega_n^2 - \omega^2)} \tag{3.8.3 c}$$

where all $\varkappa$ occurring in (3.4.4) have been set equal to zero because of the identity $f(x)\,\delta(x) = f(0)\,\delta(x)$. In particular, this implies $\omega_n^2 = f^2 + k^2/\lambda_n^2$.

The poles of the spectral functions in the ω-plane are located at $\omega = 0$ and $\omega^2 = \omega_n^2$.

Moreover, it can be shown that there are poles at $\omega^2 = f^2$ in (3.8.3b, c). This follows from the identity

$$\frac{1}{k(\omega_n^2 - \omega^2)} = -\frac{1}{(\omega^2 - f^2)}\left[\frac{1}{k} - \frac{k}{k^2 + \lambda_n^2(f^2 - \omega^2)}\right]. \tag{3.8.4}$$

Using

$$\frac{1}{\omega(\omega_n^2 - \omega^2)} = \left(\frac{1}{\omega} + \frac{\omega}{\omega_n^2 - \omega^2}\right)\frac{1}{\omega_n^2}$$

we may separate the different poles and obtain

$$p_n(\omega, k) = -\frac{u_*^2}{h_n}\frac{2\pi\, \mathrm{i}\delta(\varkappa)}{\lambda_n^2\omega_n^2}\left[\frac{1}{\omega} + \frac{\omega}{\omega_n^2 - \omega^2}\right] \tag{3.8.5 a}$$

$$u_n(\omega, k) = \frac{u_*^2}{h_n}\frac{2\pi\, \mathrm{i}\delta(\varkappa)}{\omega^2 - f^2}\left[\frac{1}{k} - \frac{k}{\lambda^2(\omega_n^2 - \omega^2)}\right] \tag{3.8.5 b}$$

$$v_n(\omega, k) = \frac{u_*^2}{h_n} 2\pi\delta(\varkappa)\left[-\frac{1}{kf\omega} - \frac{k}{\omega_n^2\lambda_n^2 f\omega} + \frac{\omega}{\omega^2 - f^2}\left(\frac{1}{kf} + \frac{k}{\omega_n^2\lambda_n^2 f}\right) + \frac{kf}{\omega_n^2\lambda_n^2}\frac{\omega}{(\omega^2 - f^2)(\omega_n^2 - \omega^2)}\right]. \tag{3.8.5 c}$$

The first order poles at the origin of the ω-plane correspond to steady motion. The first term in the bracket of (3.8.5c) yields the Ekman transport, and the second term governs a geostrophically adjusted part of the current which balances the first term of (3.8.5a).

The remaining poles at $\omega^2 = f^2$, and $\omega^2 = \omega_n^2$ correspond to inertial oscillations and Poincaré waves, respectively. Thus the analytical properties of the spectral functions can be attributed easily to the physical properties involved.

From the set (3.8.3) or, equivalently, (3.8.5) we may obtain the Eulerian spectra of u, v, and p. To this end we have to estimate the inverse Fourier transformation with respect to k and $\varkappa$, whereby the $\varkappa$-integration is trivial because of the occurrence of $\delta(\varkappa)$.

Using the integrals, see appendix (B 20) and (B 22),

$$\int_{-\infty}^{+\infty}\frac{dk}{2\pi}\frac{e^{ikx}}{\omega_n^2(k) - \omega^2} = \frac{\mathrm{i}\lambda_n}{2}\frac{e^{\mathrm{i}|x|\lambda_n\sqrt{\omega^2 - f^2}}}{\sqrt{\omega^2 - f^2}} \tag{3.8.6}$$

and

$$\int_{-\infty}^{+\infty}\frac{dk}{2\pi}\frac{-\mathrm{i}e^{ikx}}{(k - \mathrm{i}\varepsilon)(\omega_n^2 - \omega^2)} = \frac{1}{\omega^2 - f^2}\left[-\theta(x)\frac{1}{2}\,\mathrm{sgn}(x)\, e^{\mathrm{i}|x|\lambda_n\sqrt{\omega^2 - f^2}}\right] \tag{3.8.7}$$

we obtain

$$p_n(\omega, x) = \frac{u_*^2}{h_n}\frac{e^{\mathrm{i}|x|\lambda_n\sqrt{\omega^2 - f^2}}}{2\lambda_n\omega(\omega^2 - f^2)^{1/2}} \tag{3.8.8 a}$$

$$u_n(\omega, x) = \frac{u_*^2}{h_n}\frac{1}{(\omega^2 - f^2)}\left[-\theta(x) + \mathrm{sgn}(x)\frac{1}{2}e^{\mathrm{i}|x|\lambda_n\sqrt{\omega^2 - f^2}}\right] \tag{3.8.8 b}$$

$$v_n(\omega, x) = -\frac{u_*^2}{h_n}\frac{\mathrm{i}f}{\omega(\omega^2 - f^2)}\left[-\theta(x) + \operatorname{sgn}(x)\frac{1}{2}\,\mathrm{e}^{\mathrm{i}|x|\lambda_n\sqrt{\omega^2-f^2}}\right]. \tag{3.8.8c}$$

This set represents the frequency spectra of the modal components at any point x, y and may be used to estimate further spectral quantities such as power spectra. We recognize that the spectral functions are peaked at the inertial and at zero frequency. Moreover, we note that the exponentials oscillate in the inertial frequency range $\omega > f$ but they are localized around $x = 0$, i.e., at the wind edge, in the subinertial domain $\omega < f$. Thus the shape of the low frequency part of spectra varies with x.

Now we consider the response in the time domain. In order to find the inverse Fourier transformation of (3.8.8) we use some integrals and notations which will be discussed in some detail in Sections 5.5. and 5.6.

$$C_{nt}^{(1)}(|x|, t) = -\int_{-\infty}^{+\infty}\frac{\mathrm{d}\omega}{2\pi}\frac{\mathrm{e}^{\mathrm{i}|x|\lambda_n\sqrt{\omega^2-f^2}-\mathrm{i}\omega t}}{\omega(\omega^2-f^2)^{1/2}} = \int_{-\infty}^{t}\mathrm{d}t'\,\theta(t'-\lambda_n|x|)\,J_0(f\sqrt{t'^2-\lambda_n^2x^2}) \tag{3.8.9}$$

$$C_{nt}^{(2)}(|x|, t) = -\mathrm{i}\int_{-\infty}^{+\infty}\frac{\mathrm{d}\omega}{2\pi}f^2\,\frac{\mathrm{e}^{\mathrm{i}|x|\lambda_n\sqrt{\omega^2-f^2}-\mathrm{i}\omega t}}{\omega(\omega^2-f^2)}$$

$$= -\operatorname{sgn}(x)\frac{1}{\lambda_n}\frac{\partial}{\partial x}\int_{-\infty}^{t}\mathrm{d}t'(1-\cos f(t-t'))\,\theta(t'-\lambda_n|x|)\,J_0(f\sqrt{t'^2-\lambda_n^2x^2}) \tag{3.8.10}$$

$$C_{ntt}^{(2)}(|x|, t) = -\int_{-\infty}^{+\infty}\frac{\mathrm{d}\omega}{2\pi}f^2\,\frac{\mathrm{e}^{\mathrm{i}|x|\lambda_n\sqrt{\omega^2-f^2}-\mathrm{i}\omega t}}{\omega^2-f^2}$$

$$= -\operatorname{sgn}(x)\frac{f}{\lambda_n}\frac{\partial}{\partial x}\int_{-\infty}^{t}\mathrm{d}t'\,\sin f(t-t')\,\theta(t'-\lambda_n|x|)\,J_0(f\sqrt{t'^2-\lambda_n^2x^2}). \tag{3.8.11}$$

Here J_0 is the Bessel function of zeroth order. With (3.8.9) to (3.8.11) from (3.8.8) we get

$$p_n(x, t) = -\frac{u_*^2}{h_n}\frac{1}{2\lambda_n}C_{nt}^{(1)}(|x|, t) \tag{3.8.12a}$$

$$u_n(x, t) = \frac{u_*^2}{h_n f}\left[\theta(t)\,\theta(x)\sin ft - \frac{\operatorname{sgn}(x)}{2f^2}C_{ntt}^{(2)}(|x|, t)\right] \tag{3.8.12b}$$

$$v_n(x, t) = -\frac{u_*^2}{h_n f}\left[\theta(t)\,\theta(x)\,(1-\cos ft) - \frac{\operatorname{sgn}(x)}{2f^2}C_{nt}^{(2)}(|x|, t)\right]. \tag{3.8.12c}$$

Moreover, we obtain

$$w_n(x, t) = \frac{u_*^2}{h_n}\frac{1}{2\lambda_n}\theta(t-\lambda_n|x|)\,J_0(f\sqrt{t^2-\lambda_n^2x^2}) \tag{3.8.12d}$$

where the modal vertical velocity w_n is given by

$$w_n = -\frac{\partial}{\partial t}p_n\,. \tag{3.8.13}$$

If we perform the x-derivatives in (3.8.10) and (3.8.11) explicitly, from (3.8.12b, c) it follows

$$u_n(x,t) = \frac{u_*^2}{h_n f}\left[\theta(t)\,\theta(x)\sin ft - \frac{1}{2}\operatorname{sgn}(x)\sin f(t-\lambda_n x) + \frac{f\lambda_n x}{2}\int_{-\infty}^{t} dt' \sin f(t-t')\,\theta(t'-\lambda_n|x|)\frac{J_1(f\sqrt{t'^2-\lambda_n^2x^2})}{\sqrt{t'^2-\lambda_n^2x^2}}\right] \tag{3.8.14a}$$

$$v_n(x,t) = -\frac{u_*^2}{h_n f}\left\{\theta(t)\,\theta(x)(1-\cos ft) - \frac{1}{2}\operatorname{sgn}(x)\left[1-\cos f(t-\lambda_n x)\right] + f\lambda_n x\int_{-\infty}^{t} dt'(1-\cos f(t-t'))\,\theta(t'-\lambda_n|x|)\frac{J_1(f\sqrt{t'^2-\lambda_n^2x^2})}{\sqrt{t'^2-\lambda_n^2x^2}}\right\}. \tag{3.8.14b}$$

Here the relations

$$\frac{\partial}{\partial x}\theta(t'-\lambda_n|x|)\,J_0(f\sqrt{t'^2-\lambda_n^2x^2}) = -\delta(t'-\lambda_n|x|)\,\lambda_n\operatorname{sgn}(x)\,J_0(0) + \theta(t'-\lambda_n|x|)\,J_0'(f\sqrt{t'^2-\lambda_n^2x^2})\frac{(-f\lambda_n^2x)}{\sqrt{t'^2-\lambda_n^2x^2}},$$

and $J_0(0) = 1$, and $J_0'(x) = -J_1(x)$, have been used. Due to the step functions which occur in the integrands the integrals run from $\lambda_n|x|$ to t.

For large times we may use asymptotic expansions of the Bessel function and of the convolutions involving Bessel functions, in particular

$$J_0(f\sqrt{t^2-\lambda_n^2x^2}) \approx \left(\frac{2}{\pi ft}\right)^{1/2}\sin\left(ft+\frac{\pi}{4}\right) \tag{3.8.15}$$

$$C_{nt}^{(1)}(|x|,t) \approx \frac{1}{f}\,e^{-|x|/R_n} + \frac{1}{f}\left(\frac{2}{\pi ft}\right)^{1/2}\cos\left(ft+\frac{\pi}{4}\right) \tag{3.8.16}$$

$$C_{nt}^{(2)}(|x|,t) \approx e^{-|x|/R_n} - \cos ft + \left(\frac{2}{\pi ft}\right)^{1/2}\frac{x}{R_n}\cos\left(ft+\frac{\pi}{4}\right) \tag{3.8.17}$$

$$C_{ntt}^{(2)}(|x|,t) \approx f\sin ft - \left(\frac{2}{\pi ft}\right)^{1/2}\frac{xf}{R_n}\sin\left(ft+\frac{\pi}{4}\right). \tag{3.8.18}$$

Here the Rossby radius of deformation $R_n = (\lambda_n f)^{-1}$ has been introduced. As outlined in Sections 5.5. and 5.6. below the approximations (3.8.15) to (3.8.18) amount to neglecting orders of $(\lambda_n|x|/t)^2$. Because the factor x appears in (3.8.17) and (3.8.18) it needs the more time to meet the asymptotic regime the larger the distance x from the wind edge.

Now we can express the response for large times

$$p_n(x, t) \approx -\frac{u_*^2 R_n}{2h_n}\left[e^{-|x|/R_n} + \left(\frac{2}{\pi f t}\right)^{1/2} \cos\left(ft + \frac{\pi}{4}\right)\right] \tag{3.8.19 a}$$

$$u_n(x, t) \approx \frac{u_*^2}{h_n f}\left[\frac{\sin ft}{2} + \frac{x}{R_n}\left(\frac{2}{\pi f t}\right)^{1/2} \sin\left(ft + \frac{\pi}{4}\right)\right] \tag{3.8.19 b}$$

$$v_n(x, t) \approx -\frac{u_*^2}{h_n f}\left[\theta(x) - \operatorname{sgn}(x)\frac{1}{2}e^{-|x|/R_n} - \frac{\cos ft}{2} - \frac{x}{R_n}\left(\frac{2}{\pi f t}\right)^{1/2} \cos\left(ft + \frac{\pi}{4}\right)\right] \tag{3.8.19c}$$

$$w_n(x, t) \approx \frac{u_*^2}{h_n}\frac{1}{2\lambda_n}\left(\frac{2}{\pi f t}\right)^{1/2} \sin\left(ft + \frac{\pi}{4}\right) \tag{3.8.19 d}$$

where $\operatorname{sgn}(x) = 2\theta(x) - 1$ has been used.

With (3.8.12) and (3.8.19) we may describe the response scenario.

In the initial phase, just after the onset of the forcing, inertial oscillations and the Ekman transport start to act, see the first terms in (3.8.12b, c). Since these terms can be summed up according to

$$\sum_{n=0}^{\infty} \frac{1}{h_n} F_n(z) = \frac{\theta(z + H_{\mathrm{mix}})}{H_{\mathrm{mix}}}, \tag{3.8.20}$$

we see explicitly that this part of the response is confined to the upper layer. Moreover, inertial waves are generated at the wind edge, i.e., at the y-axis, which propagate parallel to the x-axis in both negative and positive direction. These waves are governed by the Bessel functions under the integral sign of the $C_n^{(1,2)}$ in (3.8.12), and they propagate their corresponding vertical mode structure whose amplitudes are weighted by the vertical eigenvalues. The phase velocities of the inertial waves are the inverse vertical eigenvalues λ_n^{-1}. This implies that the barotropic front propagates rather fast, however, with inertial

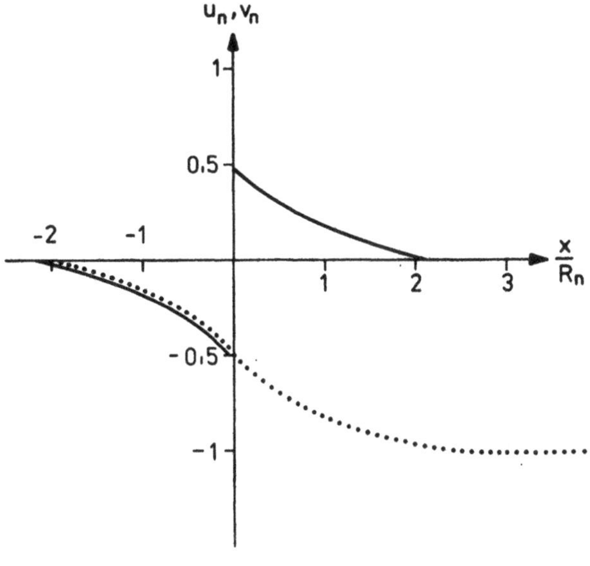

Fig. 3.8.2
Sketch of the geostrophic alongfront current (*solid line*) and of the cross-current (*dotted line*) which consists of the Ekman flux and a rectification at the wind edge

oscillations of very small amplitude, followed by the baroclinic waves with subsequently decreasing phase speeds but with much larger amplitudes.

In a sufficiently long time after a front has passed a given position, i.e., $tf \gg 1$ and $\lambda_n x/t \gg 1$, a regime governed by the asymptotic set (3.8.19) results. The inertial oscillations initially confined on the area $x > 0$ occur now in the whole ocean, however, with half amplitudes. Moreover, a wind edge trapped current follows which rectifies the Ekman flux near the line $x = 0$, see the second term in the bracket of (3.8.19c). The trapping scale is the Rossby radius $R_n = (\lambda_n f)^{-1}$. This implies the larger the mode number the closer the trapping. Obviously the rectifying current is geostrophically balanced by the pressure gradient p_x, compare the first term of (3.8.19a). The sum of Ekman flux and geostrophic current has been depicted in Fig. 3.8.2.

The inertial waves affect the whole water column from the surface to the bottom and decay algebraically like $t^{-1/2}$ even in the absence of friction. Thus the response scenario provides an example of the process of geostrophic adjustment through radiation of inertial waves. The vertical velocity is completely governed by inertial waves, hence no steady up- or downwelling occurs.

3.9. FORCING OVER A HALF PLANE, ROTATIONAL NON-DIVERGENT CASE

In this section we consider again a forcing through a wind blowing over a half plane, but with $\theta(x)$ in (3.8.1a) replaced by $\theta(y)$, i.e.,

$$X = \frac{u_*^2}{h_n}\theta(z + H_{\text{mix}})\,\theta(t)\,\theta(y) \tag{3.9.1a}$$

$$Y = 0\,. \tag{3.9.1b}$$

Now the forcing is confined to $y > 0$. The wind edge coincides with the line $y = 0$, i.e., the x-axis, see Fig. 3.9.1. Thus we deal with a rotational non-divergent wind. Again we Fourier transform (3.9.1a) with respect to x, y, and t, and expand $\theta(z + H_{\text{mix}})$ in terms of the vertical eigenfunctions $F_n(z)$. Then we get

$$X_n(\omega, \mathbf{k}) = \frac{u_*^2}{h_n}\frac{\mathrm{i}}{(\omega + \mathrm{i}\varepsilon)}\frac{-\mathrm{i}}{(\varkappa - \mathrm{i}\varepsilon)}2\pi\,\delta(k)\,. \tag{3.9.2}$$

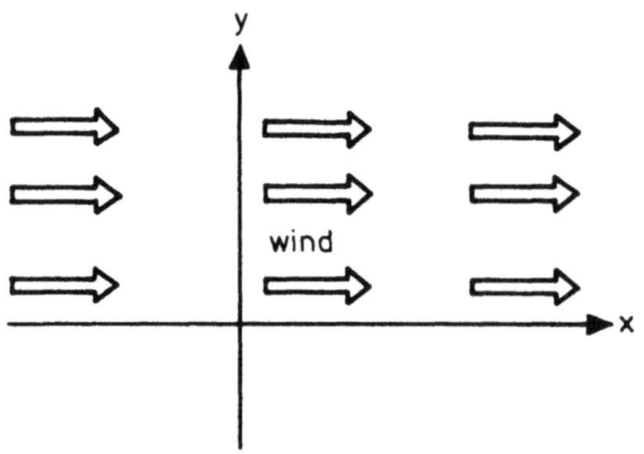

Fig. 3.9.1
Rotational non-divergent wind pattern according to (3.9.1)

Inserting (3.9.2) into the general solution set (3.4.4) it follows

$$p_n(\omega, \boldsymbol{k}) = -\frac{u_*^2}{h_n} \frac{f \cdot 2\pi\delta(k)}{\lambda_n^2\omega^2(\omega_n^2 - \omega^2)} \tag{3.9.3a}$$

$$u_n(\omega, \boldsymbol{k}) = \frac{u_*^2}{h_n} \frac{2\pi\, \mathrm{i}\delta(k)\,(\varkappa^2 - \omega^2\lambda_n^2)}{\varkappa\omega^2\lambda_n^2(\omega_n^2 - \omega^2)} \tag{3.9.3b}$$

$$v_n(\omega, \boldsymbol{k}) = -\frac{u_*^2}{h_n} \frac{f \cdot 2\pi\delta(k)}{\varkappa\omega(\omega_n^2 - \omega^2)} \tag{3.9.3c}$$

with $\omega_n^2 = f^2 + \varkappa^2/\lambda_n^2$, because of $F(k)\,\delta(k) = F(0)\,\delta(k)$.

If we discuss the set (3.9.3) in the same manner as the set (3.8.3) in the previous section the spectral function appears to have poles at $\omega = 0$, $\omega^2 = f^2$, $\omega^2 = \omega_n^2$. These poles correspond to non-oscillatory motion, inertial oscillations, and inertial waves, respectively. The striking difference between the rotational and the divergent wind case consists in the occurrence of second order poles in the origin of the ω-plane. We consider the portion of the solution due to these poles in more detail, realizing that these poles close to the origin of the ω-plane correspond in the time domain to an asymptotic expansion for large times.

For $\omega \ll f$ from (3.9.3) we find

$$p_n(\omega, \boldsymbol{k}) \approx -\frac{u_*^2}{h_n} \frac{f \cdot 2\pi\delta(k)}{\lambda_n^2\omega_n^2\omega^2} \tag{3.9.4a}$$

$$u_n(\omega, \boldsymbol{k}) \approx \frac{u_*^2}{h_n} \frac{2\pi\, \mathrm{i}\delta(k)\,\varkappa}{\lambda_n^2\omega_n^2\omega^2} \tag{3.9.4b}$$

$$v_n(\omega, \boldsymbol{k}) \approx -\frac{u_*^2}{h_n} \frac{f \cdot 2\pi\delta(k)}{\varkappa\omega_n^2\omega}\,. \tag{3.9.4c}$$

The inverse Fourier transformation of these expressions reads

$$p_n(y, t) \approx \frac{u_*^2}{h_n} \frac{t}{2\lambda_n} \mathrm{e}^{-|y|/R_n} \tag{3.9.5a}$$

$$u_n(y, t) \approx \frac{u_*^2}{h_n} t \operatorname{sgn}(y) \frac{1}{2} \mathrm{e}^{-|y|/R_n} \tag{3.9.5b}$$

$$v_n(y, t) \approx -\frac{u_*^2}{h_n f}\left[\theta(y) - \operatorname{sgn}(y)\frac{1}{2}\mathrm{e}^{-|y|/R_n}\right]. \tag{3.9.5c}$$

These expressions are sketched in Fig. 3.9.2. Introducing, moreover, the modal vertical velocity according to (3.8.13) we have for $w_n(y, t)$

$$w_n(y, t) \approx -\frac{u_*^2}{h_n} \frac{\mathrm{e}^{-|y|/R_n}}{2\lambda_n}\,. \tag{3.9.5d}$$

Thus, in the case of a rotational wind, a steady vertical current component exists. The expression (3.9.5d) implies the occurrence of downwelling in the vicinity of the wind

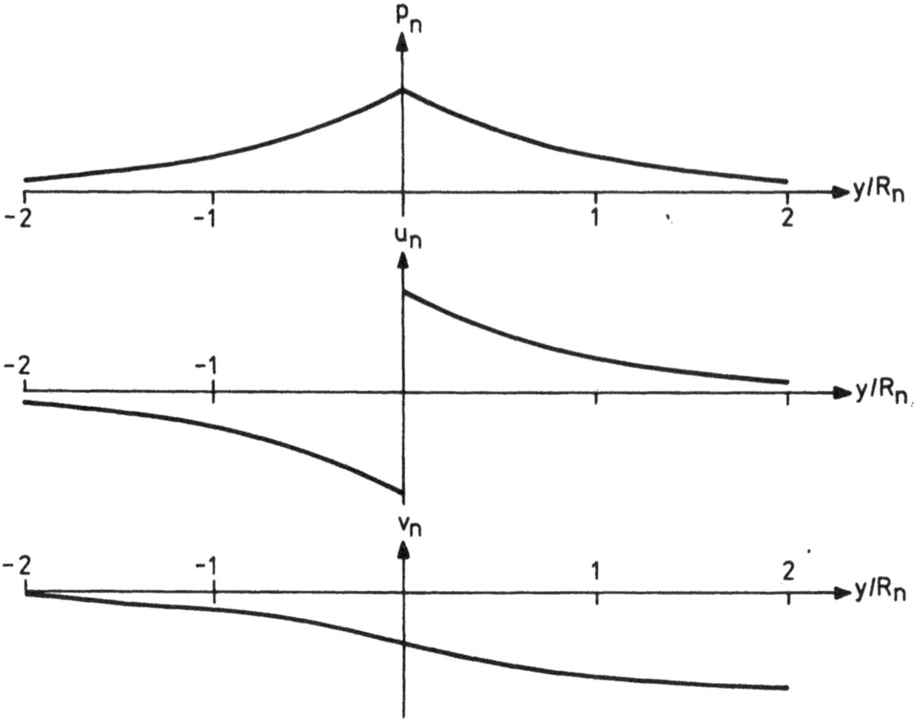

Fig. 3.9.2
A sketch of the response to the rotational wind, shown in Fig. 3.9.1, in terms of p_n (*top*), alongfront velocity u_n (*middle*) and cross-front velocity v_n (*bottom*)
Note that p_n and u_n are accelerating and geostrophically adjusted.

edge, confined to a strip of width $2R_n$. This downwelling produces a cross wind pressure gradient, see (3.9.5a), which geostrophically balances an accelerating jet trapped by the wind edge, compare (3.9.5b). The v-component according to (3.9.5c) consists in leading order of the Ekman transport and a further term which rectifies the Ekman current near the wind edge. Contrary to the divergent wind case of the previous section this rectifying current is no longer geostrophically balanced but it appears that the cross wind gradient, i.e. v_{ny}, balances the time derivative of the pressure multiplied by λ_n^2, as continuity requires, i.e.

$$v_{ny} + \lambda_n^2 p_{nt} = 0 .$$

Since p_t and w are intimately connected, e.g. (3.2.8), the part played by the Ekman current to generate vertical motion, and hence a pressure gradient, becomes quite obvious.

The accelerating jet (3.9.6b) represents a downwind current in the region $y > 0$ bordered by a countercurrent of equal magnitude in the area $y < 0$. Since the trapping scale is the Rossby radius R_n the barotropic mode will be much broader than the baroclinic portion of the jet.

On the wind edge a strong horizontal current shear exists which turns out to be proportional to a delta function. A refinement of the wind model, i.e., replacing the step simulated by $\theta(y)$ by a continuous transition, would avoid such extreme behaviour.

Nevertheless, in the transition area a large shear may be expected to exist. This implies that in the vicinity of the wind edge instabilities can occur which may generate current meanders and eddies.

Now we estimate the inverse Fourier transformation of the full expression of the set (3.9.4) and we start with the transformation with respect to the wave numbers k and $\varkappa$. Using the integrals (3.8.6) and (3.8.7) with k replaced by $\varkappa$, and utilizing, moreover, the relation

$$\int \frac{d\varkappa}{2\pi} \frac{\varkappa\, e^{i\varkappa y}}{\omega_n^2 - \omega^2} = \frac{\partial}{\partial y} \frac{1}{i} \int \frac{d\varkappa}{2\pi} \frac{e^{i\varkappa y}}{\omega_n^2 - \omega^2}$$

we obtain

$$p_n(y, \omega) = -\frac{u_*^2}{h_n} \frac{if\, e^{i|y|\lambda_n\sqrt{\omega^2-f^2}}}{2\lambda_n\omega^2 \sqrt{\omega^2 - f^2}} \tag{3.9.6 a}$$

$$u_n(y, \omega) = -\frac{u_*^2}{h_n}\left[\frac{\theta(y)}{\omega^2 - f^2} - \frac{f^2 \operatorname{sgn}(y)\, e^{i|y|\lambda_n\sqrt{\omega^2-f^2}}}{\omega^2(\omega^2 - f^2)}\right] \tag{3.9.6 b}$$

$$v_n(y, \omega) = \frac{u_*^2}{h_n} \frac{if}{\omega(\omega^2 - f^2)}\left[\theta(y) - \operatorname{sgn}(y) \frac{1}{2}\, e^{i|y|\lambda_n\sqrt{\omega^2-f^2}}\right] \tag{3.9.6 c}$$

These expressions may be used to estimate Eulerian spectra of the modal components at any fixed location (x, y). Again we recognize the poles at the inertial frequency and the localized behaviour of the exponential in the subinertial frequency range.

After inverse Fourier transformation with respect to ω we arrive at

$$p_n(y, t) = \frac{u_*^2}{h_n} \frac{f}{2\lambda_n} C_n^{(1)}(|y|, t) \tag{3.9.7 a}$$

$$u_n(y, t) = \frac{u_*^2}{h_n}\left[\theta(t)\,\theta(y) \frac{\sin ft}{f} + \frac{\operatorname{sgn}(y)}{2} C_n^{(2)}(|y|, t)\right] \tag{3.9.7 b}$$

$$v_n(y, t) = \frac{u_*^2}{h_n}\left[\theta(t)\,\theta(y)\,(1 - \cos ft) - \frac{\operatorname{sgn}(y)}{2} C_{nt}^{(2)}(|y|, t)\right] \tag{3.9.7 c}$$

where the convolutions $C_n^{(1)}$ and $C_n^{(2)}$ are given by

$$C_n^{(1)}(|y|, t) = \int_{-\infty}^{+\infty} \frac{d\omega}{2\pi i} \frac{e^{i|y|\lambda_n\sqrt{\omega^2-f^2}-i\omega t}}{\omega^2\sqrt{\omega^2 - f^2}} = \int_{-\infty}^{t} dt'(t - t')\,\theta(t' - \lambda_n|y|)\, J_0\left(f\sqrt{t'^2 - \lambda_n^2 y^2}\right) \tag{3.9.8}$$

$$\begin{aligned} C_n^{(2)}(|y|, t) &= \int_{-\infty}^{+\infty} \frac{d\omega}{2\pi} \frac{f^2\, e^{i|y|\lambda_n\sqrt{\omega^2-f^2}-i\omega t}}{\omega^2(\omega^2 - f^2)} \\ &= \frac{-\operatorname{sgn}(y)}{\lambda_n} \frac{\partial}{\partial y} \int_{-\infty}^{t} dt'\left[t - t' - \frac{\sin f(t - t')}{f}\right] \theta(t' - \lambda_n|y|)\, J_0\left(f\sqrt{t'^2 - \lambda_n^2 y^2}\right). \end{aligned} \tag{3.9.9}$$

The convolution $C_{nt}^{(2)}$ has already been used in the previous section, see (3.8.10).

The asymptotic expansions of (3.9.8) and (3.9.9) yields (compare Sections 5.5. and 5.6.)

$$C_n^{(1)}(|y|, t) \approx \frac{t}{f} e^{-|y| \lambda/R_n} + \frac{1}{f^2}\left(\frac{2}{\pi f t}\right)^{1/2} \frac{y}{R_n} \sin\left(ft + \frac{\pi}{4}\right) \tag{3.9.10}$$

$$C_n^{(2)}(|y|, t) \approx t\, e^{-|y|/R_n} - \frac{\sin ft}{f} + \frac{1}{f}\left(\frac{2}{\pi f t}\right)^{1/2} \sin\left(ft + \frac{\pi}{4}\right). \tag{3.9.11}$$

For large times, i.e., $tf \gg 1$ and $\lambda_n y/t \ll 1$, it follows from the set (3.9.7) with (3.9.10), (3.9.11) and (3.8.17)

$$p_n(y, t) \approx \frac{u_*^2}{h_n \cdot 2\lambda_n}\left[t\, e^{-|y|/R_n} + \frac{1}{f}\left(\frac{2}{\pi f t}\right)^{1/2} \sin\left(ft + \frac{\pi}{4}\right)\right] \tag{3.9.12a}$$

$$u_n(y, t) \approx \frac{u_*^2}{h_n}\left[\operatorname{sgn}(y)\, t\, e^{-|y|/R_n} + \frac{\sin ft}{2f} + \frac{y}{fR_n}\left(\frac{2}{\pi f t}\right)^{1/2} \sin\left(ft + \frac{\pi}{4}\right)\right] \tag{3.9.12b}$$

$$v_n(y, t) \approx \frac{-u_*^2}{h_n f}\left[\theta(y) - \operatorname{sgn}(y)\, e^{-|y|/R_n} - \frac{\cos ft}{2} - \frac{y}{R_n}\left(\frac{2}{\pi f t}\right)^{1/2} \cos\left(ft + \frac{\pi}{4}\right)\right] \tag{3.9.12c}$$

and with (3.8.13)

$$w_n(y, t) \approx -\frac{u_*^2}{h_n \cdot 2\lambda_n}\left[e^{-|y|/R_n} + \left(\frac{2}{\pi f t}\right)^{1/2} \cos\left(ft + \frac{\pi}{4}\right)\right]. \tag{3.9.12d}$$

By means of the sets (3.9.7) and (3.9.12) we may describe the response scenario in the manner outlined in the previous section. Again we find that pressure and current adjust toward a geostrophic balance through the radiation of inertial waves.

The main difference between the response to rotational and divergent winds consists in the different effect of the Ekman flux and the rectifying current in both cases. While in the divergent case the rectifying current balances the cross wind pressure gradient, it appears that in the rotational case these terms generate a steady vertical motion which in turn build up a cross wind pressure gradient and a geostrophically associated wind edge jet. Since friction has not been taken into account the jet current is accelerating, i.e., proportional to t.

In summary a rotational wind excites more effectively strong currents than a divergent wind. From the mathematical point of view this is due to the time derivative of the wind divergence term on the right-hand side of (3.2.10). This time derivative corresponds to a factor $-i\omega$ in the frequency domain and, hence, in the solution set of the divergent case (3.8.3) the poles are one order lower than in the rotational case (3.9.3).

Let us assume that some features of poleward parts of the trade wind systems resemble a combination of wind fields as described by (3.8.1) and (3.9.1). Then from (3.8.19) and (3.9.12) we may conclude that temporal variations of the trades generate in the surface mixed layer Ekman transports, inertial oscillations, and algebraically decaying baroclinic inertial waves which radiate away from the wind edge. Due to the smallness of the barotropic vertical eigenvalue the barotropic inertial wave is negligible.

Moreover, the geostrophic currents and countercurrents trapped at the wind edge and having a large lateral shear are likely to become unstable.

Thus, for example, the poleward boundaries of the trade wind belts may be expected to be an area of intensified eddy production. Observational evidence has been given in the Atlantic ocean by KÄSE et al. (1985).

Applications to the responses near wind edges established by the boundaries of ice covers have been discussed by several authors, see e.g. SJØBERG and MORK (1985).

3.10. FORCING OVER A STRIP

In this section we consider the example of external forcing which has been investigated by VERONIS (1956) to study the geostrophic adjustment in a two layer ocean on the f-plane, i.e., a rotational non-divergent wind blowing parallel to the x-axis over a strip of the width $2a$, see Fig. 3.10.1,

$$X = \frac{u_*^2}{H_{mix}} \theta(z + H_{mix}) \, \theta(t) \, \theta(a - |y|) \tag{3.10.1 a}$$

$$Y = 0 \,. \tag{3.10.1 b}$$

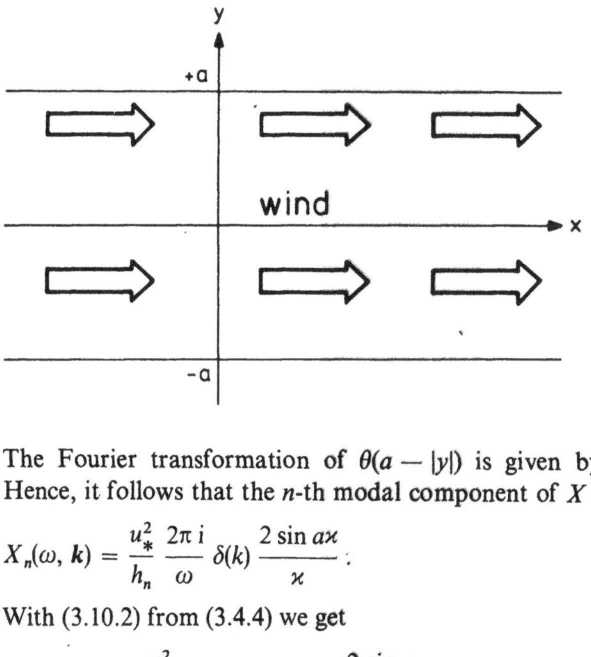

Fig. 3.10.1
The wind pattern according to (3.10.1)

The Fourier transformation of $\theta(a - |y|)$ is given by $(2 \sin a\varkappa)/\varkappa$, compare (B 23). Hence, it follows that the n-th modal component of X in the Fourier space is

$$X_n(\omega, \boldsymbol{k}) = \frac{u_*^2}{h_n} \frac{2\pi \mathrm{i}}{\omega} \delta(k) \frac{2 \sin a\varkappa}{\varkappa} \,; \tag{3.10.2}$$

With (3.10.2) from (3.4.4) we get

$$p_n(\omega, \boldsymbol{k}) = -\frac{u_*^2}{h_n} \mathrm{i} \cdot 2\pi\delta(k) \, f \frac{2 \sin a\varkappa}{\lambda_n^2 \omega^2 (\omega^2 - \omega^2)} \tag{3.10.3 a}$$

$$u_n(\omega, \boldsymbol{k}) = \frac{u_*^2}{h_n} 2\pi\delta(k) \frac{2 \sin a\varkappa}{\varkappa} \frac{(\omega^2 \lambda_n^2 - \varkappa^2)}{\omega^2 \lambda_n^2 (\omega_n^2 - \omega^2)} \tag{3.10.3 b}$$

$$v_n(\omega, \boldsymbol{k}) = \frac{-u_*^2}{h_n} \frac{2\pi\delta(k) \, f \, 2 \sin a\varkappa}{\varkappa(\omega_n^2 - \omega^2) \, \omega} \tag{3.10.3 c}$$

with $\omega_n^2 = f^2 + \varkappa^2/\lambda_n^2$. Again we recognize the occurrence of second order poles of (3.10.3a, b) in the origin of the ω-plane. If we take the identity $\sin a\varkappa = (e^{ia\varkappa} - e^{-ia\varkappa})/2i$ into account, the inverse Fourier transformation of the set (3.10.3) can readily be obtained in the same manner as in the previous section.

It follows

$$p_n(y, t) = \frac{u_*^2 f}{h_n \cdot 2\lambda_n} [C_n^{(1)}(|y + a|, t) - C_n^{(1)}(|y - a|, t)] \tag{3.10.4a}$$

$$u_n(y, t) = \frac{u_*^2}{h_n}\left[\theta(t)\,\theta(a - |y|)\frac{\sin ft}{f} - \frac{\operatorname{sgn}(y + a)}{2\lambda_n} C_n^{(2)}(|a + y|, t) + \frac{\operatorname{sgn}(y - a)}{2\lambda_n} C_n^{(2)}(|a - y|, t)\right] \tag{3.10.4b}$$

$$v_n(y, t) = \frac{-u_*^2}{h_n}\left[\theta(t)\,\theta(a - |y|)\frac{(1 - \cos ft)}{f} + \frac{\operatorname{sgn}(a + y)}{2\lambda_n f} C_{nt}^{(2)}(|a + y|, t) - \frac{\operatorname{sgn}(a - y)}{2\lambda_n f} C_{nt}^{(2)}(|a - y|, t)\right] \tag{3.10.4c}$$

where the convolutions $C_n^{(1)}$, $C_n^{(2)}$, and $C_{nt}^{(2)}$ are defined by (3.9.8), (3.9.9) and (3.8.10), respectively.

If we use the asymptotic representation of the convolutions in (3.10.4a) according to (3.9.10) we would find that the asymptotic expressions of the inertial waves in $C_n^{(1)}(|y \pm a|, t)$ cancel each other. This is due to the suppression of the phase information in that approximation. Therefore we will utilize an asymptotic formulation which conserves the phases in the arguments of the trigonometric functions, compare Section 5.5. Then we obtain for large times

$$p_n(y, t) \approx \frac{u_*^2}{h_n \cdot 2f\lambda_n}\left\{ft(e^{-|y+a|/R_n} - e^{-|y-a|/R_n}) + \left(\frac{2}{\pi f t}\right)^{1/2} \times\left[\sin\left(f\sqrt{t^2 - \lambda_n^2(y + a)^2} + \frac{\pi}{4}\right) - \sin\left(f\sqrt{t^2 - \lambda_n^2(y - a)^2} + \frac{\pi}{4}\right)\right]\right\} \tag{3.10.5a}$$

$$u_n(y, t) \approx \frac{u_*^2}{h_n \cdot 2f}\left\{ft(\operatorname{sgn}(y + a)\, e^{-|a+y|/R_n} - \operatorname{sgn}(y - a)\, e^{-|a-y|/R_n}) + \left(\frac{2}{\pi f t}\right)^{1/2}\left[\frac{y + a}{R_n}\sin\left(f\sqrt{t^2 - (y + a)^2\lambda_n^2} + \frac{\pi}{4}\right) - \frac{y - a}{R_n}\sin\left(f\sqrt{t^2 - (y - a)^2\lambda_n^2} + \frac{\pi}{4}\right)\right]\right\} \tag{3.10.5b}$$

$$v_n(y, t) \approx -\frac{u_*^2}{h_n f}\left\{\theta(a - |y|) - \operatorname{sgn}(y + a)\frac{1}{2}e^{-|y+a|/R_n} + \operatorname{sgn}(y - a)\frac{1}{2}e^{-|y-a|/R_n}\right.$$

$$-\frac{1}{2}\left(\frac{2}{\pi f t}\right)^{1/2}\left[\frac{y+a}{R_n}\cos\left(f\sqrt{t^2-(y+a)^2\lambda_n^2}+\frac{\pi}{4}\right)\right.$$

$$\left.\left.-\frac{y-a}{R_n}\cos\left(f\sqrt{t^2-(y-a)^2\lambda_n^2}+\frac{\pi}{4}\right)\right]\right\} \qquad (3.10.5\,c)$$

where the relation

$$\frac{1}{2}[\operatorname{sgn}(y+a)-\operatorname{sgn}(y-a)]=\theta(a+y)-\theta(a-y)=\theta(a-|y|)$$

has been used. Moreover, with (3.8.13) we have for the modal vertical velocity

$$w_n(y,t)\approx-\frac{u_*^2}{h_n 2\lambda_n}\left\{e^{-|y+a|/R_n}-e^{-|y-a|/R_n}+\left(\frac{2}{\pi f t}\right)^{1/2}\right.$$

$$\left.\times\left[\sin\left(f\sqrt{t^2-(a+y)^2\lambda_n^2}+\frac{\pi}{4}\right)-\sin\left(f\sqrt{t^2-(y-a)^2\lambda_n^2}+\frac{\pi}{4}\right)\right]\right\}$$

(3.10.5d)

Now we may discuss the response scenario in terms of (3.10.4) and (3.10.5). First we note that the response near the wind discontinuity line $y = -a$ behaves like the response to a rotational wind blowing over the half plane $y > 0$ which has been discussed in the preceding section. In the vicinity of the line $y = a$, however, the signs of the corresponding terms are opposite. Thus we observe downwelling at $y = -a$, but upwelling near $y = a$, compare (3.10.5d).

Immediately after the onset of the forcing inertial oscillations and Ekman transport start to act within the forcing area. Simultaneously inertial waves are generated at the wind edges which propagate in both negative and positive y-direction. Mathematically these waves are governed by the convolutions occurring in (3.10.4), in particular by the Bessel functions under the integral signs of these convolutions.

After some time, i.e., $ft \gg 1$, the asymptotic regime governed by (3.10.5) has established. Contrary to the cases of forcing over half planes in the present case the inertial oscillations have cancelled.

The response consists now of geostrophically adjusted accelerating jets trapped at the wind edges, see (3.10.5a, b), and of the Ekman transport with some rectification near the wind edges, compare (3.10.5c), and of inertial wave which radiate away and decay algebraically like $t^{-1/2}$. For the barotropic mode, i.e., $n = 0$, we obtain from (3.10.5b) under the presumption $R_0 > 2a$, i.e., the barotropic Rossby radius is larger than the width of the forcing area,

$$u_0(y,t)\approx\frac{u_*^2}{2h_0}\,t[\operatorname{sgn}(y+a)\,e^{-|a+y|/R_0}-\operatorname{sgn}(y-a)\,e^{-|a-y|/R_0}]$$

$$\approx\frac{u_*^2}{h_0}\,t\left[\theta(a-|y|)-\frac{a}{R_0}\right].$$

This represents a practically uniform downwind current within the strip, $|y| < a$, and rather weak broad countercurrents on both sides of the forcing area. The baroclinic modes, however, yield a weak current in the centre which rises sharply towards the edges, and is bordered by strong narrow countercurrents. This solution corresponds to that discussed by ROSSBY (1939) in his classical paper on the adjustment problem in which the radius of deformation was introduced.

3.11. FORCING THROUGH MOVING FRONTS

The next problem we wish to study is the response of the ocean to moving wind events such as fronts and storms. In particular we consider three explicite examples

$$X(x, y, z, t) = \frac{u_*^2}{H_{\text{mix}}} \theta(z + H_{\text{mix}}) \tilde{X}^{(j)}(x, y, t), \qquad Y = 0 \tag{3.11.1}$$

where

$$\tilde{X}^{(1)} = \frac{1}{f} \delta\left(\frac{x}{c} - t\right) \tag{3.11.2a}$$

$$\tilde{X}^{(2)} = \frac{1}{f} \delta\left(\frac{y}{c} - t\right) \tag{3.11.2b}$$

and

$$\tilde{X}^{(3)} = \theta\left(t - \frac{y}{c}\right). \tag{3.11.2c}$$

The forces given by (3.11.2a, b) correspond to moving disturbances propagating along the y-axis and the x-axis, respectively, while (3.11.2c) describes a forcing over a half plane with moving wind edge. In order to keep $X^{(1)}$ and $X^{(2)}$ dimensionless the inertial frequency has been introduced. The phase speed of the three examples is c. After Fourier transformation we get

$$X^{(1)} = \frac{(2\pi)^2}{f} \delta(\varkappa)\, \delta\left(k - \frac{\omega}{c}\right) \tag{3.11.3a}$$

$$X^{(2)} = \frac{(2\pi)^2}{f} \delta(k)\, \delta\left(\varkappa - \frac{\omega}{c}\right) \tag{3.11.3b}$$

$$X^{(3)} = \frac{(2\pi)^2 \mathrm{i}}{\omega + \mathrm{i}\varepsilon} \delta(k)\, \delta\left(\varkappa - \frac{\omega}{c}\right). \tag{3.11.3c}$$

In the following we discuss the first example in some detail and then we note only the final solution of the remaining two examples. Using the expansion of (3.11.1a) in terms of vertical eigenfunctions and the Fourier transforms according to (3.11.3a), from (3.4.4) we obtain

$$\begin{pmatrix} p_n \\ u_n \\ v_n \end{pmatrix} = \frac{u_*^2}{h_n f} (2\pi)^2\, \delta(\varkappa)\, \delta\left(k - \frac{\omega}{c}\right) \frac{1}{\gamma_n^2 \left[\omega^2 - \left(\frac{f}{\gamma_n}\right)^2\right]} \begin{pmatrix} \frac{\mathrm{i}}{c\lambda_n^2} \\ \mathrm{i}\omega \\ f \end{pmatrix} \tag{3.11.4}$$

where

$$f(k)\,\delta\left(k-\frac{\omega}{c}\right)=\delta\left(k-\frac{\omega}{c}\right)f\left(\frac{\omega}{c}\right)$$

has been used. We have introduced the quantity γ_n which is defined by

$$\gamma_n^2=\left(1-\frac{1}{c^2\lambda_n^2}\right). \tag{3.11.5}$$

Thus γ_n^2 compares the phase speed c of the wind disturbance with the phase speed λ_n^{-1} of the inertial waves. For $\gamma_n^2>0$ it follows $c>\lambda_n^{-1}$ and, hence, the forcing propagates faster than the inertial waves. Conversely, for $\gamma_n^2<0$ the inertial waves move faster than the phase of the wind. Consider first the case $\gamma_n^2>0$. The inverse Fourier transformation of the response (3.11.4) with respect to k and $\varkappa$ is trivial because of the involved delta functions. Thus, we have to estimate integrals of the type

$$\int\frac{\mathrm{d}\omega}{2\pi}\,\frac{\mathrm{e}^{-\mathrm{i}\omega\left(t-\frac{x}{c}\right)}}{(\omega+\mathrm{i}\varepsilon)^2-\frac{f^2}{\gamma_n^2}}=-\frac{\gamma_n}{f}\,\theta\left(t-\frac{x}{c}\right)\sin\left[\frac{f}{\gamma_n}\left(t-\frac{x}{c}\right)\right] \tag{3.11.6a}$$

and

$$\int\frac{\mathrm{d}\omega}{2\pi}\,\frac{\mathrm{i}\omega\,\mathrm{e}^{-\mathrm{i}\omega\left(t-\frac{x}{c}\right)}}{(\omega+\mathrm{i}\varepsilon)^2-\frac{f^2}{\gamma_n^2}}=\theta\left(t-\frac{x}{c}\right)\cos\left[\frac{f}{\gamma_n}\left(t-\frac{x}{c}\right)\right] \tag{3.116b}$$

where ω has been replaced by $\omega+\mathrm{i}\varepsilon$ in the denominators of both integrands in order to ensure causality.

With (3.11.6a, b) we arrive at

$$p_n(x,t)=\frac{u_*^2}{h_nf}\,\theta\left(t-\frac{x}{c}\right)\frac{1}{c\lambda_n^2\gamma_n^2}\cos\left[\frac{f}{\gamma_n}\left(t-\frac{x}{c}\right)\right] \tag{3.11.7a}$$

$$u_n(x,t)=\frac{u_*^2}{h_nf}\,\theta\left(t-\frac{x}{c}\right)\frac{1}{\gamma_n^2}\cos\left[\frac{f}{\gamma_n}\left(t-\frac{x}{c}\right)\right] \tag{3.11.7b}$$

$$v_n(x,t)=-\frac{u_*^2}{h_nf}\,\theta\left(t-\frac{x}{c}\right)\frac{1}{\gamma_n}\sin\left[\frac{f}{\gamma_n}\left(t-\frac{x}{c}\right)\right]. \tag{3.11.7c}$$

This describes a response in terms of a frontal edge moving with the phase speed of the wind disturbance and plane waves behind the frontal edge.

For $\gamma_n^2<0$, however, we have to deal with integrals of the type

$$\int\frac{\mathrm{d}\omega}{2\pi}\,\frac{\mathrm{e}^{-\mathrm{i}\omega\left(t-\frac{x}{c}\right)}}{\omega^2+\frac{f^2}{|\gamma_n|^2}}=\frac{1}{2}\,\frac{|\gamma_n|}{f}\,\mathrm{e}^{\frac{-f}{|\gamma_n|}\left|t-\frac{x}{c}\right|} \tag{3.11.8a}$$

and

$$\int \frac{d\omega}{2\pi} \frac{i\omega\, e^{-i\omega\left(t-\frac{x}{c}\right)}}{\omega^2 + \frac{f^2}{|\gamma_n|^2}} = \frac{1}{2}\operatorname{sgn}\left(t-\frac{x}{c}\right) e^{-\frac{f}{\gamma_n}\left|t-\frac{x}{c}\right|} \tag{3.11.8b}$$

with $|\gamma_n|^2 = -\gamma_n^2 = \frac{1}{\lambda_n^2 c^2} - 1$.

Now the integrands have poles in both the upper and the lower half of the complex ω-plane, i.e., at $\omega = \pm i f/|\gamma_n|$.

Hence, it follows that as well the integration along an infinite semicircle in the lower as in the upper half planes yield nonvanishing contributions multiplied by

$\theta\left(t-\frac{x}{c}\right)$ and $\theta\left(\frac{x}{c}-t\right)$, respectively, which may be combined to (3.11.8a, b).

Thus we obtain

$$p_n(x,t) = -\frac{u_*^2}{h_n f} \frac{\operatorname{sgn}\left(t-\frac{x}{c}\right)}{2\lambda_n^2 c\,|\gamma_n|^2} e^{-\frac{f}{|\gamma_n|}\left|t-\frac{x}{c}\right|} \tag{3.11.9a}$$

$$u_n(x,t) = -\frac{u_*^2}{h_n f} \frac{\operatorname{sgn}\left(t-\frac{x}{c}\right)}{2\,|\gamma_n|^2} e^{-\frac{f}{|\gamma_n|}\left|t-\frac{x}{c}\right|} \tag{3.11.9b}$$

$$v_n(y,t) = -\frac{u_*^2}{h_n f} \frac{1}{|\gamma_n|} e^{-\frac{f}{|\gamma_n|}\left|t-\frac{x}{c}\right|}. \tag{3.11.9c}$$

In this case the response is localized around the frontal edge and can be observed before the disturbance has passed.

Clearly, this is due to the fact that the inertial wave phase speed is now assumed to be higher than the phase speed of the wind event and so the response can propagate in advance.

Some features of the wind fields found in moderate latitudes of the globe may be modelled by (3.11.2). The disturbances and fronts occurring in these latitudes move with a speed c which is of the order of 10 m/s according to Trewartha and Horn (1980).

Compared with the propagation speeds of oceanic disturbances as given in Table 1.9.1 it appears that c is one order of magnitude smaller than the propagation speed of barotropic and one order of magnitude larger than the propagation speed of baroclinic inertial waves, respectively.

Hence, with

$\lambda_0^{-1} = \sqrt{gH} \gg c$

we obtain for the barotropic part of the response

$$p_0(x,t) = -\frac{u_*^2}{h_0} \frac{c}{2f} \operatorname{sgn}\left(t-\frac{x}{c}\right) e^{-\frac{|ct-x|}{R_0}} \tag{3.11.10a}$$

$$u_0(x, t) = -\frac{u_*^2}{h_0 f} \frac{\lambda_0^2 c^2}{2} \operatorname{sgn}\left(t - \frac{x}{c}\right) e^{-\frac{|ct-x|}{R_0}} \tag{3.11.10b}$$

$$v_0(x, t) = -\frac{u_*^2}{h_0 f} c\lambda_0 \frac{1}{2} e^{-\frac{|ct-x|}{R_0}} . \tag{3.11.10c}$$

The baroclinic modes are already given by (3.11.7).

The estimation of the response to (3.11.2b) is straightforward and we find

$$p_n(y, t) = \frac{u_*^2}{h_n f} \left\{ -\theta(\gamma_n^2) \frac{\theta\left(t - \frac{y}{c}\right)}{c\gamma_n \lambda_n^2} \sin\left[\frac{f}{\gamma_n}\left(t - \frac{y}{c}\right)\right] + \theta(-\gamma_n^2) \frac{e^{-\frac{f}{|\gamma_n|}\left|t-\frac{y}{c}\right|}}{2c\,|\gamma_n|\,\lambda_n^2} \right\} \tag{3.11.11a}$$

$$u_n(y, t) = \frac{u_*^2}{h_n f} \left\{ \theta(\gamma_n^2)\, \theta\left(t - \frac{y}{c}\right) \cos\left[\frac{f}{\gamma_n}\left(t - \frac{y}{c}\right)\right] + \theta(-\gamma_n^2) \operatorname{sgn}\left(t - \frac{y}{c}\right) \frac{1}{2} e^{-\frac{f}{|\gamma_n|}\left|t-\frac{y}{c}\right|} \right\}. \tag{3.11.11b}$$

$$v_n(y, t) = -\frac{u_*^2}{h_n f} \left\{ \frac{\theta(\gamma_n^2)}{\gamma_n} \theta\left(t - \frac{y}{c}\right) \sin\left[\frac{f}{\gamma_n}\left(t - \frac{y}{c}\right)\right] + \frac{\theta(-\gamma_n^2)}{2\,|\gamma_n|} e^{-\frac{f}{|\gamma_n|}\left|t-\frac{y}{c}\right|} \right\}. \tag{3.11.11c}$$

Moreover, we append the response to the semi-infinite forcing with moving wind edge according to (3.11.2c) or (3.11.3c)

$$\begin{aligned} p_n(y, t) = -\frac{u_*^2}{h_n} \cdot \frac{1}{fc\lambda_n^2} \Bigg\{ & \theta(\gamma_n^2)\, \theta\left(t - \frac{y}{c}\right)\left[1 - \cos \frac{f}{\gamma_n}\left(t - \frac{y}{c}\right)\right] \\ & + \theta(-\gamma_n^2)\left[\theta\left(t - \frac{y}{c}\right) - \operatorname{sgn}\left(t - \frac{y}{c}\right) \frac{1}{2} e^{-\frac{f}{|\gamma_n|}\left|t-\frac{y}{c}\right|}\right]\Bigg\} \end{aligned} \tag{3.11.12a}$$

$$u_n(y, t) = -\frac{u_*^2}{h_n f} \left\{ \theta(\gamma_n^2)\, \gamma_n \theta\left(t - \frac{y}{c}\right) \sin \frac{f}{\gamma_n}\left(t - \frac{y}{c}\right) + \theta(-\gamma_n^2)\, |\gamma_n| \frac{1}{2} e^{-\frac{f}{|\gamma_n|}\left|t-\frac{y}{c}\right|} \right\} \tag{3.11.12b}$$

$$\begin{aligned} v_n(y, t) = -\frac{u_*^2}{h_n f} \Bigg\{ & \theta\left(t - \frac{y}{c}\right) - \theta(\gamma_n^2)\, \theta\left(t - \frac{y}{c}\right) \cos\left[\frac{f}{\gamma_n}\left(t - \frac{y}{c}\right)\right] \\ & - \theta(-\gamma_n^2) \operatorname{sgn}\left(t - \frac{y}{c}\right) e^{-\frac{f}{|\gamma_n|}\left|t-\frac{y}{c}\right|} \Bigg\}. \end{aligned} \tag{3.11.12c}$$

In this case there occurs an Ekman transport behind the frontal edge, see the first term in (3.11.12c).

Contrary to the results of Section 3.9., see e.g. (3.9.5), it appears that according to (3.11.12) only the barotropic mode, i.e. $\gamma_0^2 < 0$, forms a geostrophically adjusted wind edge current. The baroclinic modes with $\gamma_n^2 > 0$ are governed by inertial waves behind the frontal edge with the somewhat enhanced frequency f/γ_n (note that $\gamma_n^2 > 0$ implies $\gamma_n < 1$). This enhancement is called "blue shift" of inertial waves. In the frame of this section the blue shift is determined by the ratio of the phase speeds λ_n^{-1} and c.

In summary, from the three examples considered it can be seen that the barotropic portion of the response of the ocean to moving wind events is characterized by a geostrophically adjusted current pattern propagating with the speed of the weather disturbances and trapped within one Rossby radius (typically 1000 km) at the atmospheric fronts. For such large scales the f-plane used throughout this chapter might be a too rough an approximation. Nevertheless, correlation between atmospheric storms and local current fluctuations in the North Atlantic have been observed by MEINCKE (1975).

Moreover, WEARN and BAKER (1980) found evidence for the response of the Antarctic Circumpolar Current to consist mainly of barotropic long gravity waves. They observed a significant coherence between local wind and the pressure at a depth of 500 m with an associated time lag of less than a day at periods of two to three days, which are the time scales of the moving atmospheric disturbances. They observed higher correlation between wind and the oceanic pressure field near the coast of Antarctica than at a site located about 600 km north of the coast. This suggests the important dynamical effects of coastal boundaries which will be discussed in Chapter 5.

The baroclinic response in midlatitudes consists mainly of algebraic decaying blue shifted inertial waves radiating behind the moving front into the interior of the ocean. The blue shift of the inertial waves given in (3.11.11) and (3.11.12) is of the order of $1.01f$, where $c = 10$ m/s and $\lambda_1 = 0.39$ s/m has been adopted. This value of the blue shift agrees fairly well with observed blue shifts of inertial waves reported by FU (1981).

The generation of baroclinic inertial waves by resting as well by moving wind patterns accounts for the global distribution of this wave type. The lack of correlation often found between the inertial waves and the local wind can be attributed to the likely large distance between the generating front of the wind pattern and the site of current observation.

CHAPTER 4

Forcing of internal waves

4.1. INTRODUCTORY REMARKS

In the preceding chapter we dealt with the response of an infinite ocean on a f-plane to wind forcing in the inertial and subinertial frequency range. High frequency motion, $3f \lesssim \omega \lesssim N$, has been filtered out by the hydrostatic approximation, i.e., neglecting w_t in (3.2.3). In the present chapter we release the hydrostatic approximation and consider the high frequency portion of the oceanic response which usually is attributed to internal waves. Those waves arise in a stable stratified fluid through the restoring force of gravity on water particles displaced from their equilibrium levels. Interfacial waves occurring between two superposed layers of different density are common, in particular at the surface of the ocean. In the continuously stratified interior of the ocean the restoring force of gravity is much weaker because of the small density differences, and consequently the periods and wavelengths of internal waves are much larger than those of surface gravity waves. Generally, waves having frequencies between the local Brunt Väisälä frequency N and the inertial frequency f are called internal waves. This implies that inertial waves discussed in the preceding chapter are included. The interval between f and N is of about 10 octaves in the ocean. Internal waves occupy an intermediate time and space scale of oceanic phenomena between small scale turbulence and geostrophic motions.

Internal waves have been observed since the early 1900's, NANSEN (1902), EKMAN (1904), and have attracted many oceanographers. However, in classical observations consisting of repeated lowerings of Nansen bottles and some current measurements internal waves could not be resolved adequately. Thus internal waves were often considered a nuisance because their complicated temporal and spatial fluctuations obscured and contaminated experimental observations. Since the 1950s, however, technology has advanced to a point where adequate internal wave instrumentation have become more available.

Theoretical investigations on internal waves preceded the observations by half a century. STOKES (1847) treated internal waves at the interface between a light fluid overlaying a heavy fluid and RAYLEIGH (1883) extended the theory to the case of vertical mode structure in continuously stratified fluids.

Historical reviews of observational and theoretical aspects of the description of internal waves may be found in the textbooks of KRAUSS (1966), PHILLIPS (1977), ROBERTS (1975), and in the survey articles of MUNK (1981) and OLBERS (1983), to mention but a few.

It is generally assumed that the internal wave region of the spectrum between the inertial frequency and the local BVF is governed primarily by linear wave dynamics. In fact, many observations fulfil certain consistency relations based on linear WKB models of free linear internal waves, e.g. VOORHIS (1968), FOFONOFF (1969) and GOULD (1971).

A strong impetus in internal wave research was made by GARRETT and MUNK (1972) who contrived, by synthesizing existing internal wave data, a universal, semi-empirical frequency-wavenumber spectrum for the internal wave field. Except for inertial-internal waves and tides this model is believed to reflect the spectral features of the internal wave climate in the deep ocean. The Garrett Munk spectrum (GM model hereafter) uses wave functions which are solutions of the free linear Boussinesq equations and the associated dispersion relation. Thus the scales are defined precisely but generation and dissipation mechanism are not taken into account. Garrett and Munk attempted to find some form of the spectrum that is at least consistent with known physical and observational constraints. The GM model, first GM 72 and, later, in the spirit of planned obsolescence GM 75 and GM 79 (GARRETT and MUNK 1975, 1979), has influenced nearly all recent activities in experimental as well as in theoretical investigations on internal waves.

The observed validity of the GM spectrum in large portions of the interior of the oceans implies the occurrence of processes affecting the spectral energy distribution of internal waves. The different processes generating internal waves are assumed to feed energy primarily into the larger scales of internal waves. Wave breaking and subsequent dissipation and mixing, however, extract energy at high wave numbers. Direct observations of internal wave breaking in the ocean were reported by WOODS (1968) and later by WOODS and WILEY (1972). Moreover, the observed fine structure of temperature and salinity found in the oceans and in lakes provides indirect evidence of such mechanisms. This has been reviewed by TURNER (1981) and MUNK (1981). Hence, the observed persistence of the GM-spectrum requires a certain non-linear redistribution of energy between the different spectral components. Such a spectral redistribution may be described by a weak non-linear interaction theory, see e.g. OLBERS (1976), MCCOMAS and MÜLLER (1981).

However, this book is devoted to the linear theory of the forcing of oceanic wave processes through wind pattern and, hence, the investigations presented below describe the first phase of the response scenario, yielding the initial spectral distributions. In later phases of the response scenario the mentioned redistribution by non-linear interaction within the internal wave spectrum may occur. Such processes, however, are beyond the scope of this book.

4.2. GENERATION OF INTERNAL WAVES

There are various possible mechanisms which may excite internal waves. Generally, under stably stratified conditions any forces that excite vertical excursions of fluid bodies will result in propagation of internal waves away from the source. Such forces may be wind stress, pressure gradients or buoyancy. Moreover, internal waves may arise from the flow of currents over topographic features, e.g. underwater sills or seamounts, which induce a significant vertical motion at depth. In particular this mechanism has turned out to be the main source of internal tides. A broad review of physical processes affecting the excitation of internal waves was given by THORPE (1975).

In the present chapter we do not attempt an exhaustive discussion of the various possibilities of forcing of internal waves. The considerations which follow are confined to the excitation of internal waves through atmospheric forces, in particular we are concerned with wind forcing, which appears to be the most significant candidate among the atmospheric forces.

This chapter is organized as follows: First we discuss the governing equations and then we calculate the appropriate Green's functions for some examples of the BVF's. We analyze the properties of the Green's functions and the corresponding internal wave dispersion relations. Finally we consider some examples of wind pattern in order to study the oceanic response in the internal wave frequency band. In particular we discuss the Eulerian spectra of the response to the external forcing.

4.3. GOVERNING EQUATIONS

We start with the linear Boussinesq equations for an infinite f-plane ocean

$$u_t - fv + p_x = X \tag{4.3.1 a}$$

$$v_t + fu + p_y = Y \tag{4.3.1 b}$$

$$w_t - b + p_z = 0 \tag{4.3.1 c}$$

$$b_t + N^2 w = 0 \tag{4.3.1 d}$$

$$u_x + v_y + w_z = 0\,. \tag{4.3.1 e}$$

We consider a flat bottom and a rigid lid condition at the surface. This implies the following boundary conditions of the vertical velocity component

$$w = 0 \quad \text{at} \quad z = 0 \quad \text{and} \quad z = -H\,. \tag{4.3.2}$$

Introducing the Fourier transformation according to

$$u(\omega, k, \varkappa, z) = \int \mathrm{d}t \int \mathrm{d}\boldsymbol{r}\; \mathrm{e}^{\mathrm{i}\omega t - \mathrm{i}\boldsymbol{k}\boldsymbol{r}}\, u(x, y, z, t)\,, \tag{4.3.3}$$

where $\boldsymbol{r} = (x, y)$ and $\boldsymbol{k} = (k, \varkappa)$, and analogous expressions for v, w, p, b, X, and Y we obtain from (4.3.1) the set

$$-\mathrm{i}\omega u - fv + \mathrm{i}kp = X \tag{4.3.4a}$$

$$-\mathrm{i}\omega v + fu + \mathrm{i}\varkappa p = Y \tag{4.3.4b}$$

$$-\mathrm{i}\omega w - b + p_z = 0 \tag{4.3.4c}$$

$$-\mathrm{i}\omega b + N^2 w = 0 \tag{4.3.4d}$$

$$\mathrm{i}ku + \mathrm{i}\varkappa v + w_z = 0\,. \tag{4.3.4e}$$

Now we may readily derive an equation for w alone. From (4.3.4c) and (4.3.4d) it follows

$$(\omega^2 - N^2)\, w + \mathrm{i}\omega p_z = 0\,. \tag{4.3.5}$$

Multiply (4.3.4a) by $\mathrm{i}k$ and (4.3.4b) by $\mathrm{i}\varkappa$ and add to obtain

$$-\mathrm{i}\omega(\mathrm{i}k + \mathrm{i}\varkappa v) - f(\mathrm{i}kv - \mathrm{i}\varkappa u) - \boldsymbol{k}^2 p = \mathrm{i}kX + \mathrm{i}\varkappa Y\,. \tag{4.3.6}$$

Multiply (4.3.4a) by $\mathrm{i}\varkappa$ and (4.3.4b) by $\mathrm{i}k$ and substract to obtain

$$\mathrm{i}\omega(\mathrm{i}kv - \mathrm{i}\varkappa u) - f(\mathrm{i}ku + \mathrm{i}\varkappa v) = \mathrm{i}\varkappa X - \mathrm{i}kY\,. \tag{4.3.7}$$

Elimination of $\mathrm{i}kv - \mathrm{i}\varkappa u$ from both equations yields

$$(\omega^2 - f^2)(\mathrm{i}ku + \mathrm{i}\varkappa v) - \mathrm{i}\omega\boldsymbol{k}^2 p = \mathrm{i}\omega(\mathrm{i}kX + \mathrm{i}\varkappa Y) + f(\mathrm{i}\varkappa X - \mathrm{i}kY)\,. \tag{4.3.8}$$

Finally we use (4.3.4e) to express the horizontal currents by w_z, differentiate the result with respect to z, and replace p_z by means of (4.3.5). Then we arrive at

$$(f^2 - \omega^2)\,w_{zz} + (\omega^2 - N^2)\,\boldsymbol{k}^2 w = \mathrm{i}\omega(\mathrm{i}kX_z + \mathrm{i}\varkappa Y_z) + f(\mathrm{i}\varkappa X_z - \mathrm{i}kY_z)\,. \tag{4.3.9a}$$

This equation forms together with the boundary conditions (4.3.2) an inhomogeneous boundary value problem for the vertical velocity. The right-hand side consists of the Fourier transform of time derivation of the wind divergence and of the vertical component of the wind curl. Thus in an infinite ocean wind induced internal waves may only be excited through variations of wind in space and time.

Let us suppose the solution of (4.3.9) with the boundary conditions (4.3.2) can be found; we show how to do this later. Then we may express the remaining observables u, v, and p in terms of w and the external forces X and Y. First, inserting (4.3.8) in (4.3.4e) yields an equation for the pressure

$$p = \mathrm{i}\,\frac{(f^2 - \omega^2)}{\boldsymbol{k}^2}\,w_z + \frac{1}{\boldsymbol{k}^2}\left[(\mathrm{i}kX + \mathrm{i}\varkappa Y) - \mathrm{i}\,\frac{f}{\omega}\,(\mathrm{i}\varkappa X - \mathrm{i}kY)\right]. \tag{4.3.9b}$$

An expression for u can be found from (4.3.7) using (4.3.4e) to eliminate v

$$u = \frac{1}{\omega\boldsymbol{k}^2}\,[(\mathrm{i}k\omega - f\varkappa)\,w_z + \mathrm{i}\varkappa^2 X - \mathrm{i}k\varkappa Y]\,. \tag{4.3.9c}$$

Finally we insert (4.3.9c) in (4.3.4e) and obtain

$$v = \frac{1}{\omega\boldsymbol{k}^2}\,[(\mathrm{i}\varkappa\omega + fk)\,w_z - \mathrm{i}k\varkappa X + \mathrm{i}\boldsymbol{k}^2 Y]\,. \tag{4.3.9d}$$

Now for any given BVF and reasonably defined external forcing the oceanic response in the internal wave band can be found by estimating w and then using (4.3.9b, c, d) to calculate u, v, and p.

We consider again the wind to act like a body force evenly distributed in an upper mixed layer of thickness H_{mix}

$$X(\omega, \boldsymbol{k}, z) = \hat{X}(\omega, \boldsymbol{k})\,\theta(z + H_{\text{mix}}) \tag{4.3.10a}$$

$$Y(\omega, \boldsymbol{k}, z) = \hat{Y}(\omega, \boldsymbol{k})\,\theta(z + H_{\text{mix}})\,. \tag{4.3.10b}$$

With (4.3.10) we may rewrite (4.3.9a)

$$w_{zz} + \frac{(N^2 - \omega^2)}{(\omega^2 - f^2)}\,\boldsymbol{k}^2 w = \frac{\delta(z + H_{\text{mix}})}{(f^2 - \omega^2)}\,[\mathrm{i}\omega(\mathrm{i}k\hat{X} + \mathrm{i}\varkappa\hat{Y}) + f(\mathrm{i}\varkappa\hat{X} - \mathrm{i}k\hat{Y})]\,. \tag{4.3.11}$$

Because of the z-derivatives of the forcing function occurring in (4.3.9a) the step functions in (4.3.10) lead to delta functions and, hence, (4.3.11) is practically a Green's function equation apart from the bracket on the right-hand side. Thus the forcing problem is formally solved if the Green's function has been determined.

In the following sections we deal with the calculation of the Green's functions corresponding to (4.3.11) for some examples of the BVF, $N^2(z)$.

4.4. GREEN'S FUNCTION FOR CONSTANT BVF

The Green's function which correspond to the boundary value problem (4.3.11) and (4.3.2) is subject to

$$G_{zz}(z, z') + \alpha^2 G(z, z') = \delta(z - z') \tag{4.4.1}$$

$$G(0, z') = G(-H, z') = 0 \tag{4.4.2}$$

$$\alpha^2 = k^2 \frac{N^2 - \omega^2}{\omega^2 - f^2}. \tag{4.4.3}$$

Generally the BVF varies with depth. This implies that for a certain frequency ω there may exist regions in the water column with both $N(z) < \omega$ and $N(z) > \omega$. The depth where $N(-z_T) = \omega$ is usually called "turning depth". Let us assume, just for definiteness, $N(z) > \omega$ for $z > -z_T$ and $N(z) < \omega$ for $z < -z_T$, see Fig. 4.4.1. Then α^2 is positive in the region $z > -z_T$ which implies an oscillatory behaviour of the Green's function. In the region $z < -z_T$, however, α^2 is negative and so the Green's function is evanescent. Thus, only in the region $z > -z_T$ real wave motion may occur. In other words the variation of the BVF in depth implies wave guide properties with regard to high frequency internal waves. We return to this phenomenon in the following sections. In this section we start with the simple case of a constant BVF.

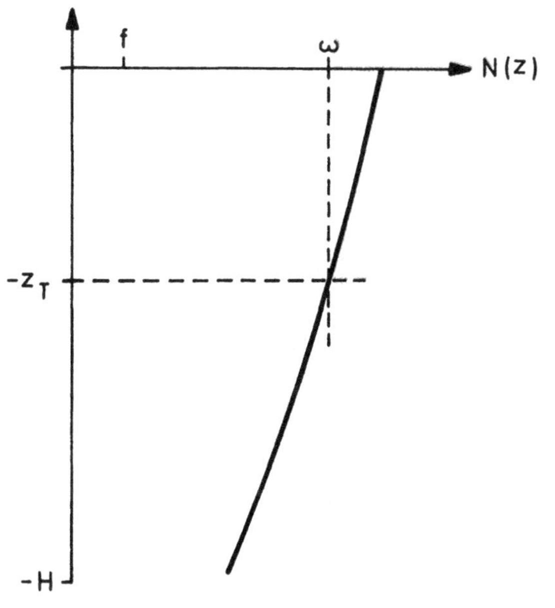

Fig. 4.4.1
A sketch of a BVF with one turning point

In that case the solution of (4.4.1) and (4.4.2) is already known from the example (2.4.29), (2.4.31). Thus we obtain

$$G(z, z') = \theta(z - z')\, G^{<}(z, z') + \theta(z' - z)\, G^{>}(z, z') \tag{4.4.4a}$$

$$G^{<}(z, z') = G^{>}(z', z) = \frac{\sin \alpha z \sin[\alpha(H + z')]}{\alpha \sin \alpha H}\,. \tag{4.4.4b}$$

Note that z and z' vary on the negative z-axis.

The equivalent bilinear expansion of G in terms of the eigenfunctions of the corresponding Sturm Liouville problem is, according to (2.4.44),

$$G(z, z') = \frac{2}{H} \sum_{n=1}^{\infty} \frac{\sin \alpha_n z \sin \alpha_n z'}{\alpha^2 - \alpha_n^2} \tag{4.4.5}$$

where

$$\alpha_n = \frac{n\pi}{H}, \qquad n = 1, 2, 3, \ldots .$$

We shortly summarize the analytical properties of $G = G(\omega, \boldsymbol{k}, z, z')$ on the complex ω-plane: First we note that G is analytic for $\alpha \to 0$, or equivalently $\omega \to N$

$$G^{>}(z, z') = \frac{z(z' + H)}{H} = G^{<}(z', z)\,. \tag{4.4.6}$$

Moreover, we observe that G has poles at $\alpha = \alpha_n = n\pi/H$. These poles define the well-known internal wave dispersion relation for constant N^2

$$\omega^2 = \omega_n^2(k) = \frac{(k^2 N^2 + \alpha_n^2 f^2)}{k^2 + \alpha_n^2}\,. \tag{4.4.7}$$

From (4.4.3) it can be seen that α is real within the internal wave frequency band, i.e., $N \gtrsim \omega \gtrsim f$, but α becomes imaginary for both frequencies larger than N and less than f. According to (4.4.4) G behaves oscillatory for real α while for imaginary α the Green's function is evanescent, i.e., localized around $z = z'$.

Since $G(z, -H_{\text{mix}})$ is closely related to w, compare (4.4.1) and (4.3.11), we may conclude that the vertical motion varies over the entire water column for frequencies within the internal wave band, but outside the internal wave band the motion appears to be confined to the vicinity of the bottom of the wind mixed layer.

Moreover, we see from (4.4.3) that for fixed wavenumber $\boldsymbol{k}$, α diminish for ω close to N, but α goes to infinity as ω approaches f. This implies that the wave function w oscillates in the vertical rapidly near the inertial frequency but slowly near the BVF.

4.5. GREEN'S FUNCTION FOR SLOWLY VARYING BVF

Now we consider BVF's which slowly vary with depth. In those cases we may use the WKB method in order to find an approximate Green's function. For $z \neq z'$ the two linearly independent solutions of (4.4.1) are in the WKB treatment

$$\frac{1}{\alpha^{1/2}(z)} \sin\left[\int_{-H}^{z} \alpha(\zeta)\, d\zeta\right] \quad \text{and} \quad \frac{1}{\alpha^{1/2}(z)} \sin\left[\int_{z}^{0} \alpha(\zeta)\, d\zeta\right].$$

The former expression vanishes at the bottom $z = -H$, and the latter one at the sea surface. The corresponding Wronskian is

$$W_r = -\sin\left[\int_{-H}^{0} \alpha(\zeta)\,d\zeta\right] \tag{4.5.1}$$

where in accordance with the WKB treatment derivatives of $\alpha(z)$ have been ignored.

Thus the Green's function may be constructed and we get

$$G(z, z') = \theta(z - z')\,G^{<}(z, z') + \theta(z' - z)\,G^{>}(z', z) \tag{4.5.2a}$$

$$G^{<}(z, z') = \frac{\sin\left[\int_{-H}^{z'} \alpha(\zeta)\,d\zeta\right] \sin\left[\int_{z}^{0} \alpha(\zeta)\,d\zeta\right]}{-\left[\alpha(z)\,\alpha(z')\right]^{1/2} \sin\left[\int_{-H}^{0} \alpha(\zeta)\,d\zeta\right]} = G^{>}(z', z)\,. \tag{4.5.2b}$$

The Green's function has poles at $\int_{-H}^{0} \alpha(\zeta)\,d\zeta = n\pi$, which implicitly define the WKB dispersion relation

$$n\pi = \frac{|\boldsymbol{k}|}{(\omega^2 - f^2)^{1/2}} \int_{-H}^{0} (N^2(z) - \omega^2)^{1/2}\,dz\,. \tag{4.5.3}$$

Let $N^2(z)$ vary between N^2_{min} and N^2_{max}. Then G oscillates for $\omega^2 < N^2_{min}$ while for $\omega^2 > N^2_{max}$ the Green's function is trapped at $z = z'$.

However, for frequencies between N_{min} and N_{max} there exist at least one turning depth $N^2(-z_T) = \omega^2$ where α^2 vanishes and consequently $G(z, z')$ has a singularity. That is the Green's function has a singularity at the turning point, which cannot be attributed to any dispersion curve, rather it indicates the breakdown of the WKB treatment at turning points.

On the other hand, the theory of linear differential equations asserts that if $\alpha(z)$ is analytic, then the exact solutions of the differential equation (4.4.1) are regular. In the next section we will show how this problem can be resolved.

4.6. UNIFORM VALID SOLUTION FOR ONE TURNING POINT

We may resolve the problem outlined at the end of the foregoing section by use of asymptotic matching to construct approximate solutions of differential equations with a turning point.

Suppose that near the turning point

$$\alpha^2(z) \approx b(z + z_T) \tag{4.6.1a}$$

with

$$b = 2N(-z_T)\,N'(-z_T)\,\frac{k^2}{\omega^2 - f^2}\,. \tag{4.6.1b}$$

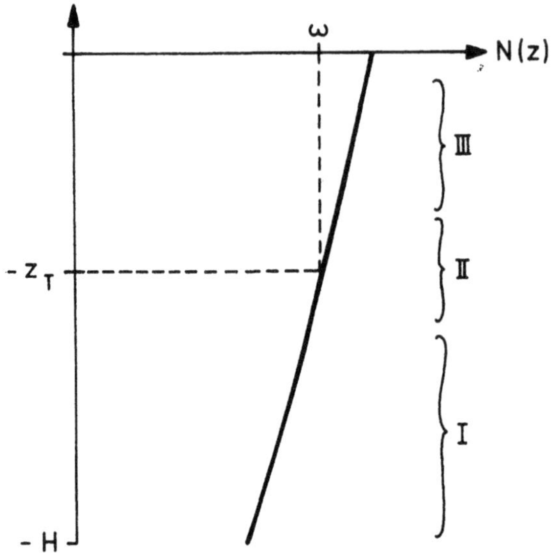

Fig. 4.6.1
Sketch of a BVF with one turning point with indication of the three different regions used for the construction of an uniformly valid approximate solution

For definiteness, we assume that α^2 has a positive slope at $z = -z_T$, and α^2 is positive above the turning point, i.e. $z > -z_T$, and negative below, compare Fig. 4.6.1. Moreover, we put $z' = -H_{mix}$ and assume the turning depth to be located below H_{mix}. Actually we need in the end $G(z, -H_{mix})$ which, apart from a factor, is proportional to the vertical velocity w, compare (4.3.11).

Our analysis of the one turning point problem proceeds as follows: We divide the z-axis between $-H_{mix}$ and $-H$ into three regions; region I with $-H \leqq z < -z_T$, region II with $z \approx -z_T$, and region III with $-z_T < z$, as indicated in Fig. 4.6.1. In regions I and III the WKB approximation is valid. In region II we find from (4.4.1) and (4.6.1) the approximate equation

$$\psi''(z) + b(z + z_T)\,\psi = 0\,. \tag{4.6.2}$$

This is the Airy equation which can be solved in terms of Airy functions. The regions I and II and the regions II and III have an overlap in which both the WKB and the Airy function approximations are valid. This enables us to match together asymptotically the solutions in the various regions. From the matching we obtain three formulas which together constitute a global approximation of the solution of (4.4.1) with $z' = -H_{mix}$.

We start our analysis of the problem in region III. The WKB approximation in this region has the form

$$\psi_{III}(z) = \tilde{A}\,\frac{e^{iQ}}{\alpha^{1/2}} + \tilde{B}\,\frac{e^{iQ}}{\alpha^{1/2}} = \frac{A\sin Q}{\alpha^{1/2}} + \frac{B\cos Q}{\alpha^{1/2}} \tag{4.6.3}$$

with

$$Q = \int_{-z_T}^{z} \alpha(\zeta)\,\mathrm{d}\zeta \tag{4.6.4 a}$$

and

$$\alpha = |\boldsymbol{k}|\left(\frac{N^2(z) - \omega^2}{\omega^2 - f^2}\right)^{1/2}. \tag{4.6.4 b}$$

We have arbitrarily chosen the lower limit of integration in (4.6.4a) to lie at the turning point. This choice is not necessary, but it simplifies expressions appearing later. In region II we have to solve the approximate equation (4.6.2). The general solution of this equation is a linear combination of Airy functions

$$\psi_{\mathrm{II}}(z) = C\,\mathrm{Ai}(-b^{1/3}(z + z_{\mathrm{T}})) + D\,\mathrm{Bi}(-b^{1/3}(z + z_{\mathrm{T}}))\,. \tag{4.6.5}$$

For $z > -z_{\mathrm{T}}$, i.e., above the turning point, the argument of the Airy functions are negative, and for $z < -z_{\mathrm{T}}$, i.e., beneath the turning points, positive. The asymptotic expansions of Ai and Bi are for negative arguments (Abramowitz and Stegun 1984)

$$\mathrm{Ai}(-b^{1/3}(z_{\mathrm{T}} - |z|)) \sim \frac{\sin\left(\frac{2}{3}\, b^{1/2}(z_{\mathrm{T}} - |z|)^{3/2} + \frac{\pi}{4}\right)}{\sqrt{\pi}\, b^{1/12}(z_{\mathrm{T}} - |z|)^{1/4}} \tag{4.6.6a}$$

$$\mathrm{Bi}(-b^{1/3}(z_{\mathrm{T}} - |z|)) \sim \frac{\cos\left(\frac{2}{3}\, b^{1/2}(z_{\mathrm{T}} - |z|)^{3/2} + \frac{\pi}{4}\right)}{\sqrt{\pi}\, b^{1/12}(z_{\mathrm{T}} - |z|)^{1/4}} \tag{4.6.6b}$$

and for positive arguments

$$\mathrm{Ai}(b^{1/3}(|z| - z_{\mathrm{T}})) \sim \frac{\exp\left[-\frac{2}{3}\, b^{1/2}(|z| - z_{\mathrm{T}})^{3/2}\right]}{2\sqrt{\pi}\, b^{1/12}(|z| - z_{\mathrm{T}})^{1/4}} \tag{4.6.7a}$$

$$\mathrm{Bi}(b^{1/3}(|z| - z_{\mathrm{T}})) \sim \frac{\exp\left[\frac{2}{3}\, b^{1/2}(|z| - z_{\mathrm{T}})^{3/2}\right]}{\sqrt{\pi}\, b^{1/12}(|z| - z_{\mathrm{T}})^{1/4}}\,. \tag{4.6.7b}$$

In region I we obtain the WKB solution

$$\psi_{\mathrm{I}} = \frac{E\,\mathrm{e}^{\tilde{Q}}}{\tilde{\alpha}^{1/2}} + \frac{F\,\mathrm{e}^{-\tilde{Q}}}{\tilde{\alpha}^{1/2}} \tag{4.6.8}$$

with the notations

$$\tilde{\alpha} = |\boldsymbol{k}|\left(\frac{\omega^2 - N^2(z)}{\omega^2 - f^2}\right)^{1/2} \tag{4.6.9a}$$

and

$$\tilde{Q}(z) = \int_{z}^{-z_T} \tilde{\alpha}(\zeta)\,\mathrm{d}\zeta\,. \tag{4.6.9b}$$

The solution ψ_{I} must vanish at $z = -H$, i.e., at the bottom. Thus we have to choose

$$E = -\frac{1}{2}\,\mathrm{e}^{-\tilde{Q}(-H)} \tag{4.6.10a}$$

$$F = \frac{1}{2}\,\mathrm{e}^{\tilde{Q}(-H)}\,. \tag{4.6.10b}$$

Now we consider the overlap region between II and I where ψ_{II} and ψ_I have a common region of validity. Inside this overlap region both functions are approximate solutions to the same differential equation and therefore they must match asymptotically.

First we consider ψ_I. In the overlap region $|z| - z_T$ is small so that $\tilde{\alpha}^2$ is approximately $b(|z| - z_T)$. Therefore

$$\tilde{\alpha}^{1/2} \sim b^{1/4}(|z| - z_T)^{1/4} \quad \text{and} \quad \tilde{Q}(z) \sim b^{1/2}\frac{2}{3}(|z| - z_T)^{3/2} . \tag{4.6.11}$$

Hence,

$$\psi_I \approx \frac{E \exp\left[b^{1/2}\frac{2}{3}(|z| - z_T)^{3/2}\right] + F \exp\left[-b^{1/2}\frac{2}{3}(|z| - z_T)^{3/2}\right]}{b^{1/4}(|z| - z_T)^{1/4}} . \tag{4.6.12}$$

Next we compare this with (4.6.5) using the asymptotic behaviour according to (4.6.7)

$$\psi_{II} \approx \frac{\frac{1}{2} C \exp\left[-b^{1/2}\frac{2}{3}(|z| - z_T)^{3/2}\right] + D \exp\left[b^{1/2}\frac{2}{3}(|z| - z_T)^{3/2}\right]}{b^{1/12}(|z| - z_T)^{1/4}} . \tag{4.6.13}$$

We observe that (4.6.12) and (4.6.13) have the same functional form and can, therefore, be matched. Requiring that both formulas match on the overlap region determines the constants C and D

$$C = \frac{2\sqrt{\pi}\, F}{b^{1/6}} \tag{4.6.14a}$$

$$D = \frac{\sqrt{\pi}\, E}{b^{1/6}} . \tag{4.6.14b}$$

Next we must match the solutions in the overlap region of II and III. In this region we have $\alpha^{1/2} \sim b^{1/4}(z_T - |z|)^{1/4}$ and

$$Q(z) \sim b^{1/2}\frac{2}{3}(z_T - |z|)^{3/2} . \tag{4.6.15}$$

From (4.6.3) and (4.6.6) it follows

$$\psi_{III} \approx \frac{A \sin\left[b^{1/2}\frac{2}{3}(z_T - |z|)^{3/2}\right] + B \cos\left[b^{1/2}\frac{2}{3}(z_T - |z|)^{3/2}\right]}{b^{1/4}(z_T - |z|)^{1/4}} . \tag{4.6.16}$$

On the other hand we obtain from (4.6.5), (4.6.6), and (4.6.14)

$$\psi_{II} \approx \frac{2F \sin\left[b^{1/2}\frac{2}{3}(z_T - |z|)^{3/2} + \frac{\pi}{4}\right] + E \cos\left[b^{1/2}\frac{2}{3}(z_T - |z|)^{3/2} + \frac{\pi}{4}\right]}{b^{1/4}(z_T - |z|)^{1/4}} . \tag{4.6.17}$$

Matching of (4.6.16) and (4.6.17) yields

$$A = \frac{\sqrt{2}}{2}(2F - E) \tag{4.6.18a}$$

$$B = \frac{\sqrt{2}}{2}(2F + E) \tag{4.6.18b}$$

where the identities $\cos \pi/4 = \sin \pi/4 = \sqrt{2}/2$, and $\sqrt{2}\sin(x + \pi/4) = \sin x + \cos x$, and $\sqrt{2}\cos(x + \pi/4) = \cos x - \sin x$ have been used.

In summary, we have found approximations to $\psi(z)$ in each of regions I, II, and III. These approximations are

$$\psi_{\mathrm{I}}(z) \approx \frac{E\,\mathrm{e}^{\tilde{Q}}}{\tilde{\alpha}^{1/2}} + \frac{F\,\mathrm{e}^{-\tilde{Q}}}{\tilde{\alpha}^{1/2}} = \frac{1}{\tilde{\alpha}^{1/2}} \sinh\left(\int_{-H}^{z} \alpha(\zeta)\,\mathrm{d}\zeta\right) \tag{4.6.19a}$$

$$\psi_{\mathrm{II}}(z) \approx \frac{\pi^{1/2}}{b^{1/6}}\left[2F\,\mathrm{Ai}(-b^{1/3}(z + z_{\mathrm{T}})) + E\,\mathrm{Bi}(-b^{1/3}(z + z_{\mathrm{T}}))\right] \tag{4.6.19b}$$

$$\psi_{\mathrm{III}}(z) \approx \frac{2F}{\alpha^{1/2}} \sin\left(Q + \frac{\pi}{4}\right) + \frac{E}{\alpha^{1/2}} \cos\left(Q + \frac{\pi}{4}\right) \tag{4.6.19c}$$

where E, F are given by (4.6.10), and b by (4.6.1a).

Using Langer's method, see e.g. BENDER and ORSZAG (1978) we may replace all three formulas (4.6.19) by a single formula which is an uniformly valid approximation to $\psi(z)$.

$$\psi(z) = \left(\frac{3}{2}Q\right)^{1/6} \frac{\pi^{1/2}}{\alpha^{1/2}}\left[2F\,\mathrm{Ai}\left(-\left(\frac{3}{2}Q\right)^{2/3}\right) + E\,\mathrm{Bi}\left(-\left(\frac{3}{2}Q\right)^{2/3}\right)\right]. \tag{4.6.20}$$

We verify this formula in all three regions. In region III Q can assumed to be large and we may use the asymptotic formulas (4.6.6) for the argument $(3/2Q)^{2/3}$, and then we obtain ψ_{III}. In the region II we expand Q around the turning point according to (4.6.15) and it follows $(3/2Q)^{2/3} \sim b^{1/2}(z + z_{\mathrm{T}})$, and $(3/2Q)^{1/6}/\alpha^{1/2} = b^{-1/6}$, and, therefore, (4.6.20) equals (4.6.19b) in region II. Finally we consider region I. Here one has to be carefully about positive and negative signs. With

$$Q(z) = -\mathrm{i}\int_{z}^{-z_T} \tilde{\alpha}(\zeta)\,\mathrm{d}\zeta = -\mathrm{i}\tilde{Q}(z)$$

and $(-\mathrm{i})^{2/3} = -1$ (note that $-\mathrm{i} = \mathrm{e}^{\mathrm{i}\cdot 3\pi/2}$, hence $(-\mathrm{i})^{2/3} = \mathrm{e}^{\mathrm{i}\pi} = \cos\pi = -1$), we find from the asymptotic representations (4.6.7)

$$\mathrm{Ai}\left(-\left(\frac{3}{2}Q\right)^{2/3}\right) \approx \frac{\mathrm{e}^{-\tilde{Q}}}{2\sqrt{\pi}\left(\frac{3}{2}\tilde{Q}\right)^{1/6}}$$

and

$$\mathrm{Bi}\left(-\left(\frac{3}{2}Q\right)^{2/3}\right) \approx \frac{\mathrm{e}^{\tilde{Q}}}{\sqrt{\pi}\left(\frac{3}{2}\tilde{Q}\right)^{1/6}}.$$

Next we estimate

$$\frac{\left(\frac{3}{2}Q\right)^{1/6}}{\alpha^{1/2}} = \frac{\left(\frac{3}{2}\tilde{Q}\right)^{1/6}}{\tilde{\alpha}^{1/2}} \frac{(-i)^{1/6}}{i^{1/2}} = \frac{\left(\frac{3}{2}\tilde{Q}\right)^{1/6}}{\tilde{\alpha}^{1/2}}$$

where $(-i)^{1/6} = (e^{i\cdot 3\pi/2})^{1/6} = e^{i\pi/4} = i^{1/2}$ was used. Then we obtain from (4.5.20) the expression (4.6.19a). Thus the equivalence of (4.6.20) and of the set (4.6.19) has been shown.

With $\psi(z)$ one solution of (4.4.1) for $z = z'$ is given. The second linear independent solution is $\varphi(z) = \frac{1}{\alpha^{1/2}} \sin\left[\int_z^0 \alpha(\zeta)\, d\zeta\right]$ which fulfills the condition $\varphi(0) = 0$. Now we may construct the Green's function $G(z, -H_{mix})$, note that $H_{mix} < z_T$ was presumed. First we estimate the Wronskian at $z = z' = -H_{mix}$

$$W_r(-H_{mix}) = \varphi'(-H_{mix})\,\psi(-H_{mix}) - \varphi(-H_{mix})\,\psi'(-H_{mix})\,.$$

From (4.6.10) follows that $F \gg E$ can be assumed, whence we may use

$$\psi(-H_{mix}) \approx \frac{2F}{\alpha^{1/2}} \sin\left(Q(-H_{mix}) + \frac{\pi}{4}\right).$$

Thus we arrive at

$$W_r \approx 2F \sin\left[\int_{z_T}^0 \alpha(\zeta)\, d\zeta + \frac{\pi}{4}\right]. \tag{4.6.21}$$

With (4.6.21) we obtain the approximate Green's function

$$G(z, -H_{mix}) = \theta(z + H_{mix})\, G^<(z, -H_{mix}) + \theta(-z - H_{mix})\, G^>(z, -H_{mix}) \tag{4.6.22a}$$

where

$$G^<(z, -H_{mix}) = -\frac{\psi(-H_{mix})}{\alpha^{1/2}(z)} \frac{\sin\left(\int_z^0 \alpha(\zeta)\, d\zeta\right)}{2F \sin\left(\int_{-z_T}^0 \alpha(\zeta)\, d\zeta + \frac{\pi}{4}\right)} = G^>(-H_{mix}, z)\,. \tag{4.6.22b}$$

The poles of the Green's function (4.6.22) are determined by the zeros of the Wronskian (4.6.21)

$$\int_{-z_T}^0 \alpha(\zeta)\, d\zeta = \left(n - \frac{1}{4}\right)\pi\,. \tag{4.6.23}$$

This defines the internal wave dispersion relation in the one turning point WKB treatment.

4.7. THE TWO TURNING POINT CASE

In the present section we will examine the problem where α^2 given by (4.4.3) has two simple turning points at $z = -z_1$, and $z = -z_2$, see Fig. 4.7.1. We introduce the notations

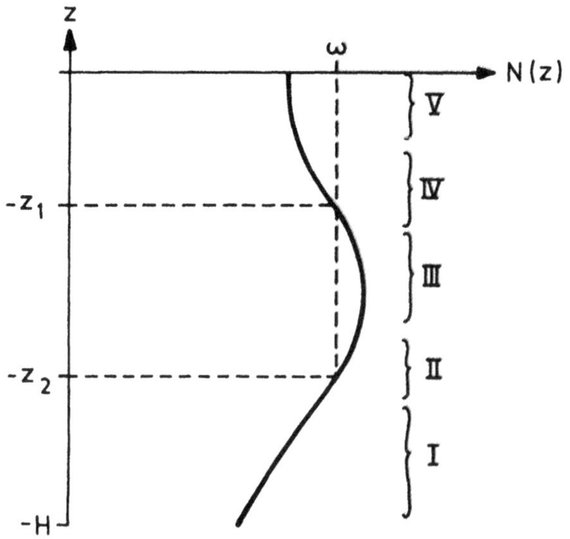

Fig. 4.7.1
Sketch of a BVF with two turning points and with indication of the five regions considered for the construction of an uniformly valid approximate solution

$$\alpha^2(z) \approx -a(z + z_1) \qquad \text{for } z \text{ near } z_1 \tag{4.7.1 a}$$

with

$$a = 2N(-z_1)\,|N'(z - z_1)|\,\frac{k^2}{\omega^2 - f^2} \tag{4.7.1 b}$$

and

$$\alpha^2(z) \approx b(z + z_2) \qquad \text{for } z \text{ near } z_2 \tag{4.7.2 a}$$

with

$$b = 2N(-z_2)\,N'(-z_a)\,\frac{k^2}{\omega^2 - f^2} \tag{4.7.2 b}$$

where $N'(-z_1) < 0$ and $N'(-z_2) > 0$ was assumed.

Again we wish to construct an approximate Green's function $G(z, z')$ which is subject to the equations (4.4.1) and (4.4.2). Let z' be located between the turning points as indicated in Fig. 4.7.1. We need two linear independent solutions of (4.4.1) for $z \neq z'$, each must fulfil one of the boundary conditions (4.4.2). We may divide the z-axis in five regions I to V, beginning from the bottom to the sea surface, as depicted in Fig. 4.7.1. For the regions I to III we use the uniform valid approximation $\psi(z)$ given by (4.6.20) which vanishes at the bottom $z = -H$. The second solution $\varphi(z)$ can be estimated in the same manner. We start with region V and find the WKB approximation

$$\varphi_{\mathrm{V}}(z) = \frac{1}{\tilde{\alpha}} [E_* e^{\tilde{Q}_*} + F_* e^{-\tilde{Q}_*}] \tag{4.7.3}$$

with

$$\tilde{\alpha} = |\boldsymbol{k}| \left(\frac{\omega^2 - N^2}{\omega^2 - f^2}\right)^{1/2} \tag{4.7.4a}$$

$$\tilde{Q}_*(z) = \int_{-z_1}^{z} \tilde{\alpha}(\zeta)\, \mathrm{d}\zeta . \tag{4.7.4b}$$

The constants E_* and F_* are chosen such that $\varphi_{\mathrm{V}}(0) = 0$, i.e.,

$$E_* = -\frac{1}{2} e^{-\tilde{Q}_*(0)} \tag{4.7.5a}$$

$$F_* = \frac{1}{2} e^{\tilde{Q}_*(0)} . \tag{4.7.5b}$$

In the region IV from (4.4.1) and (4.7.1) it follows

$$\varphi_{\mathrm{IV}}'' - a(z + z_1)\, \varphi_{\mathrm{IV}} = 0 . \tag{4.7.6}$$

The solution can be expressed in terms of Airy functions

$$\varphi_{\mathrm{IV}}(z) = \frac{\sqrt{\pi}}{\alpha^{1/2}} [2F_* \operatorname{Ai}(a^{1/3}(z + z_1)) + E_* \operatorname{Bi}(a^{1/3}(z + z_1))] . \tag{4.7.7}$$

In region III the WKB treatment yields

$$\varphi_{\mathrm{III}}(z) = \frac{1}{\alpha^{1/2}} \left[2F_* \sin\left(Q_* + \frac{\pi}{4}\right) + E_* \cos\left(Q_* + \frac{\pi}{4}\right)\right] \tag{4.7.8}$$

where

$$\alpha = |\boldsymbol{k}| \left(\frac{N^2 - \omega^2}{\omega^2 - f^2}\right)^{1/2} \tag{4.7.9a}$$

and

$$Q_* = \int_{z}^{-z_1} \alpha(\zeta)\, \mathrm{d}\zeta . \tag{4.7.9b}$$

These three solutions have been matched together asymptotically in the manner outlined in the previous section. Again we may replace all three formulas (4.7.3), (4.7.7), and (4.7.8) by a single formula

$$\varphi(z) = \left(\frac{3}{2} Q_*\right)^{1/6} \frac{\sqrt{\pi}}{\alpha^{1/2}} \left[2F_* \operatorname{Ai}\left(\left(\frac{3}{2} Q_*\right)^{2/3}\right) + E_* \operatorname{Bi}\left(\left(\frac{3}{2} Q_*\right)^{2/3}\right)\right] . \tag{4.7.10}$$

Next we estimate the Wronskian of φ and ψ. Using the fact that z' has chosen between $-z_1$ and $-z_2$ we have

$$W_r(z') \approx -4FF_* \sin\left(Q_* + \tilde{Q}_* + \frac{\pi}{2}\right) . \tag{4.7.11}$$

The Green's function is given by

$$G(z, z') = \theta(z - z')\, G^<(z, z') + \theta(z' - z)\, G^>(z, z') \tag{4.7.12a}$$

with

$$G^<(z, z') = \frac{\varphi(z)\, \psi(z')}{W_r} = G^>(z', z)\,. \tag{4.7.12b}$$

The poles of the Green's function determine the two turning point dispersion of internal waves in the WKB treatment

$$\varrho_* + \varrho_* + \frac{\pi}{2} = n\varkappa$$

or

$$\int_{-z_2}^{-z_1} \alpha(\zeta)\, d\zeta = \left(n - \frac{1}{2}\right)\pi\,. \tag{4.7.13}$$

4.8. GREEN'S FUNCTION FOR EXPONENTIAL BVF

In the preceding section we considered slowly varying BVF's without specification of the analytical form of $N(z)$. Now we deal with the important example of the exponential BVF which has been considered in Subsection 1.9.4. and which is referred to GARRETT and MUNK (1972)

$$N(z) = \theta(z + H_{mix})\, N_1 + \theta(-z - H_{mix})\, N_2\, e^{a(z + H_{mix})}\,. \tag{4.8.1}$$

First we estimate the Green's function $G(z, -H_{mix})$ governed by the equations (4.4.1), (4.4.2), and (4.8.1). In the upper layer we have with

$$\alpha_1^2 = k^2 \frac{N_1^2 - \omega^2}{\omega^2 - f^2} \tag{4.8.2}$$

$$\varphi_{zz} + \alpha_1^2 \varphi = 0\,.$$

The solution with $\varphi(0) = 0$ is

$$\varphi(z) = \sin \alpha_1 z\,. \tag{4.8.3}$$

In the lower layer it follows

$$\psi_{zz} + \frac{k^2}{\omega^2 - f^2}\left(N_2^2\, e^{2a(z + H_{mix})} - \omega^2\right)\psi = 0\,. \tag{4.8.4}$$

Introducing

$$q(z) = \frac{\beta}{\omega} N_2\, e^{a(z + H_{mix})} \tag{4.8.5a}$$

with

$$\beta^2 = \frac{k^2}{a^2\left(1 - \frac{f^2}{\omega^2}\right)} \tag{4.8.5 b}$$

equation (4.8.4) transforms into

$$q^2\psi''(q) + q\psi'(q) + (q^2 - \beta^2)\,\psi(q) = 0\,. \tag{4.8.6}$$

This is the Bessel differential equation which has the linear independent solutions $J_\beta(q)$, $J_\beta(q)$ or $J_{-\beta}(q)$ and $Y_\beta(q)$. We choose the latter pair of Bessel and Neumann functions because these solutions are also for integer β linear independent. Thus the solution with $\psi(-H) = 0$ is

$$\psi(q(z)) = J_\beta(q(z)) - \frac{J_\beta(q(-H))}{Y_\beta(q(-H))}\,Y_\beta(q(z))\,. \tag{4.8.7}$$

Next we estimate the Wronskian of φ and ψ

$$W_r = \varphi_z(z)\,\psi(q) - q_z\varphi(z)\,\psi'(q)$$

and we obtain for $z = -H_{mix}$

$$\begin{aligned} W_r = {}& \alpha_1 \cos \alpha_1 H_{mix}\left[J_\beta\left(\frac{\beta N_2}{\omega}\right) - \frac{J_\beta(q(-H_{mix}))}{Y_\beta(q(-H))}\,Y_\beta\left(\frac{\beta N_2}{\omega}\right)\right] \\ & + aN_2\,\frac{\beta}{\omega}\,\sin \alpha_1 H_{mix}\left[J'\left(\frac{\beta N_2}{\omega}\right) - \frac{J_\beta(q(-H))}{Y_\beta(q(-H))}\right] Y'_\beta\left(\frac{\beta N_2}{\omega}\right). \end{aligned} \tag{4.8.8}$$

Hence, the Green's function defined by

$$G(z, -H_{mix}) = \theta(z + H_{mix})\,G^<(z, -H_{mix}) + \theta(-z - H_{mix})\,G^>(z, -H_{mix}) \tag{4.8.9 a}$$

$$G^<(z, -H_{mix}) = \frac{\varphi(z)\,\psi(-H_{mix})}{W_r} = G^>(-H_{mix}, z) \tag{4.8.9 b}$$

is completely determined by (4.8.3), (4.8.7), and (4.8.8). The poles of $G(z, -H_{mix})$, or equivalently the zeros of the Wronskian determine the dispersion relation of internal waves for the BVF (4.8.1). In the following sections we will discuss some analytical approximations of the dispersion relation. Now we show that $\psi(z)$ and, hence, $G^>(z, -H_{mix})$ diminishes exponentially below the turning depth $z_T(\omega)$. Let z_T be determined by $\omega = N_2 e^{-az_T}$, i.e., $z_T = \frac{1}{a}\ln(\omega/N_2)$, and consider $\psi(z)$ beneath $z = z_T$.
Then we have

$$q(z) = \frac{\beta N_2}{\omega}\,e^{a(z + H_{mix})} \approx \beta\,e^{-a(|z| - z_T)} = \beta\xi \tag{4.8.10}$$

where $H_{mix} \ll |z|$ has assumed and ξ is defined by (4.8.10).

Now we use the Airy approximations of Bessel functions which are powerful for large β (ABRAMOWITZ and STEGUN 1984, p. 112)

$$J_\beta(\beta\xi) \approx \left(\frac{4\zeta}{1-\xi^2}\right)^{1/4} \frac{\mathrm{Ai}(\beta^{2/3}\zeta)}{\beta^{1/3}} \tag{4.8.11 a}$$

$$Y_\beta(\beta\xi) \approx -\left(\frac{4\zeta}{1-\xi^2}\right)^{1/4} \frac{\mathrm{Bi}(\beta^{2/3}\zeta)}{\beta^{1/3}} \tag{4.8.11 b}$$

where

$$\frac{2}{3}\zeta^{3/2} = \ln\left[\frac{1+\sqrt{1-\xi^2}}{\xi}\right] - \sqrt{1-\xi^2}\,. \tag{4.8.11 c}$$

For $|z| > z_T$ we have in leading order

$$\zeta \sim \left[\frac{3a}{2}(|z| - z_T)\right]^{2/3}$$

Thus from (4.8.11) and from the asymptotic representations (4.6.7) it follows

$$J_\beta(\beta\xi) \sim \frac{2}{\sqrt{\pi}} \frac{e^{-|k|(|z|-z_T)}}{\beta^{1/6}}\,. \tag{4.8.12}$$

The second term of (4.8.7) decreases more rapidly so that $\psi(z)$ and hence $G^>(z, -H_{mix})$ drops in fact like an exponential below a turning point. Above a turning point the oscillatory behaviour of $\psi(z)$ is quite obvious.

4.9. SOME APPROXIMATIONS OF THE DISPERSION RELATION FOR EXPONENTIAL BVF

The dispersion relations of internal waves for the exponential BVF (4.8.1) are given by the zeros of the Wronskian (4.8.8), which may be rewritten as

$$\tan(\alpha_1 H_{mix}) = -\frac{(N_1^2-\omega^2)^{1/2}}{N_2} \frac{\psi\left(\dfrac{\beta N_2}{\omega}\right)}{\psi'\left(\dfrac{\beta N_2}{\omega}\right)} \tag{4.9.1}$$

where ψ is given by (4.8.7) with $q(-H_{mix}) = \beta N_2/\omega$, and ψ' is defined by

$$\psi'\left(\frac{\beta N_2}{\omega}\right) = J'_\beta\left(\frac{\beta N_2}{\omega}\right) - \frac{J_\beta(q(-H))}{Y_\beta(q(-H))} Y'_\beta\left(\frac{\beta N_2}{\omega}\right). \tag{4.9.2}$$

Clearly, the roots of (4.9.1) can be obtained numerically. But for some regions in the $\omega - k$ plane it is possible to derive approximate dispersion relations. The following discussion is close to the paper of GARRETT and MUNK (1972).

We consider first the effect of the mixed layer depth on the dispersion. Assume $\alpha_1 H_{mix}$ to be small. Then (4.9.1) becomes

$$\frac{|\boldsymbol{k}|\, H_{\text{mix}}}{\sqrt{\omega^2 - f^2}} = -\frac{1}{N_2}\,\frac{\psi\left(\dfrac{\beta N_2}{\omega}\right)}{\psi'\left(\dfrac{\beta N_2}{\omega}\right)}\,. \tag{4.9.3}$$

We try the ansatz

$$\frac{|\boldsymbol{k}|}{\sqrt{\omega^2 - f^2}} = z_n(1 - aH_{\text{mix}}) \tag{4.9.4}$$

where $z_n N_2/a$ is the n-th zero of $\psi(\beta N_2/\omega)$, and expand ψ for small aH_{mix} in a Taylor series

$$\psi\left(\frac{\beta N_2}{\omega}\right) = \psi\left(\frac{z_n(1 - aH_{\text{mix}})\, N_2}{a}\right) \approx \psi\left(\frac{z_n N_2}{a}\right) - z_n N_2 H_{\text{mix}} \psi'\left(\frac{z_n N_2}{a}\right). \tag{4.9.5}$$

Then with (4.9.5) and (4.9.4) we obtain

$$\frac{\boldsymbol{k}}{\omega} = z_n + \mathrm{O}(aH_{\text{mix}})\,. \tag{4.9.6}$$

Since, as a rule, aH_{mix} is in fact a small quantity in the real ocean we may expect that the correction to the z_n's given by the mixed layer depth are small even if $\alpha_1 H_{\text{mix}}$ is not small. This implies that we may study the approximate equation

$$\psi\left(\frac{\beta N_2}{\omega}\right) = 0 \tag{4.9.7}$$

instead of (4.9.1) in order to find essential properties of the dispersion relation.

Next we study the effect of the ocean depth. First we assume $\omega \gg N(-H)$. Then there must exist a turning point well above the bottom, implying that according to (4.8.10) the bottom is not felt. Hence, we may replace (4.9.7) by

$$J_\beta\left(\frac{\beta N_2}{\omega}\right) = 0\,. \tag{4.9.8}$$

If in addition $\omega \ll N_2$ is assumed we have from the asymptotic expression of the Bessel function $J_\beta(z) \approx \sqrt{2/(\pi z)}\, \sin(z - \beta\pi/2 + \pi/4)$

$$\frac{\beta N_2}{\omega} \approx n\pi - \frac{\pi}{4} + \frac{\beta\pi}{2}\,. \tag{4.9.9}$$

This is called the "infinite depth approximation" of the dispersion relation.

For $N(-H) \gg \omega$ we may replace all Bessel functions occurring in (4.9.7) by their asymptotic representations and we obtain with $Y_\beta(z) \approx \sqrt{2/(\pi z)}\, \sin(z - \beta\pi/2 - \pi/4)$

$$\frac{\beta N_2}{\omega} = \frac{n\pi}{(1 - \mathrm{e}^{-aH})}\,. \tag{4.9.10a}$$

it should be noted that the replacement of $J_\beta(q(-H))$ and $Y_\beta(q(-H))$ by their asymptotics requires large mode numbers so that (4.9.10) is valid only for higher modes. A somewhat better approximation of the zeros of (4.9.7) yields

$$\frac{\beta N_2}{\omega} = \frac{n\pi}{1 - e^{-aH}} - \frac{\left(\frac{4k^2}{a^2}\right) - 1}{8n} (e^{aH} - 1) + O\left(\frac{1}{n^3\pi^3}\right), \tag{4.9.10b}$$

see ABRAMOWITZ and STEGUN (1984), p. 118. We call (4.9.10a, b) the "finite depth approximation".

It can be shown that the approximate relations (4.9.9) and (4.9.10) may also be derived from the WKB formulas (4.6.23) and (4.5.3), respectively. We start with the finite depth approximation and find from (4.5.3)

$$\frac{|k|}{\sqrt{\omega^2 - f^2}} \int_{-H}^{0} dz \sqrt{N^2(z) - \omega^2} = n\pi .$$

Inserting $N^2(z) = N_2^2 e^{2az}$ the integral yields with $N(z)/\omega = \xi$ and $dz = aN(z)\, d\xi$

$$\int_{\frac{N(-H)}{\omega}}^{\frac{N_2}{\omega}} \frac{\sqrt{\xi^2 - 1}}{a\xi} d\xi \approx \frac{N_2}{a\omega} (1 - e^{-aH}) + O\left(\frac{\omega}{N_2}\right)$$

Thus (4.9.10a) follows.

In the infinite depth approximation with $\omega \gg N(-H)$ there exist a turning point at $z = -z_T$ with $\omega = N(-z_T)$. Then from (4.6.23) it follows

$$\frac{|k|}{\sqrt{\omega^2 - f^2}} \int_{-z_T}^{0} dz \sqrt{N^2(z) - \omega^2} = \left(n - \frac{1}{4}\right) \pi .$$

The integration yields

$$\int_{1}^{\frac{N_2}{\omega}} \frac{\sqrt{\xi^2 - 1}}{a\xi} d\xi = \frac{1}{a} \sqrt{\frac{N^2}{\omega^2} - 1} - \arccos \frac{\omega}{N_2} .$$

Using $\arccos x \approx \pi/2 - x$ and $N_2^2/\omega^2 \gg 1$ we arrive at (4.9.9).

Finally we consider the large wavenumber approximation. Large horizontal wavenumbers imply ω to be near N_2, compare, for instance, (4.4.7). Thus we may assume $\omega \gg f$ and $\beta = |k|/a$, and

$$\frac{N_2}{\omega} \approx \left(1 - \frac{N_2 - \omega}{N_2}\right)^{-1} \approx 1 + \frac{N_2 - \omega}{N_2} .$$

Using the Airy approximation for Bessel functions

$$J_\beta(\beta + z\beta)^{1/3} = \frac{2^{1/3}}{\beta^{1/3}} \mathrm{Ai}(-2^{1/3}z) + O\left(\frac{1}{\beta}\right), \tag{4.9.11}$$

(ABRAMOWITZ and STEGUN 1984) with $z = (N_2 - \omega/N_2)\, \beta^{3/2}$ we find

$$\left(\frac{2k^2}{a^2}\right)^{1/3} \left(1 - \frac{\omega}{N_2}\right) = 2.34, 4.09, \ldots, \left(\frac{3\pi(4n - 1)}{8}\right)^{2/3} \tag{4.9.12}$$

where the rightmost expression of (4.9.12) follows from the relation

$$\mathrm{Ai}\,(-x) \sim \sin\left(\frac{2}{3}x^{3/2} + \frac{\pi}{4}\right)$$

with the zeros $x_n^{3/2} = (4n - 1)(3\pi/8)$.

In the WKB treatment we get to the same approximation from (4.6.23)

$$|\boldsymbol{k}| \int\limits_{-z_T}^{0} \sqrt{\frac{N^2(z)}{\omega^2} - 1}\, \mathrm{d}z = \left(n - \frac{1}{4}\right)\pi .$$

With

$\frac{N^2}{\omega^2} - 1 \approx \frac{2(N - \omega)}{N}$ the integral

$$|\boldsymbol{k}| \int\limits_{-z_T}^{0} \sqrt{2}\sqrt{1 - \frac{\omega}{N}}\, \mathrm{d}z$$

yields

$$|\boldsymbol{k}| \frac{2^{3/2}}{3a}\left(1 - \frac{\omega}{N^2}\right)^{3/2}$$

and we arrive at

$$\left(2\frac{\boldsymbol{k}^2}{a^2}\right)^{1/3}\left(1 - \frac{\omega}{N_2}\right) = \left(\frac{(4n - 1)\,3\pi}{8}\right)^{2/3}$$

which is equivalent to (4.9.12).

4.10. INTERNAL WAVE SPECTRUM OF THE VERTICAL VELOCITY, AN ILLUSTRATIVE EXAMPLE

In the preceding sections the internal wave Green's function has been calculated for some examples of the BVF. We have already outlined that if the Green's function is known the oceanic response to external forcing is completely known in Fourier space. So we find for the vertical velocity from (4.3.11) and (4.4.1) respecting the assumed structure of the forcing according to (4.3.10)

$$w(z, \boldsymbol{k}, \omega) = \frac{G(z, -H_{\mathrm{mix}})}{f^2 - \omega^2}\left[\mathrm{i}\omega(\mathrm{i}k\hat{X} + \mathrm{i}\varkappa\hat{Y}) + f(\mathrm{i}\varkappa\hat{X} - \mathrm{i}k\hat{Y})\right] . \qquad (4.10.1)$$

Moreover, we have from (4.3.9b) and (4.10.1)

$$p(z, \boldsymbol{k}, \omega) = \left[G_z(z, -H_{\mathrm{mix}}) - \theta(z + H_{\mathrm{mix}})\right] \frac{\omega(\mathrm{i}k\hat{X} + \mathrm{i}\varkappa\hat{Y}) + f(\varkappa\hat{X} - k\hat{Y})}{\omega\boldsymbol{k}^2} , \qquad (4.10.2)$$

and inserting (4.10.1) into (4.3.9c) and (4.3.9d) it follows

$$u(z, \boldsymbol{k}, \omega) = \frac{1}{\omega \boldsymbol{k}^2}\left[\theta(z + H_{\mathrm{mix}})\,(\mathrm{i}\varkappa^2 \hat{X} - \mathrm{i}k\varkappa \hat{Y}) + \frac{G_z(z, -H_{\mathrm{mix}})}{f^2 - \omega^2}\left(-\mathrm{i}(\omega^2 k^2 + \varkappa^2 f^2)\,\hat{X} + (\omega f \boldsymbol{k}^2 + \mathrm{i}\varkappa k(f^2 - \omega^2))\,\hat{Y}\right)\right] \tag{4.10.3}$$

and

$$v(z, \boldsymbol{k}, \omega) = \frac{1}{\omega \boldsymbol{k}^2}\left[\theta(z + H_{\mathrm{mix}})\,(\mathrm{i}k^2 \hat{Y} - \mathrm{i}k\varkappa \hat{X}) + \frac{G_z(z, -H_{\mathrm{mix}})}{f^2 - \omega^2}\left(-\mathrm{i}(\omega^2 \varkappa^2 + k^2 f^2)\,\hat{Y} + (\mathrm{i}\varkappa k(f^2 - \omega^2) - \omega f \boldsymbol{k}^2)\,\hat{X}\right)\right]. \tag{4.10.4}$$

These equations represent a general formulation of the internal wave response of the ocean to wind forcing in the $\omega - \boldsymbol{k}$ space. Unfortunately it appears that calculations of further expressions from these solutions, such as Eulerian spectra or representations in physical space, may lead to cumbersome integrals which require numerical methods — at the expense of losing the transparency of analytical expressions. Therefore, we consider in the following sections, where the response to some example of external forcing will be discussed, only the simple case of constant BVF which permits analytical treatment.

In order to get an impression of the gross structure of the spectrum of the vertical velocity at a fixed location (Eulerian spectrum) we start with a rather simplified heuristic case where the Green's function itselfs is the response, i.e., we set the bracket on the right-hand side of (4.3.11) equal to $-A$. That means we analyze the equation

$$w_{zz} + \frac{N^2 - \omega^2}{\omega^2 - f^2}\,\boldsymbol{k}^2 w = -\frac{A\delta(z + H_{\mathrm{mix}})}{\omega^2 - f^2} \tag{4.10.5}$$

where A is a constant of the dimension (velocity)^2. Expressing w in terms of the Green's function (4.4.5) we have

$$w = \sum_n \frac{A}{H}\,\frac{2 \sin \alpha_n z \sin \alpha_n H_{\mathrm{mix}}}{(\alpha^2 - \alpha_n^2)(\omega^2 - f^2)} = \sum_n w_n \sqrt{2} \sin \alpha_n z \tag{4.10.6}$$

with

$$w_n(\boldsymbol{k}, \omega) = \frac{B_n}{(\boldsymbol{k}^2 + \alpha_n^2)(\omega_n^2(\boldsymbol{k}) - \omega^2)} \tag{4.10.7}$$

where

$$B_n = \frac{A}{H}\sqrt{2} \sin \alpha_n H_{\mathrm{mix}} \quad \text{and} \quad \omega_n^2(\boldsymbol{k}) = \frac{\boldsymbol{k}^2 N^2 + \alpha_n^2 f^2}{\boldsymbol{k}^2 + \alpha_n^2}.$$

From the discussion of Section 2.3. we know that causality can be accounted for by replacing of ω by $\omega + \mathrm{i}\varepsilon$ in the denominator of (4.10.7). Then, using Dirac's identity (2.3.13)

$$\frac{1}{x \pm \mathrm{i}\varepsilon} = \frac{P}{x} \mp \mathrm{i}\pi\delta(x)$$

where the letter P denotes the principle value, we have from (4.10.7)

$$w_n = \frac{B_n}{\boldsymbol{k}^2 + \alpha_n^2} \left\{ \frac{P}{\omega_n^2 - \omega^2} + \frac{\mathrm{i}\pi}{2\omega_n} [\delta(\omega - \omega_n) - \delta(\omega + \omega_n)] \right\}. \tag{4.10.8}$$

The principle value contribution may be attributed to the switch-on processes while the imaginary part exhibits the pure system properties. Let us assume that the system was forced by a pulse at $t \to -\infty$, and that the switch-on process has been forgotten by the system. Then we may consider only the contribution of the imaginary part of (4.10.8). In order to estimate the Eulerian spectrum we have to perform the inverse Fourier transformation with respect to $\boldsymbol{k}$. In plane polar coordinates it follows

$$w_n(\boldsymbol{r}, \omega) = B_n \frac{\mathrm{i}\pi}{(2\pi)^2} \int_0^{2\pi} \mathrm{d}\varphi \int_0^{\infty} \mathrm{d}\,|\boldsymbol{k}|\,|\boldsymbol{k}| \frac{\mathrm{e}^{\mathrm{i}|\boldsymbol{k}|\,|\boldsymbol{r}| \cos\varphi}}{(\boldsymbol{k}^2 + \alpha_n^2) \cdot 2\omega_n} [\delta(\omega - \omega_n) - \delta(\omega + \omega_n)]. \tag{4.10.9}$$

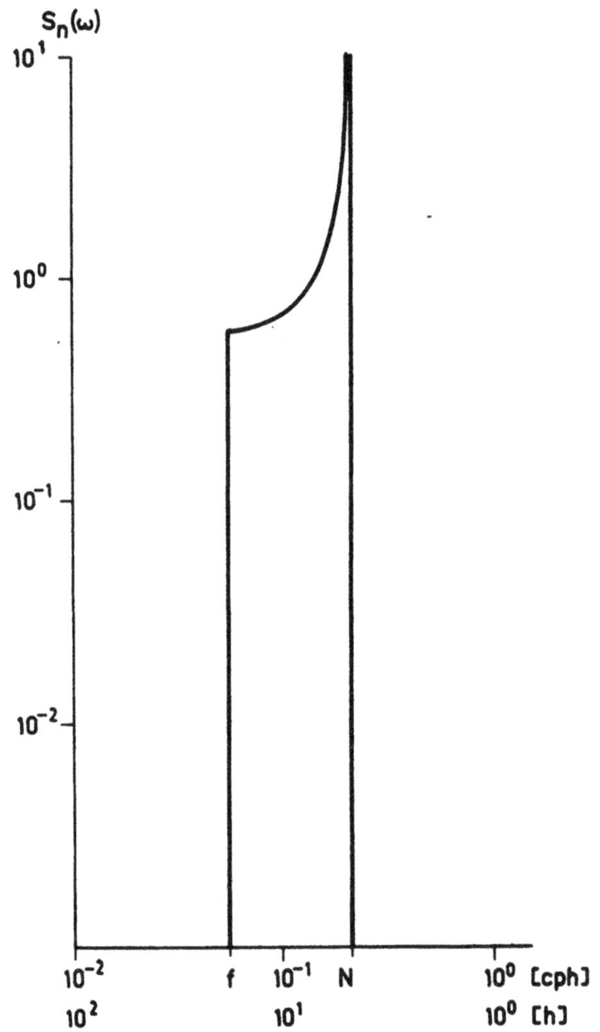

Fig. 4.10.1
Theoretically estimated spectrum of a heuristic example according to (4.10.12)

In order to do this integral we substitute the variable $s = \omega_n(\boldsymbol{k})$ with

$$\frac{\mathrm{d}s}{N^2 - s^2} = \frac{|\boldsymbol{k}|\,\mathrm{d}|\boldsymbol{k}|}{s(k^2 + \alpha_n^2)} \quad ,$$

and $\omega_n = f$ for $|\boldsymbol{k}| \to 0$, and $\omega_n = N$ for $k \to \infty$. Moreover, we use

$$\int_0^{2\pi} \frac{\mathrm{d}\varphi}{2\pi} \cdot \mathrm{e}^{\mathrm{i}\,|\boldsymbol{k}|\,|\boldsymbol{r}|\cos\varphi} = J_0(|\boldsymbol{k}|\,r)\,, \qquad r = |\boldsymbol{r}| = \sqrt{x^2 + y^2}$$

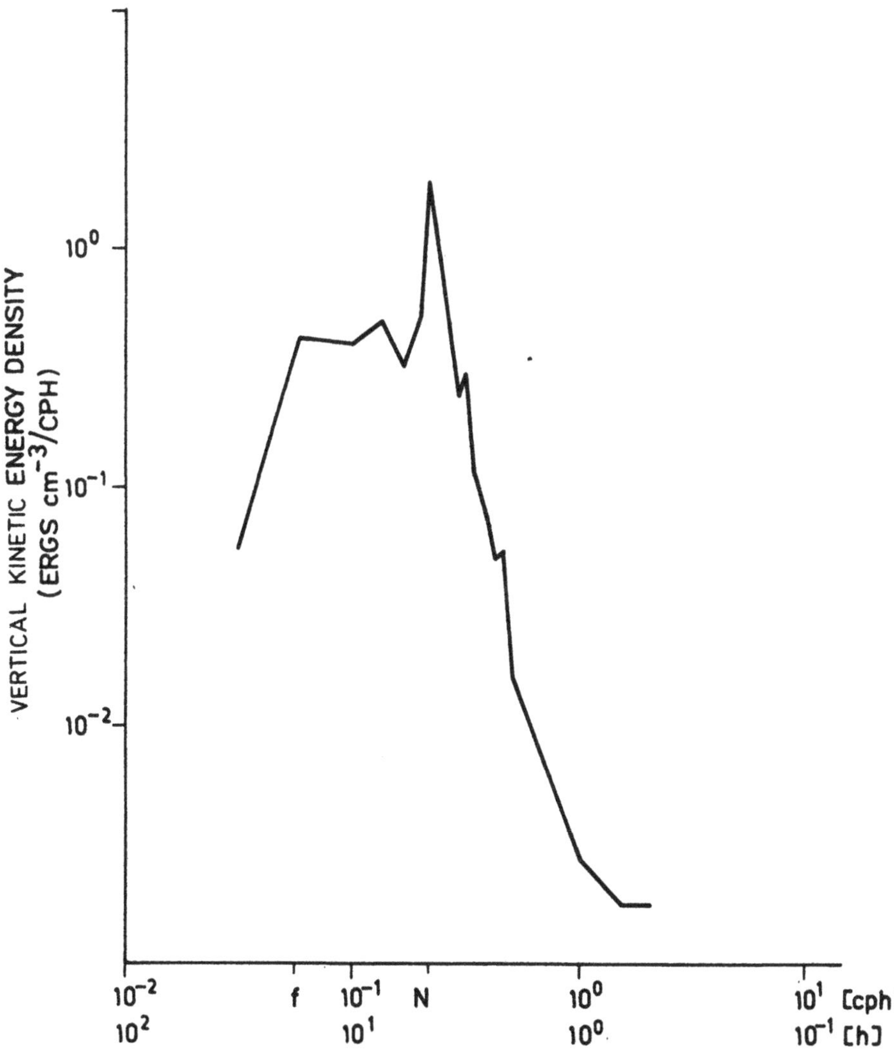

Fig. 4.10.2
Typical spectrum of the vertical velocity in the internal wave band (redrawn from GASCARD (1973), with permission from Deep Sea Research **20** (1973), Pergamon Journals Ltd.).

and have

$$w_n(\boldsymbol{r}, \omega) = B_n \frac{\mathrm{i}}{4} \int_f^N \frac{\mathrm{d}s\, J_0(k_n(s)\, r)}{N^2 - s^2} [\delta(\omega - s) - \delta(\omega + s)] \tag{4.10.10}$$

with

$$k_n(s) = \alpha_n \left(\frac{s^2 - f^2}{N^2 - s^2}\right)^{1/2}$$

Because of the delta functions this integral is easy to evaluate and we arrive at

$$w_n(\boldsymbol{r}, \omega) = \frac{\mathrm{i}}{4} B_n \operatorname{sgn}(\omega) \frac{(\theta(|\omega| - f) - \theta(|\omega| - N))}{N^2 - \omega^2} J_0(k_n(\omega)\, r) \tag{4.10.11}$$

In particular at $r = 0$ the modulus of w_n is

$$|w_n(0, \omega)| = \frac{B_n}{4} \frac{\theta(|\omega| - f) - \theta(|\omega| - N)}{N^2 - \omega^2}. \tag{4.10.12}$$

The spectrum of w_n is characterized by a limiting function which confines the spectrum to the internal wave band between f and N, and by a peak at the BVF. In Fig. 4.10.1 the structure of $|w_n(0, \omega)|$ is shown, and in Fig. 4.10.2 a measured spectrum of the vertical velocity is depicted, see GASCARD (1973). It appears that our heuristic example describes already the typical features of spectra of the vertical velocity in the internal wave range. Clearly, the sharp cut-offs at f and N appearing in (4.10.12) would be smeared out if the contribution of the switch-on processes is taken into account.

4.11. INTERNAL WAVE RESPONSE TO A WIND PULSE

Now we consider the response of the ocean to a pulse-like forcing in a more sophisticated manner than in the heuristic example discussed in the previous section. In particular, we choose

$$\hat{X}(t, \boldsymbol{r}) = \frac{u_*^2}{H_{\mathrm{mix}}} \delta(tN)\, \delta(\boldsymbol{r}/H), \qquad \hat{Y} = 0 \tag{4.11.1 a}$$

or, in the Fourier space,

$$\hat{X}(\omega, \boldsymbol{k}) = \frac{u_*^2}{H_{\mathrm{mix}}} \frac{H^2}{N}. \tag{4.11.1 b}$$

The factors N and H^{-1} were introduced arbitrarily in order to keep the delta functions dimensionless.

We wish to calculate the Eulerian spectra of the response to that forcing. To this end we first calculate the solutions of w, p, u, and v in the Fourier space, and then we transform the results in the physical space in order to find frequency spectra of the observables at any location x, y, z. We remember that the rigid lid case is considered throughout this chapter. Thus, $\sqrt{2} \sin \alpha_n z$ with $\alpha_n = (n\pi/H)$ is the n-th vertical eigenfunction of the vertical velocity for constant BVF. According to (1.9.34) we introduce the notation

$$\frac{1}{h_n} = \frac{\sqrt{2}\sin\alpha_n H_{\text{mix}}}{H_{\text{mix}}\alpha_n}. \tag{4.11.2}$$

Using (4.4.5) and (4.11.1 b) from (4.10.1) it follows

$$w(z, \boldsymbol{k}, \omega) = u_*^2 \sum_n \frac{\sqrt{2}\sin\alpha_n z}{h_n} \frac{(\mathrm{i}\varkappa f - \omega k)\,\alpha_n}{(f^2-\omega^2)(\alpha_n^2-\alpha^2)} = \sum_n w_n(\boldsymbol{k}, \omega)\sqrt{2}\sin\alpha_n z \tag{4.11.3}$$

The w_n is defined by (4.11.3). With the notations

$$\alpha^2 = \boldsymbol{k}^2 \frac{N^2-\omega^2}{\omega^2-f^2} \quad \text{and} \quad k_n^2(\omega) = \alpha_n^2 \frac{(\omega^2-f^2)}{(N^2-\omega^2)}$$

we obtain

$$w_n(\boldsymbol{k}, \omega) = \frac{-u_*^2 H^2}{h_n N} \alpha_n \frac{(\mathrm{i}\varkappa f - \omega k)}{(N^2-\omega^2)(\boldsymbol{k}^2-k_n^2)}. \tag{4.11.4}$$

In order to find the remaining observables p, u, and v we need the expression of $G_z(z, -H_{\text{mix}})$. Using (4.4.5) we have

$$\frac{1}{H_{\text{mix}}} G_z(z, -H_{\text{mix}}) = \sum_{n=1} \frac{\alpha_n^2 F_n(z)}{h_n(\alpha_n^2-\alpha^2)} \tag{4.11.5}$$

with $F_n(z) = \sqrt{2}\cos\alpha_n z$.

With (3.8.20) and (4.11.5) from (4.10.2) it follows

$$p(z, \boldsymbol{k}, \omega) = \sum_{n=1}^{\infty} p_n(\boldsymbol{k}, \omega) F_n(z) \tag{4.11.6a}$$

with

$$p_n(\boldsymbol{k}, \omega) = \frac{-u_*^2 H^2}{h_n N}\left(\mathrm{i}k + \frac{f\varkappa}{\omega}\right)\frac{1}{\boldsymbol{k}^2-k_n^2(\omega)}. \tag{4.11.6b}$$

Moreover, we obtain from (4.10.3) and (4.10.4)

$$u(z, \boldsymbol{k}, \omega) = \sum_n^{\infty} u_n F_n(z) \tag{4.11.7a}$$

with

$$u_n(\boldsymbol{k}, \omega) = \frac{u_*^2 H^2}{h_n N}\left(\varkappa^2 - \frac{\omega^2\alpha_n^2}{N^2-\omega^2}\right)\frac{\mathrm{i}}{\omega(\boldsymbol{k}^2-k_n^2)} \tag{4.11.7b}$$

and

$$v(z, \boldsymbol{k}, \omega) = \sum_n v_n F_n(z) \tag{4.11.8a}$$

with

$$v_n(\boldsymbol{k}, \omega) = \frac{-u_*^2 H^2}{h_n N}\frac{1}{\omega(\boldsymbol{k}^2-k_n^2)}\left(\mathrm{i}k\varkappa + \frac{f\alpha_n^2\omega}{N^2-\omega^2}\right). \tag{4.11.8b}$$

Next we estimate the Eulerian frequency spectra of these fields. This implies we have to evaluate the inverse Fourier transformations of the expressions (4.11.3), (4.11.6), (4.11.7), and (4.11.8). Let us introduce the integral

$$I(r) = \int \frac{\mathrm{d}k}{2\pi} \int \frac{\mathrm{d}\varkappa}{2\pi} \frac{\mathrm{e}^{ikx+i\varkappa y}}{k^2+\varkappa^2-k_n^2(\omega)} \tag{4.11.9}$$

which depends only on the modulus of $\boldsymbol{r}$. Then we may express all the Eulerian spectral components in terms of this integral, i.e.,

$$w_n(\boldsymbol{r}, \omega) = \frac{u_*^2 H^2}{h_n N} \frac{\alpha_n}{N^2-\omega^2} \left[f \frac{\partial}{\partial y} I(r) + \mathrm{i}\omega \frac{\partial}{\partial x} I(r) \right] \tag{4.11.10a}$$

$$p_n(\boldsymbol{r}, \omega) = \frac{u_*^2 H^2}{h_n N} \left[\frac{\mathrm{i}f}{\omega} \frac{\partial}{\partial y} I(r) - \frac{\partial}{\partial x} I(r) \right] \tag{4.11.10b}$$

$$u_n(\boldsymbol{r}, \omega) = \frac{u_*^2 H^2}{h_n N\, \mathrm{i}\omega} \left[\frac{\partial^2}{\partial y^2} I(r) - \frac{\alpha_n^2 \omega^2}{N^2-\omega^2} I(r) \right] \tag{4.11.10c}$$

$$v_n(\boldsymbol{r}, \omega) = \frac{u_*^2 H^2}{h_n N} \left[\mathrm{i} \frac{\partial^2}{\partial x\, \partial y} I(r) - \frac{\omega f \alpha_n^2}{N^2-\omega^2} I(r) \right] \tag{4.11.10d}$$

Let us evaluate the basic integral (4.11.9). We start with the k-integration. For $\varkappa^2 > k_n^2$ the integrand has one pole in the upper and one pole in the lower half of the complex k-plane, i.e., at $k = \pm \mathrm{i}\sqrt{\varkappa^2 - k_n^2}$. Thus we can evaluate the integral by closing the contour by a large half circle in the upper half plane for positive x, and in the lower half plane for negative x. From the residue theorem it follows

$$\int \frac{\mathrm{d}k}{2\pi} \frac{\mathrm{e}^{ikx}}{k^2+\varkappa^2-k_n^2} = \mathrm{i}\left[\theta(x) \frac{\mathrm{e}^{-\sqrt{\varkappa^2-k_n^2}\,x}}{2\mathrm{i}\sqrt{\varkappa^2-k_n^2}} + \theta(-x) \frac{\mathrm{e}^{\sqrt{\varkappa^2-k_n^2}\,x}}{2\mathrm{i}\sqrt{\varkappa^2-k_n^2}} \right]$$

$$= \frac{\mathrm{e}^{-\sqrt{\varkappa^2-k_n^2}\,|x|}}{2(\varkappa^2-k_n^2)^{1/2}} \quad \text{for} \quad \varkappa^2 > k_n^2 . \tag{4.11.11}$$

For $\varkappa^2 < k_n^2$ the integrand has poles on the k-axis, i.e., on the integration path. We know, however, that causality requires the replacement of ω by $\omega + \mathrm{i}\varepsilon$, implying $k_n(\omega + \mathrm{i}\varepsilon) \approx k_n(\omega) + \mathrm{i}\varepsilon$. Thus the poles are shifted into the complex plane, i.e., $k = \pm(\sqrt{k_n^2 - \varkappa^2} + \mathrm{i}\varepsilon)$, and we use again the residue theorem to obtain

$$\int \frac{\mathrm{d}k}{2\pi} \frac{\mathrm{e}^{ikx}}{k^2-(k_n^2-\varkappa^2)} = \mathrm{i}\left[\theta(x) \frac{\mathrm{e}^{\mathrm{i}\sqrt{k_n^2-\varkappa^2}\,x}}{2\sqrt{k_n^2-\varkappa^2}} + \theta(-x) \frac{\mathrm{e}^{-\mathrm{i}\sqrt{k_n^2-\varkappa^2}\,x}}{2\sqrt{k_n^2-\varkappa^2}} \right]$$

$$= \frac{\mathrm{i}}{2} \frac{\mathrm{e}^{\mathrm{i}\sqrt{k_n^2-\varkappa^2}\,|x|}}{\sqrt{k_n^2-\varkappa^2}} \quad \text{for} \quad \varkappa^2 < k_n^2 . \tag{4.11.12}$$

Now we may add (4.11.11) and (4.11.12) and have

$$I(r) = \int_0^\infty \frac{d\varkappa}{2\pi} \cos \varkappa y \left\{ \theta(\varkappa^2 - k_n^2) \frac{e^{-\sqrt{\varkappa^2 - k_n^2}|x|}}{(\varkappa^2 - k_n^2)^{1/2}} \right.$$

$$\left. + \frac{\theta(k_n^2 - \varkappa^2)}{(k_n^2 - \varkappa^2)^{1/2}} \left[i \cos(\sqrt{k_n^2 - \varkappa^2}\, x) - \sin(\sqrt{k_n^2 - \varkappa^2}\,|x|) \right] \right\}. \quad (4.11.13)$$

This integral can be found in standard tables, e.g. OBERHETTINGER (1957), and we arrive at

$$I(r) = \frac{i}{4} H_0^{(1)}(k_n r), \qquad r = (x^2 + y^2)^{1/2} \quad (4.11.14)$$

where $H_0^{(1)}(z) = J_0(z) + iY_0(z)$ is the Hankel function of the first kind. Moreover, we have

$$\frac{\partial}{\partial x} I(r) = -\frac{ik_n x}{4r} H_1^{(1)}(k_n r) \quad (4.11.15a)$$

$$\frac{\partial}{\partial y} I(r) = -\frac{ik_n y}{4r} H_1^{(1)}(k_n r) \quad (4.11.15b)$$

and with $\frac{d}{dz} H_1^{(1)}(z) = H_0^{(1)}(z) - \frac{1}{z} H_1^{(1)}(z)$

$$\frac{\partial^2}{\partial x \partial y} I(r) = \frac{-ik_n}{4} \left[-\frac{2xy}{r^3} H_1^{(1)}(k_n r) + \frac{xy}{r^2} k_n H_0^{(1)}(k_n r) \right] \quad (4.11.15c)$$

$$\frac{\partial^2 I(r)}{\partial y^2} = -\frac{ik_n}{4} \left[\frac{x^2 - y^2}{r^3} H_1^{(1)}(k_n r) + \frac{y^2}{r^2} k_n H_0^{(1)}(k_n r) \right]. \quad (4.11.15d)$$

By means of these expressions all terms appearing in the set (4.11.10) are known and we have explicitly

$$w_n(\mathbf{r}, \omega) = \frac{u_*^2 k_n \alpha_n H^2}{4h_n(N^2 - \omega^2)\, rN} (ify - \omega x)\, H_1^{(1)}(k_n r) \quad (4.11.16a)$$

$$p_n(\mathbf{r}, \omega) = \frac{u_*^2 k_n H^2}{4h_n \omega r N} (fy + i\omega x)\, H_1^{(1)}(k_n r) \quad (4.11.16b)$$

$$u_n(\mathbf{r}, \omega) = \frac{-u_*^2 H^2}{4h_n \omega N} \left[k_n \frac{(x^2 - y^2)}{r^3} H_1^{(1)}(k_n r) + k_n^2 \frac{y^2}{r^2} H_0^{(1)}(k_n r) + \frac{\omega^2 \alpha_n^2}{N^2 - \omega^2} H_0^{(1)}(k_n r) \right] \quad (4.11.16c)$$

$$v_n(\mathbf{r}, \omega) = \frac{-u_*^2 H^2}{4h_n \omega N} \left[k_n \frac{2xy}{r^3} H_1^{(1)}(k_n r) - k_n^2 \frac{xy}{r^2} H_0^{(1)}(k_n r) + \frac{if\omega \alpha_n^2}{N^2 - \omega^2} H_0^{(1)}(k_n r) \right]. \quad (4.11.16d)$$

These expressions represent the Eulerian frequency spectra of the n-th modal components of the dynamical fields. Let r be large and using the asymptotic formula of the Hankel function for large arguments

$$H_m^{(1)}(z) \approx \left(\frac{2}{\pi z}\right)^{1/2} e^{i\left(z - m\frac{\pi}{2} - \frac{\pi}{4}\right)}$$

then we may rewrite the set (4.11.16) in the following form

$$w_n(\boldsymbol{r}, \omega) \approx \frac{u_*^2 \alpha_n H^2}{4 h_n N} \frac{i f y - \omega x}{N^2 - \omega^2} \sqrt{\frac{2k_n}{\pi r}} \frac{1}{r} e^{i\left(k_n r - \frac{3\pi}{4}\right)} \tag{4.11.17a}$$

$$p_n(\boldsymbol{r}, \omega) \approx \frac{u_*^2 H^2 (f y + i \omega x)}{4 h_n N \omega} \sqrt{\frac{2k_n}{\pi r}} e^{i\left(k_n r - \frac{3\pi}{4}\right)} \tag{4.11.17b}$$

$$u_n(\boldsymbol{r}, \omega) \approx \frac{-u_*^2 H^2}{4 h_n N} \left[-i \frac{x^2 - y^2}{r^3} k_n^{1/2} + \frac{y^2}{r^2} k_n^{3/2} + \frac{\alpha_n^2 \omega^2}{N^2 - \omega^2} k_n^{-1/2} \right] \sqrt{\frac{2}{\pi r}} e^{i\left(k_n r - \frac{\pi}{4}\right)}. \tag{4.11.17c}$$

$$v_n(\boldsymbol{r}, \omega) \approx \frac{-u_*^2 H^2}{4 h_n N} \sqrt{\frac{2}{\pi r}} \left[-i \frac{2xy}{r^3} k_n^{1/2} - \frac{xy}{r^2} k_n^{3/2} + \frac{i \omega f \alpha_n^2}{N^2 - \omega^2} k_n^{-1/2} \right] e^{i\left(k_n r - \frac{\pi}{4}\right)}. \tag{4.11.17d}$$

This is valid for large r and, moreover, ω larger than f, say $\omega > 2f$, otherwise k_n would tend to zero.

We observe that all expressions of (4.11.17) have poles at the BVF but with different orders. The strongest peak occurs in the vertical velocity, $w_n \sim (N^2 - \omega^2)^{-5/4}$, p_n has the weakest peak with the strength $(N^2 - \omega^2)^{-1/4}$, and the horizontal components possess leading peaks of the order $(N^2 - \omega^2)^{-3/4}$. The Hankel functions act as limiting functions, which oscillate for frequencies below N and decrease exponentially for frequencies above the BVF.

In all spectral expressions the horizontal coordinates occur at several places and, consequently, the spectra are neither homogeneous nor isotropic.

Next we consider the spectral functions near the inertial frequency, i.e., for small k_n. Then we may use the small argument approximation of the Hankel functions which are dominated by the singular behaviour of the Neumann functions Y_0 and Y_1, i.e.,

$$H_0^{(1)}(z) \approx \frac{2i}{\pi} \left(\frac{1}{2} \ln z + C \right) \tag{4.11.18a}$$

and

$$H_1^{(1)}(z) \approx \frac{-2i}{z\pi} \tag{4.11.18b}$$

with $C = 0.5772\ldots$ being Euler's constant. For ω close to f we may express k_n by

$$k_n^2 = \frac{1}{R_n^2} \left(\frac{\omega^2}{f^2} - 1 \right) \tag{4.11.19}$$

where $\lambda_n = \alpha_n / N$ is the n-th vertical eigenvalue in the hydrostatic approximation and $R_n = 1/(\lambda_n f)$ is the n-th baroclinic Rossby radius of deformation. Then from (4.11.16) and (4.11.18) it follows

$$w_n(\boldsymbol{r}, \omega) = \frac{u_*^2 H^2}{h_n N} \frac{1}{2\pi r} \frac{1}{N R_n} \left(\frac{y}{r} + \frac{i\omega}{f} \frac{x}{r} \right) \tag{4.11.20a}$$

$$p_n(\mathbf{r}, \omega) = \frac{u_*^2 H^2}{h_n N} \frac{1}{2\pi r} \left(\frac{x}{r} - \mathrm{i} \frac{y}{r} \frac{f}{\omega} \right) \tag{4.11.20b}$$

$$u_n(\mathbf{r}, \omega) = -\frac{u_*^2 H^2}{h_n N} \frac{\mathrm{i}}{2\pi\omega} \left\{ \frac{(y^2 - x^2)}{4} + \frac{y^2}{r^2 R_n^2} \left(\frac{\omega^2}{f^2} - 1 \right) \left[\ln \frac{r}{2R_n} + C + \frac{1}{2} \ln \left(\frac{\omega^2}{f^2} - 1 \right) \right] \right\} \tag{4.11.20c}$$

$$v_n(\mathbf{r}, \omega) = \frac{u_*^2 H^2}{h_n N} \frac{\mathrm{i}}{2\pi\omega} \left\{ \frac{2xy}{\omega} + \frac{xy}{r^2 R^2} \left(\frac{\omega^2}{f^2} - 1 \right) \left[\ln \frac{r}{2R_n} + C + \frac{1}{2} \ln \left(\frac{\omega^2}{f^2} - 1 \right) \right] \right.$$
$$\left. - \frac{\mathrm{i}\omega}{f} \frac{1}{R_n^2} \left[\ln \frac{r}{2R_n} + C + \frac{1}{2} \ln \left(\frac{\omega^2}{f^2} - 1 \right) \right] \right\}. \tag{4.11.20d}$$

It appears that the spectra of w_n and p_n are smooth in the vicinity of the inertial frequency, while the horizontal components u_n and v_n have a logarithmic inertial peak, see the rightmost terms in (4.11.20c, d).

4.12. SPECTRA OF THE INTERNAL WAVE RESPONSE TO DIVERGENT, AND TO ROTATIONAL WIND

In this section we consider the response of the ocean in the internal wave band to two examples of external forcing, which have already been analyzed in the hydrostatic approximation in Section 3.8. and 3.9.

In particular we consider a divergent wind

$$\hat{X} = \frac{u_*^2}{H_{\text{mix}}} \theta(t)\,\theta(x)\,, \qquad \hat{Y} = 0 \tag{4.12.1a}$$

and a rotational wind

$$\hat{X} = \frac{u_*^2}{H_{\text{mix}}} \theta(t)\,\theta(y)\,, \qquad \hat{Y} = 0\,. \tag{4.12.1b}$$

In the Fourier space we obtain respectively

$$\hat{X} = \frac{u_*^2}{H_{\text{mix}}} 2\pi\delta(\varkappa) \frac{\mathrm{i}}{\omega + \mathrm{i}\varepsilon} \frac{-\mathrm{i}}{k - \mathrm{i}\varepsilon}, \qquad \hat{Y} = 0 \tag{4.12.2a}$$

and

$$\hat{X} = \frac{u_*^2}{H_{\text{mix}}} 2\pi\delta(k) \frac{\mathrm{i}}{\omega + \mathrm{i}\varepsilon} \frac{-\mathrm{i}}{\varkappa - \mathrm{i}\varepsilon}, \qquad \hat{Y} = 0\,. \tag{4.12.2b}$$

The positive infinitesimal ε gives the prescription how to go around poles on the real axis when evaluating Fourier integrals. Bearing this in mind there is no need to write the ε explicitly in all formulas. The corresponding solutions follow from (4.10.1) to (4.10.4) with (4.4.5). Using, moreover, (3.8.20), (4.11.2), and (4.11.5) it follows

$$w = \sum_n w_n \sqrt{2} \sin \alpha_n z \tag{4.12.3}$$

$$(p, u, v) = \Sigma\, (p_n, u_n, v_n)\, F_n(z) \tag{4.12.4}$$

where in the divergent case (4.12.2a)

$$\begin{pmatrix} w_n(\boldsymbol{k},\omega) \\ \\ p_n(\boldsymbol{k},\omega) \\ \\ u_n(\boldsymbol{k},\omega) \\ \\ v_n(\boldsymbol{k},\omega) \end{pmatrix} = \frac{u_*^2}{h_n}\frac{2\pi\delta(\varkappa)}{k^2-k_n^2} \begin{pmatrix} -\dfrac{\alpha_n}{N^2-\omega^2} \\ \\ -\dfrac{\mathrm{i}}{\omega} \\ \\ -\dfrac{\mathrm{i}\alpha_n^2}{k(N^2-\omega^2)} \\ \\ -\dfrac{f\alpha_n^2}{\omega k(N^2-\omega^2)} \end{pmatrix} \tag{4.12.5a}$$

and in the rotational case (4.12.2b)

$$\begin{pmatrix} w_n(\boldsymbol{k},\omega) \\ \\ p_n(\boldsymbol{k},\omega) \\ \\ u_n(\boldsymbol{k},\omega) \\ \\ v_n(\boldsymbol{k},\omega) \end{pmatrix} = \frac{u_*^2}{h_n}\frac{2\pi\delta(k)}{\varkappa^2-k_n^2} \begin{pmatrix} \dfrac{\mathrm{i}f}{\omega(N^2-\omega^2)} \\ \\ -\dfrac{f}{\omega^2} \\ \\ \mathrm{i}\varkappa-\dfrac{\mathrm{i}\omega^2\alpha_n^2}{(N^2-\omega^2)} \\ \\ -\dfrac{f^2\alpha_n}{\omega\varkappa(N^2-\omega^2)} \end{pmatrix} \tag{4.12.5b}$$

Next we invert Fourier transforms with respect to k and $\varkappa$ in order to find the Eulerian frequency spectra. We consider the integral

$$\hat{I}(x) = \int_{-\infty}^{+\infty} \frac{\mathrm{d}k}{2\pi}\frac{\mathrm{e}^{\mathrm{i}kx}}{k^2-k_n^2} \tag{4.12.6}$$

This integral can be taken from (4.11.12) with $\varkappa = 0$, and we find

$$\hat{I}(x) = \frac{\mathrm{i}}{2}\frac{1}{k_n}\mathrm{e}^{\mathrm{i}k_n|x|}. \tag{4.12.7}$$

Clearly, this integral can also be done with k and x replaced by $\varkappa$ and y, respectively. Moreover, using the identity

$$\frac{1}{k}\frac{1}{k^2-k_n^2} = -\frac{1}{k_n^2}\left(\frac{1}{k}-\frac{k}{k^2-k_n^2}\right)$$

we get

$$\begin{aligned}\int_{-\infty}^{+\infty}\frac{\mathrm{d}k}{2\pi}\frac{\mathrm{e}^{\mathrm{i}kx}}{(k-\mathrm{i}\varepsilon)(k^2-k_n^2)} &= -\frac{\mathrm{i}}{k_n^2}\left[\theta(x)+\frac{\partial}{\partial x}\hat{I}(x)\right] \\ &= -\frac{\mathrm{i}}{k_n^2}\left[\theta(x)-\operatorname{sgn}(x)\frac{1}{2}\mathrm{e}^{\mathrm{i}k_n|x|}\right].\end{aligned} \tag{4.12.8}$$

With (4.12.7) and (4.12.8) all integrals appearing in the inverse Fourier integrals of (4.12.5a, b) are calculated and we obtain the Eulerian spectra in the divergent case

$$w_n(x, \omega) = -\frac{u_*^2}{h_n}\frac{\mathrm{i}e^{\mathrm{i}k_n|x|}}{[(\omega^2 - f^2)(N^2 - \omega^2)]^{1/2}} \tag{4.12.9a}$$

$$p_n(x, \omega) = \frac{u_*^2}{h_n}\left(\frac{N^2 - \omega^2}{\omega^2 - f^2}\right)^{1/2}\frac{e^{\mathrm{i}k_n x}}{2\omega\alpha_n} \tag{4.12.10a}$$

$$u_n(x, \omega) = \frac{u_*^2}{h_n}\frac{1}{(\omega^2 - f^2)}\left[-\theta(x) + \operatorname{sgn}(x)\frac{1}{2}e^{\mathrm{i}k_n|x|}\right] \tag{4.12.11a}$$

$$v_n(x, \omega) = \frac{u_*^2}{h_n}\frac{\mathrm{i}f}{(\omega^2 - f^2)\,\omega}\left[\theta(x) - \operatorname{sgn}(x)\frac{1}{2}e^{\mathrm{i}k_n|x|}\right] \tag{4.12.12a}$$

and, analogously, in the rotational case

$$w_n(y, \omega) = -\frac{u_*^2}{h_n}\frac{f\,e^{\mathrm{i}k_n|y|}}{2\omega[(\omega^2 - f^2)(N^2 - \omega^2)]^{1/2}} \tag{4.12.9b}$$

$$p_n(y, \omega) = -\frac{u_*^2}{h_n}\left(\frac{N^2 - \omega^2}{\omega^2 - f^2}\right)^{1/2}\frac{\mathrm{i}f\,e^{\mathrm{i}k_n|y|}}{\omega^2\alpha_n} \tag{4.12.10b}$$

$$u_n(y, \omega) = \frac{u_*^2}{h_n}\frac{1}{\omega^2 - f^2}\left[-\theta(y) + f^2\operatorname{sgn}(y)\frac{1}{2\omega^2}e^{\mathrm{i}k_n|y|}\right] \tag{4.12.11b}$$

$$v_n(y, \omega) = \frac{u_*^2}{h_n}\frac{\mathrm{i}f}{\omega(\omega^2 - f^2)}\left[\theta(y) - \operatorname{sgn}(y)\frac{1}{2}e^{\mathrm{i}k_n|y|}\right] \tag{4.12.12b}$$

The w_n-spectra have in both cases peaks of the same order at the BVF and at the inertial frequency, see (4.12.9a, b), while in the spectra of p_n, u_n, and v_n only at the inertial peaks occur. The exponentials $\exp(\mathrm{i}k_n|x|)$ and $\exp(\mathrm{i}k_n|y|)$ serve in both cases as limiting functions. For $\omega > N$, in fact, k_n is imaginary and, therefore, the spectra diminish exponentially above the BVF.

For ω close to f we may replace k_n by $R_n^{-1}[(\omega^2/f^2) - 1]^{1/2}$ and we obtain the results of Sections 3.8 and 3.9, compare (3.8.8) and (3.9.6), i.e., in the subinertial frequency range the spectra drop rapidly outside a strip of the width of $2R_n$, while within that strip the low frequency motion in terms of geostrophic currents and Ekman transport occurs. Moreover, this demonstrates that the hydrostatic approximation amounts to filtering out the high frequencies, close to the BVF, as mentioned in Section 3.2.

Comparing these results with those of the foregoing section we get an impression of the sensitivity of the spectral shape regarding the structure of the external forcing considered.

4.13. FORCING THROUGH A MOVING FRONT

The last example we wish to analyze in this chapter is that of a front which propagates with phase speed c in the y-direction,

$$\hat{X}(x, y, t) = \frac{u_*^2}{H_{\text{mix}}}\,\theta\left(t - \frac{y}{c}\right), \qquad \hat{Y} = 0 . \tag{4.13.1}$$

This is a rotational non-divergent forcing. We confine the analysis to the estimation of the vertical velocity. In particular we calculate the Eulerian spectrum of w and discuss the pattern of the vertical velocity in the time domain.

By Fourier transforming we get

$$\hat{X}(\omega, k, \varkappa) = \frac{u_*^2}{H_{\text{mix}}} \cdot 2\pi\,\delta(k)\,\frac{2\pi i}{\omega}\,\delta\left(\varkappa - \frac{\omega}{c}\right).$$

With (4.10.1), (4.4.5) and (4.11.2) the solution reads

$$w(\omega, \boldsymbol{k}, z) = u_*^2 f c (2\pi)^2\,\delta(k)\,\delta\left(\varkappa - \frac{\omega}{c}\right) \sum_n \frac{\alpha_n}{h_n}\,\frac{\sqrt{2}\sin\alpha_n z}{\omega^4 + \omega^2(c^2\alpha_n^2 - N^2) - c^2\alpha_n^2 f^2} \tag{4.13.2}$$

where properties of the delta function were used, see Appendix A. The expression (4.13.2) has poles at

$$\omega^2 = \omega_{n_{1,2}}^2 = \frac{N^2}{2}\left\{(1 - c^2\lambda_n^2) \pm \left[(1 - c^2\lambda_n^2) + 4c^2\lambda_n^2\left(\frac{f}{N}\right)^2\right]^{1/2}\right\} \tag{4.13.3}$$

where $\lambda_n = \alpha_n/N = n\pi/HN$ is the vertical eigenvalue in the hydrostatic approximation, see Subsection 1.9.2. We may distinguish three cases:

Firstly, there may exists a certain $n = n^*$ where $c\lambda_{n^*} = 1$. Then (4.12.3) has the roots

$$\omega_{n^*_{1,2}}^2 = \pm N f . \tag{4.13.4}$$

Secondly, for $c\lambda_n^2 \neq 1$ we use the smallness of f^2/N^2 and expand the square root in a Taylor series to obtain

$$\omega_{n_{1,2}}^2 = \frac{N^2}{2}\left\{(1 - c^2\lambda_n^2) \pm (1 - c^2\lambda_n^2)\left[1 - \frac{2\lambda_n^2 c^2}{(1 - c^2\lambda_n^2)^2}\,\frac{f^2}{N^2} + \mathrm{O}\left(\frac{f^4}{N^4}\right)\right]\right\}. \tag{4.13.5}$$

Then the approximate solutions are

$$\omega_{n_1}^2 = N^2(1 - c^2\lambda_n^2) \tag{4.13.6a}$$

$$\omega_{n_2}^2 = -\frac{\lambda_n^2 c^2 f^2}{1 - c^2\lambda_n^2}\,. \tag{4.13.6b}$$

Thirdly, for $\lambda_1^2 c^2 > 1$, implying $\lambda_n^2 c^2 > 1$ for all n, we have an imaginary ω_{n_1} and a real ω_{n_2}. This is the case of a fast moving front since c exceeds λ_n^{-1}. In the opposite case of a slowly moving front there is a finite number of modes, say $n \leqq \bar{n}$, where $\lambda_n^2 c^2 < 1$, that means ω_{n_1} is real and ω_{n_2} is imaginary. For $n > \bar{n}$, however, it follows that $\lambda_n^2 c^2 > 1$ as in the former case.

Next we calculate the Eulerian spectrum of w. That is we have to invert the Fourier transforms (4.13.2) concerning the wavenumbers. Because of the involved delta functions we get

$$w(\omega, \boldsymbol{r}, z) = \int \frac{d\boldsymbol{k}}{2\pi} e^{i\boldsymbol{k}\boldsymbol{r}} w(\omega, \boldsymbol{k}, z)$$

$$= u_*^2 \sqrt{2cf}\, e^{i\frac{\omega}{c}y} \sum_{n=1}^{\infty} \frac{\alpha_n}{h_n} \frac{\sin \alpha_n z}{\omega^4 + \omega^2(c\alpha_n^2 - N^2) - c^2\alpha_n^2 f^2}. \tag{4.13.7}$$

For $\lambda_1^2 c^2 > 1$, i.e., a fast moving front, $w(\omega)$ has according to (4.13.6b) peaks at a somewhat enhanced inertial frequency, whereby the "blue shift" is given by the factor $\lambda_n^2 c^2/(c^2\lambda_n^2 - 1)$ which approaches 1 for large n. This amounts to broadening of the inertial peak.

For a slowly moving front, i.e., $\lambda_n^2 c^2 < 1$ for $n \leqq \bar{n}$, we have in addition a peak at the somewhat reduced BVF, i.e., at $\omega^2 = N^2(1 - c^2\lambda_n^2)$. If there exist a n^* with $c^2\lambda_{n^*}^2 = 1$ there occurs a further peak at the frequency $(fN)^{1/2}$.

Contrary to the examples considered in the foregoing sections the spectrum (4.13.7) does not contain a limiting function which concentrates the spectrum more or less sharply on the band between f and N.

The Fourier transform of (4.13.7) yields

$$w(x, y, z, t) = u_*^2 c \sqrt{2} \left\{ \sum_{n=1}^{\bar{n}} \frac{\alpha_n}{h_n} \frac{\sin \alpha_n z}{N^2(1 - c^2\lambda_n)} \left[-\theta\left(t - \frac{y}{c}\right) \frac{f \sin\left(N\sqrt{1 - c^2\lambda_n^2}\left(t - \frac{y}{c}\right)\right)}{N\sqrt{1 - c^2\lambda_n^2}} \right.\right.$$

$$\left. + \frac{\sqrt{1 - c_2\lambda_n^2}}{2c\lambda_n} e^{-\frac{cf\lambda_n}{\sqrt{1-c^2\lambda_n^2}}\left|t - \frac{y}{c}\right|} \right] + \sum_{n=\bar{n}+1}^{\infty} \frac{\alpha_n}{h_n} \frac{\sin \alpha_n y}{N^2(c^2\lambda_n^2 - 1)}$$

$$\times \left[-\theta\left(t - \frac{y}{c}\right) \frac{\sqrt{c^2\lambda_n^2 - 1}}{c\lambda_n} \sin\left(\frac{cf\lambda_n\left(t - \frac{y}{c}\right)}{\sqrt{\lambda_n^2 c^2 - 1}}\right) \right.$$

$$\left.\left. - \frac{f}{2N} \frac{e^{-N\sqrt{\lambda_n^2 c^2 - 1}\left|t - \frac{y}{c}\right|}}{\sqrt{\lambda_n^2 c^2 - 1}} \right] \right\}. \tag{4.13.8}$$

In the case of a slowly moving front we observe that oscillations with frequencies close to f and N occur behind the frontal edge. For a fast moving front we have $\bar{n} = 0$ and only the second sum in (4.13.8) remains. In addition to the wave processes there are some non-oscillatory vertical motions localized at the frontal edge with the scales $(f\lambda_n)^{-1}(\sqrt{1 - c^2\lambda_n^2})$ for $n \leqq \bar{n}$ and $c/N(\lambda_n^2 c^2 - 1)^{-1}$ for $n > \bar{n}$.

Finally we mention that if a $n = n^*$ exists with $c\lambda_{n^*} = 1$, e.g., $n^* = \bar{n} + 1$, then the second sum in (4.13.8) runs from $n = \bar{n} + 2$ to infinity, and we have to add the term

$$-u_*^2 c \sqrt{2} \frac{\alpha_{n*}}{h_{n*}} \frac{f \sin(\alpha_{n*} z)}{(Nf)^{3/2}} \left[\theta\left(t - \frac{y}{c}\right) \sin\left(\sqrt{Nf}\left(t - \frac{y}{c}\right)\right) + \frac{1}{2} e^{-\sqrt{Nf}\left|t - \frac{y}{c}\right|} \right]$$

That is, a further vibration with the frequency $\sqrt{Nf}$ occurs.

CHAPTER 5

5. Forcing of a semi-infinite ocean

5.1. FORMULATION OF THE PROBLEM

In this chapter we return to the hydrostatic approximation, and consider a flat bottomed ocean bounded by a straight coast. Again we consider an f-plane ocean and analyze the response pattern of this model to some examples of external forcing. Such problems have been studied by several authors: CREPON (1967) investigated the response of a barotropic ocean, MILLIOT and CREPON (1981), CREPON and RICHEZ (1982) considered a two layer ocean, and KUNDU et al. (1983), and KUNDU (1984) dealt with a continuously stratified ocean.

We will present a systematic account on the basis of the Green's function technique in order to give a unified representation of the theory. We consider again a continuously stratified ocean, characterized by a BVF $N(z)$. We start with the Boussinesq equations (3.2.1) to (3.2.7) and expand the pressure and the horizontal velocities in terms of a set of vertical eigenfunctions $F_n(z)$. As it was outlined in Subsection 1.9.1 it depends on the assumed BVF which functional form the $F_n(z)$'s have. We may use one of the examples discussed in Subsections 1.9.2.—1.9.4.

With

$$(u, v, p) = \sum_n (u_n, v_n, p_n) F_n(z) \tag{5.1.1}$$

and, using the same expansion of the forcing functions,

$$(X, Y) = \sum_n (X_n, Y_n) F_n(z) \tag{5.1.2}$$

we find the following set of equations

$$u_{nt} - fv_n + p_{nx} = X_n \tag{5.1.3a}$$

$$v_{nt} + fu_n + p_{ny} = Y_n \tag{5.1.3b}$$

$$u_{nx} + v_{ny} + \lambda_n^2 p_{nt} = 0 . \tag{5.1.3c}$$

Again we assume that the external forcing starts to act at $t = 0$ on a fluid at rest. This implies the initial conditions

$$u_n = v_n = p_n = 0 \quad \text{for} \quad t < 0\,. \tag{5.1.4}$$

Moreover, we have the boundary condition that the cross-shore current vanishes at the coast. Let the coast be located along the x-axis, i.e., the ocean is confined to the half space $y > 0$. Then we have

$$v_n(x, 0, t) = 0\,. \tag{5.1.5}$$

In addition we postulate that v_n remains finite for large distances from the coast, i.e.,

$$v_n(x, y, t) < \infty \quad \text{for} \quad y \to \infty\,. \tag{5.1.6}$$

Since these boundary conditions are rather simple it seems appropriate to build up a solution theory on the basis of the Green's function which corresponds to v_n. First we derive from the set (5.1.3) an equation for v_n alone. In the following we drop the vertical mode index n.

Inserting (5.1.3a) into (5.1.3b) we have

$$v_{tt} + f^2 v = fp_x - p_{yt} - fX + Y_t\,. \tag{5.1.7}$$

Using (5.1.3c) we find

$$fp_{xt} - p_{ytt} = \frac{1}{\lambda^2}(v_{tyy} + u_{txy} - fu_{xx} - fv_{xy})\,. \tag{5.1.8}$$

Let $\partial/\partial y$ act on (5.1.3a) and $\partial/\partial x$ on (5.1.3b) and substract

$$u_{ty} - v_{tx} - f(u_x + v_y) = X_y - Y_x\,. \tag{5.1.9}$$

Next we differentiate this with respect to x in order to eliminate u in (5.1.8) and insert the result into (5.1.7). Then we obtain

$$v_{ttt} + f^2 v_t - \frac{1}{\lambda^2}\Delta v_t = Y_{tt} - \frac{1}{\lambda^2} Y_{xx} - fX_t + \frac{1}{\lambda^2} X_{xy} \tag{5.1.10a}$$

where $\Delta = \partial^2/\partial x^2 + \partial^2/\partial y^2$ is the horizontal Laplacian.

Equations (5.1.10) form together with the boundary conditions (5.1.5) and (5.1.6) an inhomogeneous boundary value problem which can be solved with the aid of our Green's function technique. If v is evaluated we can use the equations

$$u_{tt} - \frac{1}{\lambda^2} u_{xx} = fv_t + \frac{1}{\lambda^2} v_{xy} + X_t \tag{5.1.10b}$$

and

$$p_{tt} - \frac{1}{\lambda^2} p_{xx} = -\frac{1}{\lambda^2}(fv_x + v_{yt} + X_x) \tag{5.1.10c}$$

which can readily be obtained from (5.1.3a, c) in order to express u and p in terms of the external forcing and of v.

Now we Fourier transform the set (5.1.10) according to

$$v(\omega, k, y) = \int_{-\infty}^{+\infty} \mathrm{d}x \int_0^{\infty} \mathrm{d}t\; \mathrm{e}^{i\omega t - ikx}\, v(x, y, t) \tag{5.1.11}$$

and analogous formulas for u and p, and we obtain

$$v_{yy} + \alpha^2 v = \frac{\mathrm{i}(\omega^2\lambda^2 - k^2)}{\omega} Y + \left(\lambda^2 f + \frac{k}{\omega}\frac{\partial}{\partial y}\right) X \tag{5.1.12a}$$

$$u = \frac{\mathrm{i}\omega}{\omega^2\lambda^2 - k^2}\left[\lambda^2 X + \left(\lambda^2 f - \frac{k}{\omega}\frac{\partial}{\partial y}\right) v\right] \tag{5.1.12b}$$

$$p = \frac{\mathrm{i}k}{\omega^2\lambda^2 - k^2}\left[X + \left(f - \frac{\omega}{k}\frac{\partial}{\partial y}\right) v\right] \tag{5.1.12c}$$

where $\alpha^2 = \lambda^2(\omega^2 - f^2) - k^2$. This set will be our starting point for the solution theory which will be elaborated in the following sections. The initial conditions (5.1.4) will be respected by the replacement of ω by $\omega + \mathrm{i}\varepsilon$ if inverse Fourier transformations will be calculated. Clearly, the boundary conditions (5.1.5), (5.1.6) are not affected by the Fourier transformation (5.1.11).

5.2. THE GREEN'S FUNCTION AND THE FORMAL SOLUTION OF THE PROBLEM

We define the Green's function corresponding to the v component of the current by replacing the right-hand side of (5.1.12a) by the delta function $\delta(y - y')$, i.e.,

$$\left(\frac{\partial^2}{\partial y^2} + \alpha^2\right) G(y, y') = \delta(y - y') \tag{5.2.1a}$$

and we choose the boundary conditions

$$G(0, y') = 0 \tag{5.2.1b}$$

$$G(y, y') < \infty \quad \text{for} \quad y \to \infty \tag{5.2.1c}$$

where

$$\alpha^2 = \lambda^2(\omega^2 - f^2) - k^2 \,. \tag{5.2.2}$$

The estimation of $G(y, y')$ is standard. First we determine the two linear independent solutions for $y \neq y'$. For $\alpha^2 > 0$ the solutions are oscillatory, e.g., $\sin \alpha y$ and $\mathrm{e}^{\pm \mathrm{i}\alpha y}$, and for $\alpha^2 < 0$, i.e., $\alpha = \mathrm{i}\,|\alpha|$ where $|\alpha|$ is the modulus of α, we have $\sinh |\alpha|\, y$ and $\mathrm{e}^{-|\alpha| y}$. The former vanishes at the coast and the latter diminishes for large y. Hence, we may take for all α the pair $\sin \alpha y$ and $\mathrm{e}^{\mathrm{i}\alpha y}$, with the Wronskian $-\alpha$, to construct the Green's function

$$G(y, y') = \theta(y - y')\, G^{>}(y, y') + \theta(y' - y)\, G^{<}(y, y') \tag{5.2.3a}$$

with

$$G^{>}(y, y') = \frac{-\mathrm{e}^{\mathrm{i}\alpha y} \sin \alpha y'}{\alpha} = G^{<}(y', y) \,. \tag{5.2.3b}$$

An equivalent expression is given by

$$G(y, y') = \frac{1}{2\mathrm{i}\alpha}\left(\mathrm{e}^{\mathrm{i}\alpha|y - y'|} - \mathrm{e}^{\mathrm{i}\alpha(y + y')}\right). \tag{5.2.4}$$

We notice that for $\alpha \to 0$ the Green's function is regular

$$G(y, y')_{\alpha \to 0} \to \frac{1}{2}(|y - y'| - (y + y')) .$$

Thus, $G(y, y')$ has no poles. The only singularities are two branch points in the ω-plane at $\omega = \pm\sqrt{f^2 + k^2/\lambda^2}$. Moreover, the Green's function oscillates with y for real α and $G(y, y')$ is localized at $y = y'$ for imaginary α.

Introducing the notations

$$G * X = \int_0^\infty \mathrm{d}y'\, G(y, y')\, X(y') , \qquad G * Y = \int_0^\infty \mathrm{d}y'\, G(y, y')\, Y(y') , \tag{5.2.5}$$

we may express the solution of (5.1.12a) in terms of $G(y, y')$ and the external forcing by means of source representation

$$v(\omega, k, y) = \frac{\mathrm{i}}{\omega}(\omega^2\lambda^2 - k^2)\, G * Y + \lambda^2 f G * X + \frac{k}{\omega} G * X_{y'} . \tag{5.2.6a}$$

With (5.2.6a) we obtain from (5.1.12b, c)

$$u(\omega, k, y) = -f\lambda^2 G * Y + \frac{k}{\omega}\frac{\partial}{\partial y} G * Y + \frac{\mathrm{i}}{\omega^2\lambda^2 - k^2}$$
$$\times \left[\omega\lambda^2 X + \omega\lambda^4 f^2 G * X + kf\lambda^2 G * X_{y'} - \frac{\partial}{\partial y}\left(\lambda^2 fkG * X + \frac{k^2}{\omega} G * X_{y'} \right) \right] \tag{5.2.6b}$$

$$p(\omega, k, y) = -\frac{fk}{\omega} G * Y + \frac{\partial}{\partial y} G * Y + \frac{ik}{\omega^2\lambda^2 - k^2}$$
$$\times \left[kX - \omega\lambda^2 f \frac{\partial}{\partial y} G * X - k \frac{\partial}{\partial y} G * X_{y'} + k\lambda^2 f^2 G * X + \frac{k^2 f}{\omega} G * X_{y'} \right] . \tag{5.2.6c}$$

This is already the general formal solution of the problem considered. Contrary to (5.2.6a) the expressions of u and p, i.e. (5.2.6b, c) have poles at $\omega = \pm k/\lambda$, which correspond to the ordinary dispersion relation of non-dispersive waves, and which we will call Kelvin poles. Those poles do not occur in the solution for an unbounded ocean, compare (3.4.3). Hence, we may argue that the appearence of these poles is a direct consequence of the existence of the coast. We study this in detail in the following section.

Clearly, we could use a Green's function associated with another dynamical variable. We shortly discuss the Green's function which corresponds to the pressure p.

From (5.1.3) an equation of p is readily derived

$$p_{ttt} + f^2 p_t - \frac{1}{\lambda^2}\Delta p_t = -\frac{1}{\lambda^2}(X_{xt} - fX_y + Y_{yt} + fY_x) . \tag{5.2.7}$$

Fourier transformation concerning x and t yields

$$p_{yy} + \alpha^2 p = \frac{\mathrm{i}}{\omega}(\omega kX - fX_y - \mathrm{i}\omega Y_y + \mathrm{i}kfY) . \tag{5.2.8}$$

The associated Green's function $K(y, y')$ is subject to

$$K_{yy}(y, y') + \alpha^2 K(y, y') = \delta(y - y') . \tag{5.2.9}$$

The boundary condition at $y = 0$ is, however, more complicated than (5.2.1b).
First, from (5.1.7) it follows

$$fp_x - p_{yt} = fX - Y_t \quad \text{for} \quad y = 0 . \tag{5.2.10}$$

For large y we postulate

$$p < \infty \quad \text{for} \quad y \to \infty . \tag{5.2.11}$$

This implies the choice

$$K_y + \frac{kf}{\omega} K = 0 \quad \text{for} \quad y = 0 , \tag{5.2.12a}$$

$$K = 0 \quad \text{for} \quad y \to \infty . \tag{5.2.12b}$$

For $y \neq y'$ the two linear independent solutions of (5.2.9) are $\cos \alpha y - kf/\omega\alpha \sin \alpha y$ and $e^{i\alpha y}$ with the Wronskian $i\alpha + kf/\omega$.

Thus we get

$$K^{>}(y, y') = \left(\cos \alpha y' - \frac{kf}{\omega\alpha} \sin \alpha y'\right) \frac{e^{i\alpha y}}{i\alpha + \dfrac{kf}{\omega}} = K^{<}(y', y) \tag{5.2.13}$$

or equivalently

$$K^{>}(y, y') = \frac{\omega}{\omega^2 - f^2} \frac{kf - i\omega\alpha}{(\lambda^2\omega^2 - k^2)} \left(\cos \alpha y' - \frac{kf}{\alpha\omega} \sin \alpha y'\right) e^{i\alpha y'} = K^{<}(y, y) . \tag{5.2.14}$$

Notably this Green's function involves explicitly the Kelvin poles. Comparing (5.2.14) with (5.2.3b) it turns out that the Green's function associated with v is in fact more convenient to deal with. This is the only reason for using $G(y, y')$ throughout this chapter.

5.3. THE FORCED KELVIN WAVE

In this section we investigate which kind of motion is associated with the poles of the alongshore velocity and of the pressure at $\omega^2 = k^2/\lambda^2$. First we observe that these poles only occur in terms associated with X. This implies the corresponding motion is independent of the forcing Y which acts perpendicular to the coast. Thus, in this section we may put $Y = 0$, just for simplicity.

Using the identities

$$\frac{k}{\omega^2 - \left(\dfrac{k}{\lambda}\right)^2} = \frac{\lambda}{2}\left(\frac{1}{\omega - \dfrac{k}{\lambda}} - \frac{1}{\omega + \dfrac{k}{\lambda}}\right) \tag{5.3.1a}$$

$$\frac{\omega}{\omega^2 - \left(\dfrac{k}{\lambda}\right)^2} = \frac{1}{2}\left(\frac{1}{\omega - \dfrac{k}{\lambda}} - \frac{1}{\omega + \dfrac{k}{\lambda}}\right) \tag{5.3.1b}$$

$$\frac{1}{\omega}\frac{1}{\omega^2-\left(\frac{k}{\lambda}\right)^2}=-\frac{\lambda^2}{k^2}\left[\frac{1}{\omega}-\frac{1}{2}\left(\frac{1}{\omega-\frac{k}{\lambda}}+\frac{1}{\omega+\frac{k}{\lambda}}\right)\right] \tag{5.3.1c}$$

we may rewrite (5.2.6b) and obtain

$$u(\omega,k,y)=\frac{\mathrm{i}}{\omega}G_y*X_{y'}+\frac{\mathrm{i}}{2}\left[\frac{1}{\left(\omega-\frac{k}{\lambda}\right)}\left(X-\frac{G_y*X}{R}+\frac{G*X}{R^2}+\frac{G*X_{y'}}{R}-G_y*X_{y'}\right)\right.$$
$$\left.+\frac{1}{\omega+\frac{k}{\lambda}}\left(X+\frac{G*X}{R^2}-\frac{G*X_{y'}}{R}+\frac{G_y*X}{R}-G_y*X_{y'}\right)\right]$$

where the Rossby radius $R=(\lambda f)^{-1}$ was introduced.

With the aid of

$$X=X*\delta(y-y')=\int_0^\infty \mathrm{d}y'\;X(y')\,\delta(y-y')$$

and

$$G*X_{y'}=\int_0^\infty \mathrm{d}y'\;G(y,y')\frac{\partial}{\partial y'}X(y')=-\int_0^\infty \mathrm{d}y'\;\frac{\partial}{\partial y'}G(y,y')\,X(y')=-G_{y'}*X \tag{5.3.2}$$

we arrive at

$$u(\omega,k,y)=\frac{\mathrm{i}}{\omega}G_y*X_{y'}$$
$$+\frac{\mathrm{i}X}{2\left(\omega-\frac{k}{\lambda}\right)}*\left[\delta(y-y')+\frac{G}{R^2}+G_{yy'}-\frac{1}{R}(G_{y'}+G_y)\right]$$
$$+\frac{\mathrm{i}X}{2\left(\omega+\frac{k}{\lambda}\right)}*\left[\delta(y-y')+\frac{G}{R^2}+G_{yy'}+\frac{1}{R}(G_{y'}+G_y)\right]. \tag{5.3.3}$$

The contributions of the poles at $\omega=\pm k/\lambda$ are given by the corresponding residues, i.e., the values of the terms in the brackets at $\omega=k/\lambda$ and $\omega=-k/\lambda$, respectively. With $\alpha=\mathrm{i}f\lambda=\mathrm{i}/R$ for $\omega=\pm k/\lambda$ it follows from (5.2.4)

$$G(y,y')|_{\omega^2\lambda^2=k^2}=-\left(\frac{1}{R}\right)\left(\mathrm{e}^{-\frac{1}{R}|y-y'|}-\mathrm{e}^{-\frac{1}{R}(y+y')}\right) \tag{5.3.4}$$

and from (5.2.1a)

$$G_{yy}=\delta(y-y')+\frac{G}{R^2}.$$

We readily verify that

$$G_{y'} + G_y = -e^{-(y+y')/R} \tag{5.3.5}$$

and

$$G_{yy} + G_{yy'} = \frac{1}{R}\, e^{-\frac{1}{R}(y+y')} \tag{5.3.6}$$

Thus only one of the poles, namely $\omega = k/\lambda$ has a non-vanishing residue and contributes with

$$u_{\text{Kelvin}} = \frac{i}{\left(\omega - \dfrac{k}{\lambda}\right)} X * \frac{1}{R}\, e^{-\frac{1}{R}(y+y')} \tag{5.3.7}$$

where $X(\omega, k, y) = X(k/\lambda, k, y)$ has to be taken. We use the subscript "Kelvin" because we will show that this expression describes in fact a Kelvin wave. From (5.3.7) we see that, firstly, only a wave with positive phase velocity exists, i.e., the signal propagates with the coast to the right, and secondly, the motion is coastally trapped with the trapping scale R. The latter statement follows from the finiteness of the wind force X for large y, hence, because of the exponential the integral in (5.3.7) will be diminish for large y.

In the same manner we obtain from (5.2.6c)

$$p(\omega, k, y) = -ifg * X_y + \frac{i}{2\lambda\left(\omega - \dfrac{k}{\lambda}\right)} X * \left[\delta(y - y') + \frac{G}{R^2} + G_{yy'} - \frac{1}{R}(G_y + G_{y'})\right]$$

$$- \frac{i}{2\lambda\left(\omega + \dfrac{k}{\lambda}\right)} X * \left[\delta(y - y') + \frac{G}{R^2} + G_{yy'} + \frac{1}{R}(G_y + G_{y'})\right]. \tag{5.3.8}$$

Again we consider the residues at the poles at $\omega = \pm k/\lambda$, and we find that only the pole at $\omega = k/\lambda$ yields a non-zero contribution

$$p_{\text{Kelvin}} = -\frac{i}{\lambda\left(\omega - \dfrac{k}{\lambda}\right)} X * \frac{1}{R}\, e^{-\frac{1}{R}(y+y')}. \tag{5.3.9}$$

Obviously u_{Kelvin} and p_{Kelvin} are in geostrophical balance

$$\frac{\partial}{\partial y} p_{\text{Kelvin}} = -\frac{i}{\lambda\left(\omega - \dfrac{k}{\lambda}\right)} X * e^{-\frac{1}{R}(y+y')} = -f u_{\text{Kelvin}}.$$

Thus this contribution possesses in fact all properties of a Kelvin wave; that is a coastal trapped wave which propagates with the coast to the right, which has a vanishing cross-shore component, and where the alongshore current and the pressure gradient perpendicular to the coast are geostrophically adjusted. Notably, all these properties follow from the general formal solution (5.2.6), and from the Green's function without further specification of the forcing.

5.4. LARGE SCALE FORCING AND COASTAL JETS

In addition to the Kelvin wave discussed in the foregoing section we study now a further coastal effect, the formation of coastal jets. This notation was introduced by CHARNEY (1955). We start with a rather simple wind field given by a constant large scale alongshore wind switched on at $t = 0$, i.e.,

$$X = \frac{u_*^2}{H_{\text{mix}}} \theta(z + H_{\text{mix}})\, \theta(t)\,, \qquad Y = 0\,. \tag{5.4.1}$$

Expansion in terms of vertical eigenfunctions, and Fourier transformation with respect to x and t yield

$$X_n(\omega, k) = \frac{u_*^2}{h_n} \frac{\mathrm{i}}{\omega + \mathrm{i}\varepsilon}\, 2\pi\, \delta(k)\,, \qquad Y = 0\,. \tag{5.4.2}$$

Using the identity $f(k)\,\delta(k) = f(0)\,\delta(k)$ from the set (5.2.6) it follows

$$v = \frac{u_*^2}{h} \frac{2\pi\,\mathrm{i}\delta(k)}{f\omega R^2}\, G * 1 \tag{5.4.3a}$$

$$u = \frac{-u_*^2}{h} \frac{2\pi\,\delta(k)}{\omega^2} (1 + \lambda^2 f^2 G * 1) \tag{5.4.3b}$$

$$p = \frac{u_*^2}{h} \frac{f \cdot 2\pi\delta(k)}{\omega^2} \frac{\partial}{\partial y} G * 1 \tag{5.4.3c}$$

$$w = \frac{u_*^2}{h} \frac{\mathrm{i}f \cdot 2\pi\delta(k)}{\omega} \frac{\partial}{\partial y} G * 1 \tag{5.4.3d}$$

where the modal vertical velocity is given by $w = \mathrm{i}\omega p$, see formula (3.8.13). Again we suppress the index n and the infinitesimal $\mathrm{i}\varepsilon$. First we estimate the integral $G * 1$ which is according to (5.2.5)

$$G * 1 = \int_0^\infty \mathrm{d}y'\, G(y, y') = \int_0^y \mathrm{d}y'\, G^>(y, y') + \int_y^\infty \mathrm{d}y'\, G^>(y, y')\,. \tag{5.4.4}$$

In order to calculate this integrals we consider a somewhat more complicated expression, $G * \theta(l - y')$, which we deal with in Section 5.7., and which yields $G * 1$ for $l \to \infty$. Here we have to be careful about the cases $y > l$ and $y < l$. It follows

$$G * \theta(l - y') = \theta(y - l) \int_0^l \mathrm{d}y'\, G^>(y, y') + \theta(l - y) \left[\int_0^y \mathrm{d}y'\, G^>(y, y') + \int_y^l \mathrm{d}y'\, G^<(y, y') \right].$$

With (5.2.3b) we have

$$\int_0^y \mathrm{d}y'\, G^>(y, y') = \frac{1}{\alpha^2} \mathrm{e}^{\mathrm{i}\alpha y} (\cos \alpha y - 1)\,,$$

and

$$\int_y^l \mathrm{d}y'\, G^<(y, y') = \frac{\mathrm{i}}{\alpha^2} \sin \alpha y (\mathrm{e}^{\mathrm{i}\alpha l} - \mathrm{e}^{\mathrm{i}\alpha y})\,,$$

and hence

$$G * \theta(l - y') = \frac{1}{\alpha^2}\left[\theta(l - y) - e^{i\alpha y} + \frac{1}{2}\left(e^{i\alpha(y+l)} + \operatorname{sgn}(y - l)\, e^{i\alpha|y+l|}\right)\right]. \tag{5.4.5}$$

Now let l go to infinity. For $\alpha^2 < 0$ each of the two rightmost terms of (5.4.5) diminishes exponentially, while for $\alpha^2 > 0$ the two terms are rapidly oscillating functions whose phase difference tends to zero for infinite l, and hence the terms cancel each other.

Thus we obtain

$$G * 1 = \frac{1}{\alpha^2}\left(1 - e^{i\alpha y}\right). \tag{5.4.6}$$

With (5.4.6) and with $\alpha^2 = \lambda^2(\omega^2 - f^2)$, note that k vanishes because of the delta function appearing in (5.4.3), we arrive at

$$u(\omega, k, y) = \frac{u_*^2}{h} \cdot 2\pi\delta(k)\left(\frac{-1}{\omega^2 - f^2} + \frac{f^2\, e^{i\lambda\sqrt{\omega^2 - f^2}\, y}}{\omega^2(\omega^2 - f^2)}\right) \tag{5.4.7a}$$

$$v(\omega, k, y) = \frac{u_*^2}{h} \cdot 2\pi\delta(k)\, \frac{if}{\omega(\omega^2 - f^2)}\left(1 - e^{i\lambda\sqrt{\omega^2 - f^2}\, y}\right) \tag{5.4.7b}$$

$$p(\omega, k, y) = \frac{u_*^2}{h}\, \frac{2\pi\delta(k)\, f}{\lambda\, i\omega^2(\omega^2 - f^2)^{1/2}}\, e^{i\lambda\sqrt{\omega^2 - f^2}\, y} \tag{5.4.7c}$$

$$w(\omega, k, y) = \frac{u_*^2}{h} \cdot 2\pi\delta(k)\, \frac{f}{\omega}\, \frac{e^{i\lambda\sqrt{\omega^2 - f^2}\, y}}{\lambda(\omega^2 - f^2)^{1/2}}. \tag{5.4.7d}$$

From (5.4.7) the corresponding Eulerian frequency spectra are easily to obtain because of the delta function which makes the Fourier transformation with respect to k trivial. The spectral functions have peaks at the inertial frequency and are increasing for low frequencies.

Next we consider the behaviour of these functions near the origin of the ω-plane where the dynamical fields have poles of different order. The contributions from these poles correspond to non-oscillatory motion in the time domain. The residues are readily obtained

$$u \approx \frac{u_*^2}{h}\, \frac{(-1)}{\omega^2} \cdot 2\pi\delta(k)\, e^{-\frac{y}{R}} \tag{5.4.8a}$$

$$v \approx \frac{u_*^2}{h} \cdot 2\pi\delta(k)\, \frac{-i}{\omega f}\left(1 - e^{-\frac{y}{R}}\right) \tag{5.4.8b}$$

$$p \approx \frac{-u_*^2}{h}\, \frac{2\pi\delta(k)}{\lambda\omega^2}\, e^{-\frac{y}{R}} \tag{5.4.8c}$$

$$w \approx \frac{-u_*^2}{h}\, 2\pi\delta(k)\, \frac{i}{\lambda\omega}\, e^{-\frac{y}{R}}. \tag{5.4.8d}$$

In the time domain this yields

$$u(t, y) \approx \frac{u_*^2}{h} t \, e^{-\frac{y}{R}} \tag{5.4.9 a}$$

$$v(t, y) \approx -\frac{u_*^2}{hf} \left(1 - e^{-\frac{y}{R}}\right) \tag{5.4.9 b}$$

$$p(t, y) \approx \frac{u_*^2}{h\lambda} t \, e^{-\frac{y}{R}} \tag{5.4.9 c}$$

$$w(t, y) \approx \frac{-u_*^2}{h\lambda} e^{-\frac{y}{R}} . \tag{5.4.9 d}$$

Thus the non-oscillatory part of the response consists of an accelerating coastally trapped alongshore current with the trapping scale R, i.e., the Rossby radius. This type of motion is called "coastal jet", the notation was introduced by CHARNEY (1955). The jet is geostrophically balanced by the pressure gradient perpendicular to the coast. The cross component v consists of the Ekman transport to the right of wind and a rectifying term which ensures v to vanish at the coast. The vertical component shows that the jet is associated with coastal downwelling. The complete results follow after summations of the modal components (5.4.9) according to (5.1.1).

Using, for example, a constant BVF, with $F_n(z)$ and $1/h_n$ according to (1.9.25) and (1.9.34), respectively, we can perform the mode summations for the velocity components, i.e.,

$$\begin{aligned} u(y, z, t) &= u_*^2 t \sum_{n=0}^{\infty} \frac{1}{h_n} e^{-\frac{y}{R_n}} F_n(z) \\ &= u_*^2 t \left\{ \frac{1}{H} e^{-\frac{y}{R_0}} + \frac{1}{\pi H_{\text{mix}}} \left[\arctan\left(\frac{\sin \frac{\pi}{H}(H_{\text{mix}} + z)}{e^{\frac{y}{R_1}} - \cos \frac{\pi}{H}(H_{\text{mix}} + z)} \right) \right. \right. \\ &\quad \left. \left. + \arctan\left(\frac{\sin \frac{\pi}{H}(H_{\text{mix}} - z)}{e^{\frac{y}{R_1}} - \cos \frac{\pi}{H}(H_{\text{mix}} - z)} \right) \right] \right\} \end{aligned} \tag{5.4.10 a}$$

and analogously

$$\begin{aligned} v(y, z, t) &= -\frac{u_*^2}{f} \left\{ \frac{\theta(H_{\text{mix}} + z)}{H_{\text{mix}}} - \frac{1}{H} e^{-\frac{y}{R_0}} - \frac{1}{\pi H_{\text{mix}}} \left[\arctan\left(\frac{\sin \frac{\pi}{H}(H_{\text{mix}} + z)}{e^{\frac{y}{R_1}} - \cos \frac{\pi}{H}(H_{\text{mix}} + z)} \right) \right. \right. \\ &\quad \left. \left. + \arctan\left(\frac{\sin \frac{\pi}{H}(H_{\text{mix}} - z)}{e^{\frac{y}{R_1}} - \cos \frac{\pi}{H}(H_{\text{mix}} - z)} \right) \right] \right\} . \end{aligned} \tag{5.4.10 b}$$

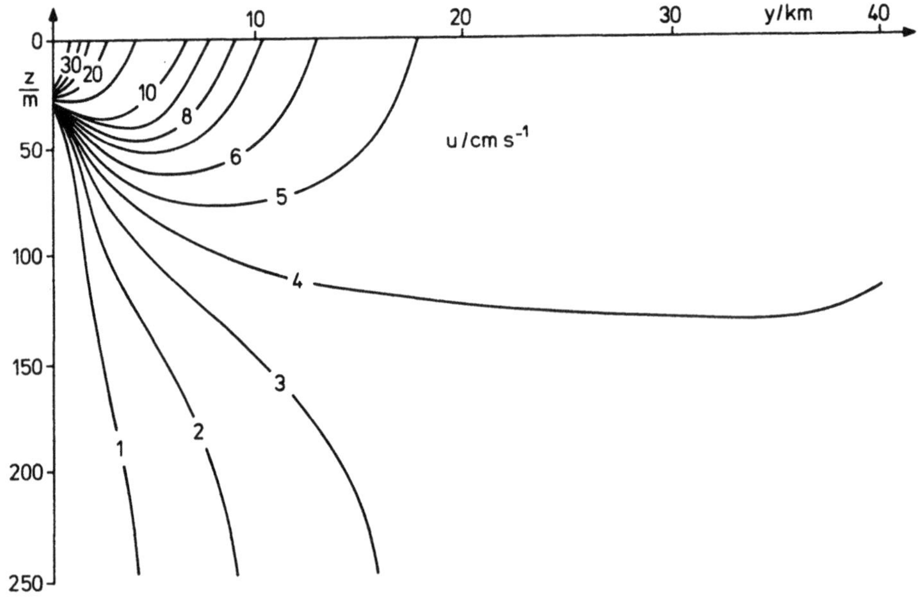

Fig. 5.4.1
Contours of the velocity of the coastal jet in cm/s
The adopted numerical values of the involved parameters are listed in the text.

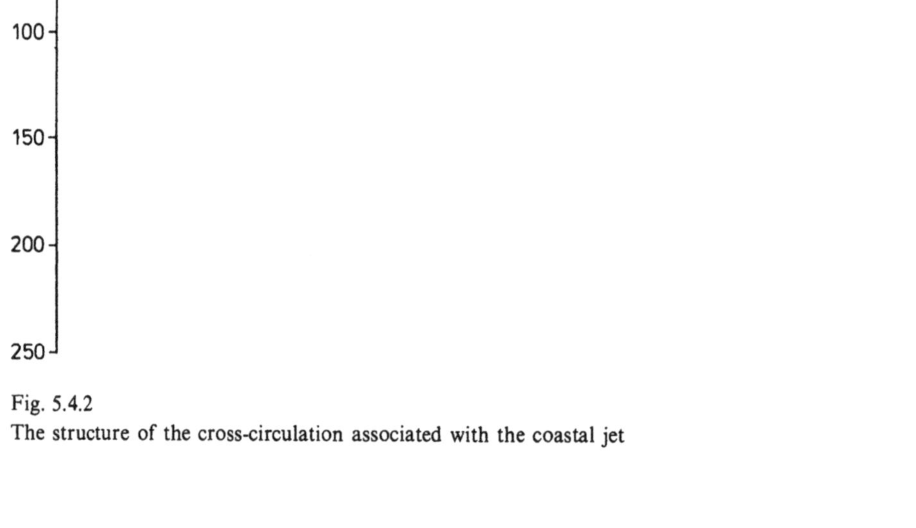

Fig. 5.4.2
The structure of the cross-circulation associated with the coastal jet

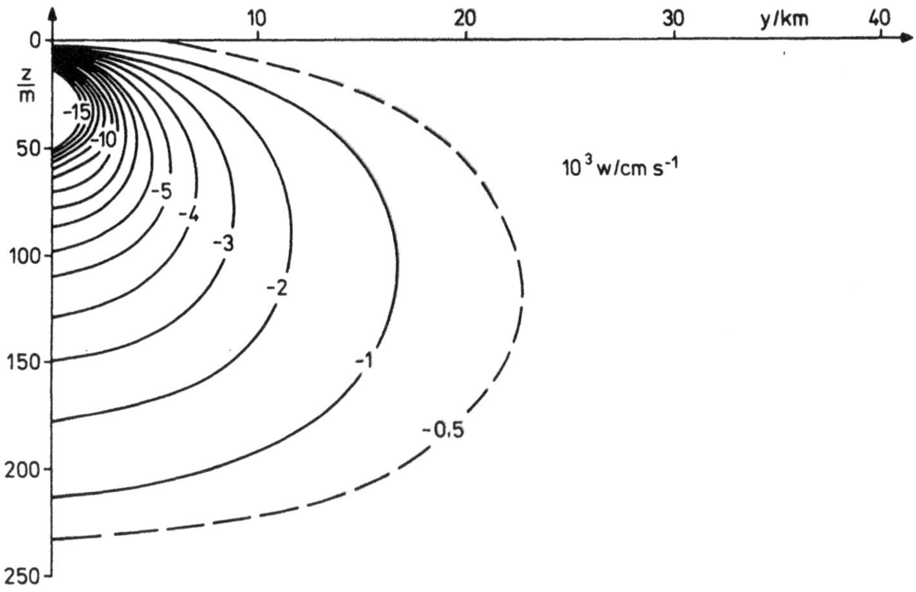

Fig. 5.4.3
Contours of the vertical velocity associated with the coastal jet

Moreover, with (1.8.11 a)

$$w(y, z, t) = u_*^2 \sum_{n=0}^{\infty} w_n(y, t) \frac{F_n'(z)}{N^2},$$

we get

$$w(y, z, t) = -u_*^2 \left\{ -\lambda_0 \, e^{-\frac{y}{R_0}} \left(1 + \frac{z}{H}\right) \right.$$

$$\left. + \frac{1}{2\pi N H_{mix}} \ln \left[\frac{\cosh\left(\frac{y}{R_1}\right) - \cos \frac{\pi}{H}(H_{mix} - z)}{\cosh\left(\frac{y}{R_1}\right) - \cos \frac{\pi}{H}(H_{mix} + z)} \right] \right\}. \tag{5.4.11}$$

Adopting the numerical values $u_*^2 = 1 \text{ cm}^2 \text{ s}^{-2}$, $H = 250$ m, $H_{mix} = 25$ m, $f = 10^{-4} \text{ s}^{-1}$, $N = 10^{-2} \text{ s}^{-1}$, $t = 10^5$ s, and $R_1 = 8$ km, the resulting contours of the velocity components are shown in Figs. 5.4.1 to 5.4.3.

Next we wish to transform the complete set (5.4.7) in the $t - y$ domain. To this end we use the convolutions $C^{(1)}$ and $C^{(2)}$ according to (5.5.1) and (5.6.1). The next two sections are devoted to the investigation of these integrals. Here we use the results in advance and we get

$$u(y, t) = \frac{u_*^2}{h} \left[\theta(t) \frac{\sin ft}{f} + C^{(2)}(y, t) \right] \tag{5.4.12 a}$$

$$v(y, t) = \frac{u_*^2}{h}\left[\frac{-\theta(t)}{f} + \theta(t)\frac{\cos ft}{f} + \frac{1}{f}C_t^{(2)}(y, t)\right] \tag{5.4.12b}$$

$$p(y, t) = \frac{u_*^2 f}{h\lambda}C^{(1)}(y, t) \tag{5.4.12c}$$

$$w(y, t) = \frac{u_*^2 f}{h\lambda}C_t^{(1)}(y, t)\,. \tag{5.4.12d}$$

With the aid of (5.5.6), (5.5.5), (5.6.4a), and (5.6.4b), and writing the vertical mode index explicitly, we obtain

$$u_n(y, t) = \frac{u_*^2}{h_n}\left\{\theta(t)\frac{\sin ft}{f} + \theta(t-\lambda_n y)\left[(t-\lambda_n y) - \frac{\sin f(t-\lambda_n y)}{f}\right]\right.$$
$$\left. - \int_{-\infty}^{t} dt'\left[t - t' - \frac{\sin f(t-t')}{f}\right]\frac{y}{R_n}\theta(t'-\lambda_n y)\frac{J_1(f\sqrt{t'^2-\lambda_n^2 y^2})}{\sqrt{t'^2-\lambda_n^2 y^2}}\right\} \tag{5.4.13a}$$

$$v_n(y, t) = \frac{u_*^2}{h_n}\left\{-\frac{\theta(t)}{f}(1-\cos ft) + \theta(t-\lambda_n y)\left[1 - \frac{\cos f(t-\lambda_n y)}{f}\right]\right.$$
$$\left. - \int_{-\infty}^{t} dt'\frac{1-\cos f(t-t')}{f}\frac{y}{R_n}\theta(t'-\lambda_n y)\frac{J_1(f\sqrt{t'^2-\lambda_n^2 y^2})}{\sqrt{t'^2-\lambda_n^2 y^2}}\right\} \tag{5.4.13b}$$

$$p_n(y, t) = \frac{u_*^2}{h_n}\frac{f}{\lambda_n}\int_{-\infty}^{t} dt'(t-t')\,J_0(f\sqrt{t'^2-\lambda_n^2 y^2})\,\theta(t'-\lambda_n y) \tag{5.4.13c}$$

$$w_n(y, t) = \frac{u_*^2 f}{h_n\lambda_n}\int_{-\infty}^{t} dt'\,\theta(t'-\lambda_n y)\,J_0(f\sqrt{t'^2-\lambda_n^2 y^2})\,. \tag{5.4.13d}$$

Using the asymptotic expressions of $C^{(1)}$, $C^{(2)}$ according to (5.5.11), (5.6.15) which applies for large times t we find

$$u_n(y, t) = \frac{u_*^2}{h_n}\left[t\,e^{-\frac{y}{R_n}} + \frac{y}{R_n}\sqrt{\frac{2}{\pi ft}}\,\frac{\sin\left(ft + \frac{\pi}{4}\right)}{}\right] \tag{5.4.14a}$$

$$v_n(y, t) = \frac{u_*^2}{h_n}\left[-\frac{1}{f} + \frac{1}{f}e^{-\frac{y}{R_n}} + \frac{y}{R_n}\sqrt{\frac{2}{\pi ft}}\,\frac{\cos\left(ft + \frac{\pi}{4}\right)}{f}\right] \tag{5.4.14b}$$

$$p_n(y, t) = \frac{u_*^2}{h_n}\left[\frac{t}{\lambda_n}e^{-\frac{y}{R_n}} - R_n\sqrt{\frac{2}{\pi ft}}\sin\left(ft + \frac{\pi}{4}\right)\right] \tag{5.4.14c}$$

$$w_n(y, t) = \frac{u_*^2}{h_n}\left[-\frac{1}{\lambda_n} e^{-\frac{y}{R_n}} + \lambda_n \sqrt{\frac{2}{\pi f t}} \sin\left(ft + \frac{\pi}{4}\right)\right]. \tag{5.4.14d}$$

This asymptotic set represents an accurate solution of the Boussinesq equations in the version (5.1.3) and with the right-hand side given by (5.4.1), i.e., a large scale alongshore forcing, as long as orders of $y^2 t^{-3/2}$ can be neglected.

On the basis of these solution sets we may discuss some aspects of the response scenario. First we remember that the complete response is governed by the vertical mode summation according to (5.1.1). The sums of the non-oscillatory terms of (5.4.14a, b, d) are given by (5.4.10a, b) and (5.4.11), respectively. Moreover, we notice that

$$\sum_{n=0}^{\infty} \frac{F_n(z)}{h_n} = \frac{\theta(z + H_{\mathrm{mix}})}{H_{\mathrm{mix}}}.$$

We start with the response in the time domain $t \lesssim f^{-1}$. According to (5.4.13a) the alongshore velocity consists of inertial oscillations confined to the upper layer and of a series of inertial waves, which appear as pure inertial waves, (second term in (5.4.13a)), and under the integral sign of $C^{(2)}$ in terms of Bessel functions. The inertial waves propagate off shore with the phase speeds λ_n^{-1}. The barotropic mode propagates rather fast with $1/\lambda_0 = \sqrt{gH}$ and then the baroclinic modes follow with subsequently decreasing phase velocity.

The velocity perpendicular to the coast, v, consists of the Ekman drift and inertial oscillations, both confined to the upper layer, and of inertial waves which also radiate off shore.

The solutions of p and w contain only the off-shore radiating inertial waves, governed by the Bessel functions in the convolution terms, see (5.4.13c, d). Thus we confirm the result of Chapter 3 that pure inertial oscillations may only occur in the horizontal current components but not in the pressure and in the vertical velocity.

Next let us consider a certain distance y from the coast with $y \ll R_0$. Before the front of the barotropic wave reaches the point y the solutions of u and v are governed by pure inertial oscillation confined to the mixed layer. We consider, for instance, the off-shore component v,

$$v(y, z, t) = -\frac{1 - \cos ft}{f} \sum_{n=0}^{\infty} \frac{u_*^2}{h_n} F_n(z) = -\frac{1 - \cos ft}{H_{\mathrm{mix}}} \frac{u_*^2}{H_{\mathrm{mix}}} \theta(z + H_{\mathrm{mix}}).$$

Behind the rapidly propagating barotropic front we have approximately

$$v(y, z, t) = -\frac{1 - \cos ft}{f} \sum_{n=1}^{\infty} \frac{u_*^2}{h_n} F_n(z) = -\frac{1 - \cos ft}{f} \frac{u_*^2}{H_{\mathrm{mix}}}\left[\theta(z + H_{\mathrm{mix}}) - \frac{H_{\mathrm{mix}}}{H}\right]$$

where terms of the order y/R_0 were neglected. This kind of motion consists of a slab-like inertial oscillation of amplitude $u_*^2(H - H_{\mathrm{mix}})/(fHH_{\mathrm{mix}})$ in the upper layer and of an oppositely directed subsurface oscillation of smaller amplitude $u_*^2/(fH)$.

After the passage of the $n = 1$ front it follows

$$v(y, z, t) = -\frac{1-\cos ft}{f}\frac{u_*^2}{H_{\mathrm{mix}}}\left[\theta(z + H_{\mathrm{mix}}) - \frac{H_{\mathrm{mix}}}{H}\right]$$

$$+\frac{u_*^2}{h_1 f}F_1(z)\left\{1 - \cos\left(ft - \frac{y}{R_1}\right) - \frac{y}{R_1}\int_{\lambda_1 y}^{t} dt'[1 - \cos f(t-t')]\frac{J_1(f\sqrt{t'^2 - \lambda_1^2 y^2})}{\sqrt{t'^2 - \lambda_1^2 y^2}}\right\}$$

The second term having the shape of the first baroclinic mode $F_1(z)$ now contributes inertial waves at all depths. Due to the large contribution of the first term in the upper layer the inertial oscillations in the mixed layer are still predominant. But the subsequent arrival of higher modes would tend to weaken such dominance, while the inertial waves penetrate vertically with more and more modes. In the end the inertial oscillations are completely compensated and the regime governed by the asymptotic expression (5.4.14b) established. Thus it remains the Ekman transport and the rectifying term, while the inertial waves decay algebraically.

Analogously, we could discuss the time history of the remaining dynamical fields. The geostrophically adjusted non-oscillatory motion establish through the radiation of inertial waves.

In order to show the complete response scenario in particularly the transients in a more thorough manner the convolutions $C^{(1,2)}$ must be evaluated numerically. This was undertaken by KUNDU et al. (1983).

5.5. DISCUSSION OF THE CONVOLUTION $C^{(1)}$

In this section we discuss the properties of the convolution $C^{(1)}$ which is defined by

$$C^{(1)}(y, t) = \int_{-\infty}^{+\infty} \frac{d\omega}{2\pi i}\frac{e^{i\sqrt{\omega^2 - f^2}\lambda y - i\omega t}}{\omega^2\sqrt{\omega^2 - f^2}}. \tag{5.5.1}$$

Let us first consider the integral

$$C_{tt}^{(1)} = -\int_{-\infty}^{+\infty} \frac{d\omega}{2\pi i}\frac{e^{i\sqrt{\omega^2 - f^2}\lambda y - i\omega t}}{\sqrt{\omega^2 - f^2}}. \tag{5.5.2}$$

Due to the requirement of causality we shift the two branch points at $\omega = \pm f$ into the lower half plane of the complex ω-plane by adding $i\varepsilon$ to ω. For $\omega \to \infty$ the f in the exponential of (5.5.2) can be neglected and the exponential becomes $\exp[i\omega(\lambda y - t)]$. For $t < \lambda y$, i.e., $\lambda y - t > 0$, the contour for the ω-integral can thus be completed by an

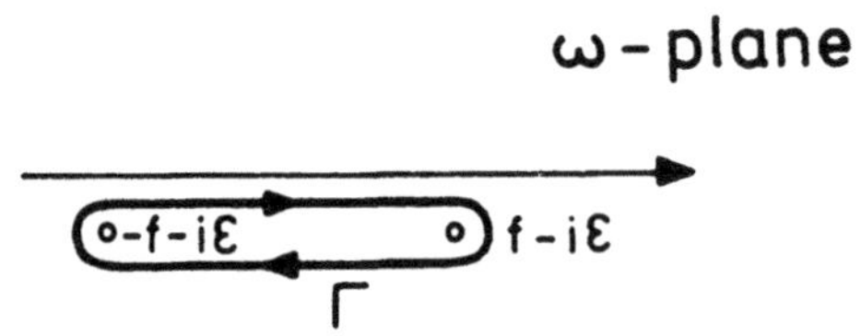

Fig. 5.5.1
The contour Γ used for the calculation of $C_{tt}^{(1)}$

infinite semicircle in the upper half plane, and since there are no singularities the integral vanishes. For $t - \lambda y > 0$ we can complete the contour by an infinite semicircle in the lower half plane, and since there are the branch points the integral can expected to be non-zero there. Now let us take the branch cuts along the joining line $-f$ and $+f$, and then use Cauchy's theorem to pull the contour in until it is just wrapped around the cut $(-f, f)$, see Fig. 5.5.1. Then we have

$$C_{tt}^{(1)} = -\int_{\Gamma} \frac{d\omega}{2\pi i} \frac{e^{i\sqrt{\omega^2 - f^2}\lambda y - i\omega t}}{\sqrt{\omega^2 - f^2}}.$$

If $\sqrt{\omega^2 - f^2}$ is positive above the cut, it is negative below, and it follows

$$C_{tt}^{(1)} = -\int_{-f}^{+f} \frac{d\omega}{2\pi i} \frac{e^{i\sqrt{\omega^2 - f^2}\lambda y - i\omega t}}{\sqrt{\omega^2 - f^2}} + \int_{+f}^{-f} \frac{d\omega}{2\pi i} \frac{e^{-i\sqrt{\omega^2 - f^2}\lambda y - i\omega t}}{\sqrt{\omega^2 - f^2}}.$$

Hence,

$$C_{tt}^{(1)} = \frac{2}{\pi} \int_0^f d\omega \frac{\cosh(\sqrt{f^2 - \omega^2}\,\lambda y)}{\sqrt{f^2 - \omega^2}} \cos \omega t.$$

This integral can be found in standard integral tables (e.g. OBERHETTINGER 1957, p. 38). Thus we arrive at

$$C_{tt}^{(1)} = \theta(t - \lambda y)\, J_0(f \sqrt{t^2 - \lambda^2 y^2}). \tag{5.5.3}$$

Next we consider

$$C_t^{(1)} = -\int_{-\infty}^{+\infty} \frac{d\omega}{2\pi i} \frac{i}{\omega} \frac{e^{i\sqrt{\omega^2 - f^2}\lambda y - i\omega t}}{\sqrt{\omega^2 - f^2}}. \tag{5.5.4}$$

Taking into account that $i/(\omega + i\varepsilon)$ is the Fourier transform of the step function $\theta(t)$ we may express $C_t^{(1)}$ by means of a convolution of θ and $C_{tt}^{(1)}$, i.e.,

$$C_t^{(1)} = \int_{-\infty}^{+\infty} dt'\, \theta(t - t')\, C_{t't'}^{(1)},$$

and, hence,

$$C_t^{(1)} = \int_{-\infty}^{t} dt'\, \theta(t' - \lambda y)\, J_0(f \sqrt{t'^2 - \lambda^2 y^2}). \tag{5.5.5}$$

Moreover, using the fact that $-1/(\omega + i\varepsilon)^2$ is the Fourier transform of $t\theta$ we have

$$C^{(1)} = \int_{-\infty}^{t} dt'(t - t')\, \theta(t' - \lambda y)\, J_0(f \sqrt{t'^2 - \lambda^2 y^2}) \tag{5.5.6}$$

Now we are going to estimate the asymptotics of $C^{(1)}$, which has turned out to be of some importance to analyze oceanic response pattern. In order to find an asymptotic expression of $C^{(1)}$ for large times we note first that the integrand of (5.5.1) oscillates wildly for large t, hence the largest contribution arises from small ω.

We introduce dimensionless variables $t' = tf$, $\omega' = \omega/f$, and $\xi = \lambda y/t$ and rewrite (5.5.1)

$$C^{(1)} = \frac{1}{f^2}\int\limits_{-\infty}^{+\infty} \frac{d\omega'}{2\pi i}\, \frac{e^{-it'\left(\omega'-\xi\sqrt{\omega'^2-1}\right)}}{\omega'^2\sqrt{\omega'^2-1}}\,. \tag{5.5.7}$$

Now let $u = \omega' - \xi\sqrt{\omega'^2 - 1}$, and let the inverse transformation be $\omega' = (u - \xi\sqrt{u^2 - \gamma^2})/\gamma^2$ where $\gamma^2 = 1 - \xi^2$. Then the integral becomes

$$C^{(1)} = -\frac{\gamma^4}{f^2}\int\limits_{-\infty}^{+\infty} \frac{du}{2\pi i}\, \frac{e^{-it'u}}{\sqrt{u^2-\gamma^2}\,(u-\xi\sqrt{u^2-\gamma^2})^2}\,. \tag{5.5.8}$$

The integrand has a pole at $u = i\xi$, which corresponds to the pole of the integrand of (5.5.7) in the origin of the ω'-plane. The contribution of this pole can be evaluated simply

$$\frac{-2\pi i}{2\pi i}\,\mathrm{Res}\, \frac{e^{-it'\left(\omega'-\xi\sqrt{\omega'^2-1}\right)}}{f^2\omega'^2\sqrt{\omega'^2-1}}\Bigg|_{\omega'=0} = \frac{t}{f}\, e^{-\frac{y}{R}}\,. \tag{5.5.9}$$

Moreover, in (5.5.8) there are two branch points at $u = \pm\gamma$. Let the cuts go to infinity parallel to the imaginary axis. Then we may close the contour for the u-integration by an infinite semicircle in the lower half plane, and use Cauchy's theorem to deform the contour until it is wrapped around the two cuts as indicated in Fig. 5.5.2.

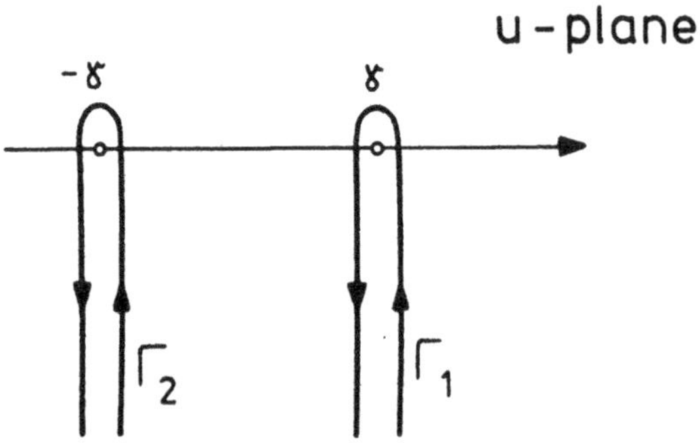

Fig. 5.5.2
The contours Γ_1 and Γ_2 used in (5.5.10)

Then it follows

$$C^{(1)}(y, t) = \frac{t}{f}\, e^{-\frac{y}{R}} + \int\limits_{\Gamma_1} \frac{du}{2\pi i}(\ldots.) + \int\limits_{\Gamma_2} \frac{du}{2\pi i}(\ldots.) \tag{5.5.10}$$

where the parentheses indicate the integrand of (5.5.8). We consider first the second term. Let $s = \sqrt{u - \gamma}$ and expand the integrand around the branch point $u = \gamma$. Then we have

$$\frac{1}{s\sqrt{s^2 + 2\gamma}\,(s^2 + \gamma - \xi s\sqrt{s^2 + 2\gamma})^2} = \frac{1}{\gamma^2\sqrt{2\gamma}\, s}(1 + O(s)) \approx \frac{1}{\gamma^2\sqrt{2\gamma}\sqrt{u-\gamma}}$$

and the integral along Γ_1 yields

$$\frac{\gamma^{3/2}}{f^2\sqrt{2}}\int_{\Gamma_1}\frac{du}{2\pi i}\frac{e^{-it'u}}{\sqrt{u-\gamma}}=\frac{\gamma^{3/2}e^{-i\gamma t'}}{f^2\sqrt{2}\,\pi i}\left[-\int_{\infty}^{0}\frac{dR\,e^{-t'R}}{2\sqrt{i}\sqrt{R}}+\int_{0}^{+\infty}\frac{dR\,e^{-t'R}}{2\sqrt{i}\sqrt{R}}\right]$$

$$=\left(\frac{\gamma}{i}\right)^{3/2}\frac{e^{-i\gamma t'}}{f^2\sqrt{2\pi t}}.$$

Here the transformation $u-\gamma=Re^{i\varphi}$ was made, and the integral

$$\int_{-\infty}^{+\infty}dx\,e^{-x^2t'}=\sqrt{\frac{\pi}{t'}}$$

has been used. Note that

$$\sqrt{u-\gamma}=\sqrt{R}\,e^{i\frac{\varphi}{2}}$$

equals $-i\sqrt{iR}$ at the right-hand side of the cut and $i\sqrt{iR}$ at the left-hand side, as indicated in Fig. 5.5.3. In the third term of (5.5.10) we expand the integrand around the branch point at $u=-\gamma$. With $s=\sqrt{u+\gamma}$ we find

$$\frac{1}{s\sqrt{s^2-2\gamma}\,(s^2-\gamma-\xi s\sqrt{s^2-2\gamma})^2}\cdot\frac{1+O(s)}{\sqrt{-2\gamma}\,\gamma^2 s}\approx\frac{1}{i\gamma^2\sqrt{2\gamma(u+\gamma)}}$$

and the integral along Γ_2 yields approximately

$$\frac{\gamma^{3/2}}{f^2\sqrt{2}\,i}\int_{\Gamma_2}\frac{du}{2\pi i}\frac{e^{it'u}}{\sqrt{u+\gamma}}=\frac{\gamma^{3/2}e^{i\gamma t'}}{f^2\sqrt{2}\,\pi i^2}\left[-\int_{\infty}^{0}\frac{dR\,e^{-t'R}}{2\sqrt{i}\sqrt{R}}+\int_{0}^{\infty}\frac{dR\,e^{-t'R}}{2\sqrt{i}\sqrt{R}}\right]$$

$$=-\frac{\gamma^{3/2}e^{i\gamma t'}}{\sqrt{i}\,f^2\sqrt{2\pi t'}}$$

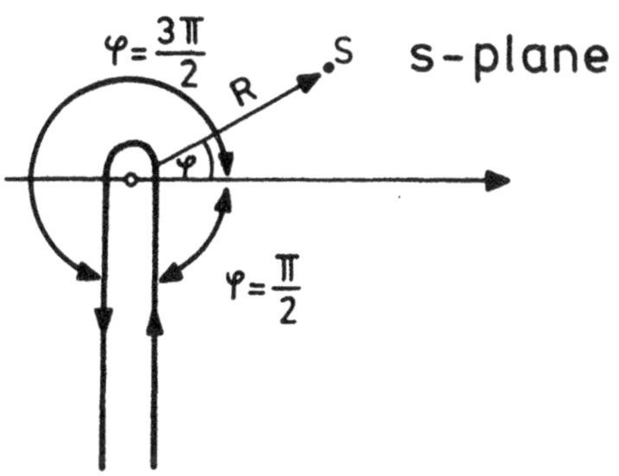

Fig. 5.5.3
Plane polar coordinates centered at the branch point

where the transformation

$$u + \gamma = R e^{i\varphi}$$

was used. With

$$1/\sqrt{i} = \sqrt{i}/i\,, \qquad \sqrt{i} = e^{i\pi/4}\,,$$

and

$$\gamma^{3/2} \approx 1 - \frac{2}{3}\left(\frac{\lambda y}{t}\right)^2$$

we obtain

$$C^{(1)}(y,t) \approx \frac{t}{f}\, e^{-\frac{y}{R}} - \frac{1}{f^2}\sqrt{\frac{2}{\pi f t}}\,\sin\left(f\sqrt{t^2 - \lambda^2 y^2} + \frac{\pi}{4}\right) \tag{5.5.11a}$$

where orders of $\lambda^2 y^2/t^2$ have been neglected, except in the argument of the sine. Thus to this approximation the phase of the wave process is conserved. We call this "phase conserving approximation". If we expand also the sine term we lose the phase information and have

$$C^{(1)}(y,t) \approx \frac{t}{f}\, e^{-\frac{y}{R}} - \frac{1}{f^2}\sqrt{\frac{2}{\pi f t}}\,\sin\left(f t + \frac{\pi}{4}\right). \tag{5.5.11b}$$

To the same approximation we obtain

$$C_t^{(1)}(y,t) \approx \frac{1}{f}\, e^{-\frac{y}{R}} - \frac{1}{f}\sqrt{\frac{2}{\pi f t}}\,\cos\left(f\sqrt{t^2 - y^2\lambda^2} + \frac{\pi}{4}\right), \tag{5.5.12}$$

and

$$C_{tt}^{(1)}(y,t) \approx \sqrt{\frac{2}{\pi f t}}\,\sin\left(f\sqrt{t^2 - y^2\lambda^2} + \frac{\pi}{4}\right). \tag{5.5.13}$$

5.6. DISCUSSION OF THE CONVOLUTION $C^{(2)}$

We define $C^{(2)}$ by the Fourier integral

$$C^{(2)}(y,t) = \int_{-\infty}^{+\infty} \frac{d\omega}{2\pi}\,\frac{f^2\, e^{i\sqrt{\omega^2 - f^2}\,\lambda y - i\omega t}}{\omega^2(\omega^2 - f^2)}. \tag{5.6.1}$$

An equivalent formulation is given by

$$C^{(2)}(y,t) = \frac{f^2}{i\lambda}\frac{\partial}{\partial y}\int_{-\infty}^{+\infty} \frac{d\omega}{2\pi}\,\frac{e^{i\sqrt{\omega^2 - f^2}\,\lambda y - i\omega t}}{\omega^2(\omega^2 - f^2)\sqrt{\omega^2 - f^2}}. \tag{5.6.2}$$

Since $f^2/[\omega^2(\omega^2 - f^2)]$ is the Fourier transform of $t - (\sin ft)/f$ we find with the aid of (5.5.2) and (5.5.3) the following convolution

$$C^{(2)}(y,t) = -\frac{1}{\lambda}\frac{\partial}{\partial y}\int_{-\infty}^{t} dt'\left(t - t' - \frac{\sin f(t-t')}{f}\right)\theta(t' - \lambda y)\, J_0\!\left(f\sqrt{t'^2 - \lambda^2 y^2}\right). \tag{5.6.3a}$$

In a similar manner we obtain from (5.6.1) and (5.6.3)

$$C_t^{(2)}(y,t) = f^2 \int_{-\infty}^{+\infty} \frac{d\omega\, e^{i\sqrt{\omega^2-f^2}\lambda y - i\omega t}}{2\pi\, i\omega(\omega^2-f^2)}$$

$$= -\frac{1}{\lambda}\frac{\partial}{\partial y}\int_{-\infty}^{t} dt'[1-\cos f(t-t')]\,\theta(t'-\lambda y)\,J_0(f\sqrt{t'^2-\lambda^2y^2}) \qquad (5.6.3\text{ b})$$

and

$$C_{tt}^{(2)}(y,t) = -f^2 \int_{-\infty}^{+\infty} \frac{d\omega}{2\pi}\frac{e^{i\sqrt{\omega^2-f^2}\lambda y - i\omega t}}{\omega^2-f^2}$$

$$= -\frac{f}{\lambda}\frac{\partial}{\partial y}\int_{-\infty}^{t} dt'\sin f(t-t')\,\theta(t'-\lambda y)\,J_0(f\sqrt{t'^2-\lambda^2y^2}). \qquad (5.6.3\text{ c})$$

Performing the y-derivate explicitly we have with

$$-\frac{1}{\lambda}\frac{\partial}{\partial y}\theta(t-\lambda y)\,J_0(f\sqrt{t^2-\lambda^2y^2}) = \delta(t-\lambda y)\,J_0(0) + \theta(t-\lambda y)\frac{y}{R}\frac{J_0'(f\sqrt{t^2-\lambda^2y^2})}{\sqrt{t^2-\lambda^2y^2}}$$

and with $J_0(0) = 1$, $J_0'(x) = -J_1(x)$

$$C^{(2)}(y,t) = \left[t-\lambda y - \frac{\sin f(t-\lambda y)}{f}\right]\theta(t-\lambda y)$$

$$-\frac{y}{R}\int_{-\infty}^{t} dt'\left(t-t'-\frac{\sin f(t-t')}{f}\right)\theta(t'-\lambda y)\frac{J_1(f\sqrt{t'^2-\lambda^2y^2})}{\sqrt{t'^2-\lambda^2y^2}} \qquad (5.6.4\text{ a})$$

$$C_t^{(2)}(y,t) = \theta(t-\lambda y)[1-\cos f(t-\lambda y)]$$

$$-\frac{y}{R}\int_{-\infty}^{t} dt'[1-\cos f(t-t')]\,\theta(t'-\lambda y)\frac{J_1(f\sqrt{t'^2-\lambda^2y^2})}{\sqrt{t'^2-\lambda^2y^2}}. \qquad (5.6.4\text{ b})$$

$$C_{tt}^{(2)}(y,t) = f\theta(t-\lambda y)\sin f(t-\lambda y)$$

$$-\frac{y}{R}\int_{-\infty}^{t} dt'\sin f(t-t')\,\theta(t'-\lambda y)\frac{J_1(f\sqrt{t'^2-\lambda^2y^2})}{\sqrt{t'^2-\lambda^2y^2}}. \qquad (5.6.4\text{ c})$$

Next we wish to derive asymptotic expressions of $C^{(2)}$ in the manner outlined in the previous section. The first part of the following discussion is close to the method outlined by VERONIS (1956).

We start with the expression (5.6.1) and note first that the contribution of the pole in the origin of the ω-plane yields

$$f^2\int\frac{d\omega}{2\pi}\frac{e^{-i\omega t}\,e^{-\frac{y}{R}}}{\omega^2(-f^2)} \to t\,e^{-\frac{y}{R}}. \qquad (5.6.5)$$

Next we introduce the dimensionless variables $t' = tf$, $\omega' = \omega/f$, and make the transformation

$$u = \omega' - \xi \sqrt{\omega'^2 - 1} \quad \text{with} \quad \xi = f\lambda \frac{y}{t'},$$

and choose

$$\omega' = \frac{u - \xi \sqrt{u^2 - \gamma^2}}{\gamma^2}$$

with

$$\gamma^2 = 1 - \xi^2 .$$

Then from (5.6.1) it follows

$$C^{(2)} = -\frac{\gamma^6}{f} \int_{-\infty}^{+\infty} \frac{du}{2\pi} \frac{e^{-it'u}}{\sqrt{u^2 - \gamma^2}(u - \xi\sqrt{u^2 - \gamma^2})^2 (u\xi - \sqrt{u^2 - \gamma^2})} \tag{5.6.6}$$

where

$$\frac{\partial \omega'}{\partial u} = -\frac{\xi u - \sqrt{u^2 - \gamma^2}}{\gamma^2 \sqrt{u^2 - \gamma^2}} \quad \text{and} \quad \omega'^2 - 1 = \frac{(u\xi - \sqrt{u^2 - \gamma^2})^2}{\gamma^4}$$

has been used.

We notice that the integrand has a second order pole at $u = i\xi$, two simple poles at $u = \pm 1$, and branch points at $u = \pm\gamma$. The pole at $u = i\xi$ in the u-plane corresponds to the pole in the origin of the ω-plane whose contribution is already given by (5.6.5).

For large t', or equivalently small ξ the branch points nearly coincide with the poles at $u = \pm 1$. Therefore we expand the integrand around these poles. We proceed as follows: First we complete the integration path by an infinite semicircle in the lower half plane of the complex u-plane, and take the cuts from $+\gamma$ to $-i\infty + \gamma$ and from $-\gamma$ to $-i\infty - \gamma$. Then we use Cauchy's theorem to deform the contour until it is wrapped around the cuts as shown in Fig. 5.6.1.

Now we consider the integral along Γ_1. Let $s = \sqrt{u - \gamma}$ and $s_0 = \sqrt{1 - \gamma}$ and expand the integrand around s and $s - s_0$. Then with

$$u\xi - \sqrt{u^2 - \gamma^2} = \frac{-2\gamma^2(s - s_0)}{\sqrt{1 + \gamma}} [1 + O(s - s_0)^2],$$

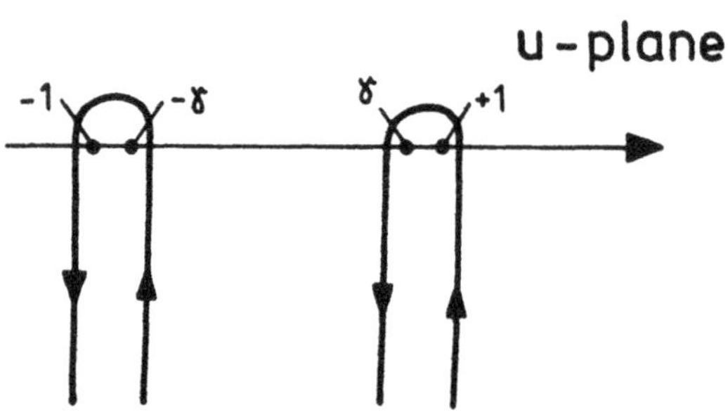

Fig. 5.6.1
The contour in the u-plane used for the calculation of (5.6.6)

$$\sqrt{u^2 - \gamma^2} = s\sqrt{1+\gamma}\,[1 + O(s - s_0)]\,,$$

and

$$u - \xi\sqrt{u^2 - \gamma^2} = \gamma^2[1 + O(s - s_0)]\,,$$

we obtain

$$\frac{\gamma^6}{f}\int\limits_{\Gamma_1}\frac{du}{2\pi}\,\frac{e^{-it'u}}{2\gamma^6\sqrt{u-\gamma}\,(\sqrt{u-\gamma}-\sqrt{1-\gamma})}\,[1 + O(s - s_0)]\,.$$

Introducing polar coordinates $u - \gamma = R\,e^{i\varphi}$, see Fig. 5.5.3, this integral becomes

$$\frac{e^{-it'\gamma}}{f\cdot 4\pi}\left[\int\limits_{\infty}^{0}\frac{dR\,e^{-t'R}}{\sqrt{R}\,(\sqrt{R}-\sqrt{i}\,\sqrt{1-\gamma})} - \int\limits_{0}^{\infty}\frac{dR\,e^{-t'R}}{\sqrt{R}\,(-\sqrt{R}-\sqrt{i}\,\sqrt{1-\gamma})}\right]$$

with $\sqrt{R} = r$ follows

$$\frac{-e^{-it'\gamma}}{f\cdot 2\pi}\int\limits_{-\infty}^{+\infty}\frac{dr\,e^{-t'r^2}}{r-\sqrt{i}\,\sqrt{1-\gamma}} = -\frac{e^{-it'\gamma}}{f\cdot 2\pi}\,I(\sqrt{i}\,\sqrt{1-\gamma})\,. \qquad (5.6.7)$$

Analogously we obtain the contribution of (5.6.6) along Γ_2. Let $s = \sqrt{u+\gamma}$ and $s_0 = \sqrt{-1+\gamma} = i\sqrt{1-\gamma}$, and expand the integrand arounds s and $s - s_0$. Then it follows

$$\frac{\gamma^6}{f}\int\limits_{\Gamma_2}\frac{du}{2\pi}\,\frac{e^{-it'u}}{i^2\gamma^2\sqrt{u+\gamma}\,(\sqrt{u+\gamma}-i\sqrt{1-\gamma})\,\gamma^4}\,,$$

where

$$u\xi - \sqrt{u^2-\gamma^2} = -\frac{2\gamma^2\,i}{\sqrt{1+\gamma}}\,(\sqrt{u+\gamma}-i\sqrt{1-\gamma})\,[1 + O(s - s_0)]\,,$$

$$\sqrt{u^2-\gamma^2} = i\sqrt{u+\gamma}\,\sqrt{1+\gamma}\,[1 + O(s - s_0)]\,,$$

and

$$u - \xi\sqrt{u^2-\gamma^2} = -\gamma^2[1 + O(s - s_0)]$$

have been used. With the transformation $u + \gamma = R\,e^{i\varphi}$ the integral becomes

$$\frac{-e^{it'\gamma}}{f\cdot 4\pi}\left[\int\limits_{\infty}^{0}\frac{dR\,e^{-t'R}}{\sqrt{R}\,(\sqrt{R}-i\sqrt{i}\,\sqrt{1-\gamma})} - \int\limits_{0}^{\infty}\frac{dR\,e^{-t'R}}{\sqrt{R}\,(-\sqrt{R}-i\sqrt{i}\,\sqrt{1-\gamma})}\right]$$

$$= \frac{e^{it'\gamma}}{f\cdot 2\pi}\,I(i\sqrt{i}\,\sqrt{1-\gamma})\,. \qquad (5.6.8)$$

With (5.6.5), (5.6.7), and (5.6.8) we find

$$C^{(2)} \approx t\,e^{-\frac{y}{R}} + \frac{1}{f \cdot 2\pi}\left[e^{it'\gamma} I(i\sqrt{i}\sqrt{1-\gamma}) - e^{-it'\gamma} I(\sqrt{i}\sqrt{1-\gamma})\right]. \tag{5.6.9}$$

Now we are left with the problem of the estimation of the integral $I(z)$

$$I(z) = \int_{-\infty}^{+\infty} dr \frac{e^{-t'r^2}}{r-z} \tag{5.6.10}$$

with $z = \sqrt{i}\sqrt{1-\gamma}$ and with $z = i\sqrt{i}\sqrt{1-\gamma}$. Note that z goes to zero for $\gamma \to 1$, or, equivalently, for $\xi \to 0$. For $z = i\varepsilon$, with $\varepsilon \to +0$, we have with the aid of Dirac's relation (2.3.13)

$$I(0) = \int_{-\infty}^{+\infty} dr \frac{e^{-t'r^2}}{r - i\varepsilon} = P \int_{-\infty}^{+\infty} dr \frac{e^{-t'r^2}}{r} + i\pi = i\pi . \tag{5.6.11}$$

The principle value integral vanishes since the integrand is odd.

We consider now

$$\frac{d}{dz} I(z) = \int_{-\infty}^{+\infty} dr\, e^{-r^2 t'} \frac{d}{dz}\frac{1}{r-z} = -\int_{-\infty}^{+\infty} dr\, e^{-r^2 t'} \frac{d}{dr}\frac{1}{r-z}$$

$$= -\int_{-\infty}^{+\infty} dr\, 2rt' \frac{e^{-r^2 t'}}{r-z} = -2t'\sqrt{\frac{\pi}{t'}} - 2t' z\, I(z)$$

where

$$\frac{r}{r-z} = 1 + \frac{z}{r-z} \quad \text{and} \quad \int_{-\infty}^{+\infty} dr\, e^{-r^2 t'} = \sqrt{\frac{\pi}{r}},$$

have been used. After multiplication by the factor $e^{z^2 t'}$ this differential equation can be put in the form

$$\frac{d}{dz}[e^{z^2 t'} I(z)] = -2\sqrt{\pi t'}\, e^{z^2 t'} .$$

Integrating from 0 to z and using the initial value (5.6.11) this becomes

$$e^{z^2 t'} I(z) - i\pi = -2\sqrt{\pi t'} \int_0^z dz'\, e^{z'^2 t'} = 2i\sqrt{\pi} \int_0^{iz\sqrt{t'}} dx\, e^{-x^2}, \tag{5.6.12}$$

i.e.,

$$I(z) = i\pi e^{-z^2 t'}[1 + \text{erf}(iz\sqrt{t'})]$$

with

$$\text{erf}(x) = \frac{2}{\sqrt{\pi}} \int_0^x dy\, e^{-y^2} = \frac{2}{\sqrt{\pi}}\left(x - \frac{x^3}{1 \cdot 3} + \frac{x^5}{2 \cdot 5} + - \ldots\right).$$

The error function erf(x) of a complex variable can be expressed in terms of Fresnel's integral

$$C(x) = \sqrt{\frac{2}{\pi}} \int\limits_0^x \cos y^2 \, dy \quad \text{and} \quad S(y) = \sqrt{\frac{2}{\pi}} \int\limits_0^x \sin y^2 \, dy \,,$$

i.e.,

$$\operatorname{erf}(\mathrm{i}\sqrt{\mathrm{i}}\, x) = -\sqrt{\frac{2}{\mathrm{i}}}\,[C(x) + \mathrm{i}S(x)]\,,$$

and

$$\operatorname{erf}(-\sqrt{\mathrm{i}}\, x) = -\sqrt{2\mathrm{i}}\,[C(x) - \mathrm{i}\, S(x)]\,.$$

With $x = \sqrt{t'(1-\gamma)}$ we arrive at

$$I(\sqrt{\mathrm{i}}\sqrt{1-\gamma}) = \mathrm{i}\pi\, \mathrm{e}^{-\mathrm{i}(1-\gamma)t'} \left\{1 - 2\,\mathrm{e}^{-\mathrm{i}\frac{\pi}{4}}[C(x) + \mathrm{i}S(x)]\right\}$$

and

$$I(\mathrm{i}\sqrt{\mathrm{i}}\sqrt{1-\gamma}) = \mathrm{i}\pi\, \mathrm{e}^{\mathrm{i}(1-\gamma)t'} \left\{1 - 2\,\mathrm{e}^{\mathrm{i}\frac{\pi}{4}}[C(x) - \mathrm{i}S(x)]\right\}.$$

Then it follows from (5.6.9)

$$C^{(2)}(y,t) = t\,\mathrm{e}^{-\frac{y}{R}} - \frac{\sin ft}{f} + \sqrt{\frac{2}{f}}\left[\sin\left(ft + \frac{\pi}{4}\right) C(\sqrt{t(1-\gamma)}) - \cos\left(ft + \frac{\pi}{4}\right) S(\sqrt{t(1-\gamma)})\right]. \quad (5.6.13)$$

Next we utilize the sum representation of the Fresnel's integral (see e.g. Gradstein, Ryshik 1971, p. 945)

$$S(x) = \sqrt{\frac{2}{\pi}}\left[\sin x^2 \sum_{n=0}^{\infty} \frac{(-1)^n\, 2^{2n} x^{4n+1}}{(4n+1)!!} - \cos x^2 \sum_{n=0}^{\infty} \frac{(-1)^n\, 2^{2n+1} x^{4n+3}}{(4n+1)!!}\right]$$

and

$$C(x) = \sqrt{\frac{2}{\pi}}\left[\sin x^2 \sum_{n=0}^{\infty} \frac{(-1)^n\, 2^{n+1} x^{4n+3}}{(4n+3)!!} + \cos x^2 \sum_{n=0}^{\infty} \frac{(-1)^n\, 2^{2n} x^{4n+1}}{(4n+1)!!}\right].$$

For large times we have $x = \sqrt{t(1-\gamma)} \approx y\lambda f/(2ft)^{1/2}$ and thus only the first term of each of the sums must be taken into account. Obviously the remaining sums converge rapidly and yield only small contribution to the leading terms.

Then we may use elementary trigonometric identities to find

$$C^{(2)}(y,t) \approx t\,\mathrm{e}^{-\frac{y}{R}} - \frac{\sin ft}{f} + \frac{y}{Rf}\left(\frac{2}{\pi f t}\right)^{1/2} \sin\left(f\sqrt{t^2 - \lambda^2 y^2} + \frac{\pi}{4}\right). \quad (5.6.14)$$

This is the phase conserving approximation to $C^{(2)}$. Clearly, if we are not interested in the phase information we may also expand the sine term and have

$$C^{(2)}(y, t) \approx t\, e^{-\frac{y}{R}} - \frac{\sin ft}{f} + \frac{y}{Rf'}\left(\frac{2}{\pi f t}\right)^{1/2} \sin\left(ft + \frac{\pi}{4}\right). \tag{5.6.15}$$

From (5.6.14) the asymptotic expressions of $C_t^{(2)}$ and $C_{tt}^{(2)}$ are readily obtained

$$C_t^{(2)}(y, t) \approx e^{-\frac{y}{R}} - \cos ft + \frac{y}{R}\left(\frac{2}{\pi f t}\right)^{1/2} \cos\left(f\sqrt{t^2 - \lambda^2 y^2} + \frac{\pi}{4}\right) \tag{5.6.16}$$

$$C_{tt}^{(2)}(y, t) \approx f \sin ft - \frac{yf}{R}\left(\frac{2}{\pi f t}\right)^{1/2} \sin\left(f\sqrt{t^2 - \lambda^2 y^2} + \frac{\pi}{4}\right). \tag{5.6.17}$$

5.7. ALONGSHORE FORCING CONFINED TO A COASTAL STRIP

The next problem we analyze is the case of a alongshore wind blowing over a coastal strip of width l. The forcing function is given by

$$X(x, y, z, t) = \frac{u_*^2}{H_{\text{mix}}}\, \theta(z + H_{\text{mix}})\, \theta(t)\, \theta(l - y)\,, \qquad Y = 0\,. \tag{5.7.1}$$

Near the wind edge, i.e., at $y = l$, the forcing resembles that considered in Section 3.9. The new feature, however, is now the assumed existence of a coast, see Fig. 5.7.1.

After Fourier transformation with respect to x and t, and expansion in terms of vertical eigenfunctions $F_n(z)$ we obtain from (5.7.1)

$$X_n(\omega, k, y) = \frac{u_*^2}{h_n} \frac{\mathrm{i} \cdot 2\pi\delta(k)}{\omega}\, \theta(l - y)\,. \tag{5.7.2}$$

According to the set (5.2.6) the dynamical response to the forcing (5.7.2) is given by (we drop the index n)

$$u(\omega, k, y) = -\frac{u_*^2}{h} \frac{2\pi\delta(k)}{\omega^2}\, [\theta(l - y) + f^2\lambda^2 G * \theta)] \tag{5.7.3a}$$

$$v(\omega, k, y) = \frac{2\pi\delta(k)}{\omega}\, \lambda^2 f \mathrm{i}\, G * \theta \tag{5.7.3b}$$

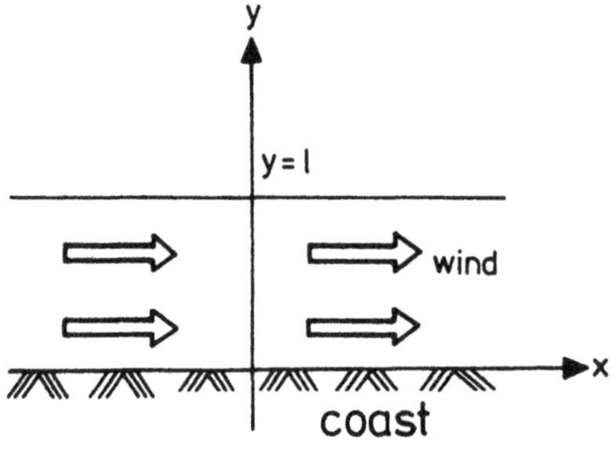

Fig. 5.7.1
Wind blowing alongshore over a coastal strip

$$p(\omega, k, y) = \frac{u_*^2}{h} \frac{2\pi\,\delta(k)}{\omega^2} f \frac{\partial}{\partial y} G * \theta\,. \tag{5.7.3c}$$

The integral $G * \theta$ was already evaluated in Section 5.4., namely (5.4.5)

$$G * \theta = \frac{1}{\alpha^2}\left[\theta(l-y) - \mathrm{e}^{\mathrm{i}\alpha y} + \frac{1}{2}\left(\mathrm{e}^{\mathrm{i}\alpha(y+l)} + \operatorname{sgn}(y-l)\,\mathrm{e}^{\mathrm{i}\alpha|y-l|}\right)\right].$$

The solution set exhibits some interesting features: Firstly, it has poles at $\omega = \pm f$ which correspond to inertial oscillations, and secondly for $\alpha^2 > 0$ the convolution $G * \theta$ oscillates over the whole basin while for $\alpha^2 < 0$ the solution appears to be trapped at the coast, i.e., at $y = 0$, and at the wind edge, i.e., at $y = l$.

Inserting (5.4.5) into the set (5.7.3) we find the Eulerian frequency spectra at any position y. Note that the inverse Fourier transformation with respect to k absorbs only the factor $2\pi\,\delta(k)$

$$u(\omega, y) = -\frac{u_*^2}{h}\left[\frac{\theta(l-y)}{\omega^2 - f^2} + \frac{f^2}{\omega^2(\omega^2 - f^2)}\left(-\mathrm{e}^{\mathrm{i}\lambda\sqrt{\omega^2-f^2}\,y} + \frac{1}{2}\,\mathrm{e}^{\mathrm{i}\sqrt{\omega^2-f^2}\,\lambda(y+l)} + \frac{1}{2}\operatorname{sgn}(y-l)\,\mathrm{e}^{\mathrm{i}\sqrt{\omega^2-f^2}\,\lambda|y-l|}\right)\right] \tag{5.7.4a}$$

$$v(\omega, y) = \frac{u_*^2}{h}\,\frac{\mathrm{i}f}{\omega(\omega^2 - f^2)}\left[\theta(l-y) - \mathrm{e}^{\mathrm{i}\sqrt{\omega^2-f^2}\,\lambda y} + \frac{1}{2}\,\mathrm{e}^{\mathrm{i}\sqrt{\omega^2-f^2}\,\lambda(y+l)} + \frac{1}{2}\operatorname{sgn}(y-l)\,\mathrm{e}^{\mathrm{i}\sqrt{\omega^2-f^2}\,\lambda|y-l|}\right] \tag{5.7.4b}$$

$$p(\omega, y) = \frac{u_*^2}{h}\,\frac{f}{\mathrm{i}\lambda\omega^2(\omega^2 - f^2)^{1/2}}\left[\mathrm{e}^{\mathrm{i}\sqrt{\omega^2-f^2}\,\lambda y} - \frac{1}{2}\,\mathrm{e}^{\mathrm{i}\lambda\sqrt{\omega^2-f^2}\,(y+l)} - \frac{1}{2}\,\mathrm{e}^{\mathrm{i}\sqrt{\omega^2-f^2}\,\lambda|y-l|}\right]. \tag{5.7.4c}$$

The Fourier inversion with respect to ω can be written in terms of the convolutions $C^{(1)}$ and $C^{(2)}$, see (5.5.1) and (5.6.1)

$$u(y, t) = \frac{u_*^2}{h}\left\{\theta(l-y)\frac{\sin ft}{f} + C^{(2)}(y, t) - \frac{1}{2}\left[C^{(2)}(y+l, t) + \operatorname{sgn}(y-l)\,C^{(2)}(|y-l|, t)\right]\right\} \tag{5.7.5a}$$

$$v(y, t) = \frac{u_*^2}{h}\left\{-\frac{\theta(l-y)}{f}(1 - \cos ft) + C_t^{(2)}(y, t) - \frac{1}{2}\left[C_t^{(2)}(y+l, t) + \operatorname{sgn}(y-l)\,C_t^{(2)}(|y-l|, t)\right]\right\} \tag{5.7.5b}$$

$$p(y, t) = \frac{u_*^2}{h}\,\frac{f}{\lambda}\left\{C^{(1)}(y, t) - \frac{1}{2}\left[C^{(1)}(y+l, t) + C^{(1)}(|y-l|, t)\right]\right\} \tag{5.7.5c}$$

For large times we may use the asymptotic representations of the convolutions, i.e., (5.5.11a), (5.6.14), and (5.6.15), and we arrive at

$$u(y,t) = \frac{u_*^2}{h}\left\{t\,e^{-\frac{y}{R}} - \frac{t}{2}\,e^{-\frac{(y+l)}{R}} + \operatorname{sgn}(y-l)\,e^{-\frac{|y-l|}{R}} + \frac{1}{2}\left(\frac{2}{\pi f t}\right)^{1/2}\left[\frac{y}{R}\sin\left(f\sqrt{t^2-\lambda^2 y^2}+\frac{\pi}{4}\right) - \frac{(y+l)}{2R}\sin\left(f\sqrt{t^2-\lambda^2(y+l)^2}+\frac{\pi}{4}\right) - \frac{(y-l)}{2R}\sin\left(f\sqrt{t^2-(y-l)^2\lambda^2}+\frac{\pi}{4}\right)\right]\right\} \tag{5.7.6a}$$

$$v(y,t) = \frac{u_*^2}{hf}\left\{-\theta(l-y) + e^{-\frac{y}{R}} - \frac{1}{2}\left(e^{-\frac{(y+l)}{R}} + \operatorname{sgn}(y-l)\,e^{-\frac{|y-l|}{R}}\right) + \left(\frac{2}{\pi f t}\right)^{1/2}\left[\frac{y}{R}\cos\left(f\sqrt{t^2-y^2\lambda^2}+\frac{\pi}{4}\right) - \frac{(y+l)}{2R}\cos\left(f\sqrt{t^2-\lambda^2(y+l)^2}+\frac{\pi}{4}\right) - \frac{(y-l)}{2R}\cos\left(f\sqrt{t^2-\lambda^2(y-l)^2}+\frac{\pi}{4}\right)\right]\right\} \tag{5.7.6b}$$

$$p(y,t) = \frac{u_*^2}{h\lambda}\left\{t\,e^{-\frac{y}{R}} - \frac{t}{2}\left(e^{-\frac{(y+l)}{R}} + \operatorname{sgn}(y-l)\,e^{-\frac{|y-l|}{R}}\right) - \frac{1}{f}\left(\frac{2}{\pi f t}\right)^{1/2} \times\left[\sin\left(f\sqrt{t^2-y^2\lambda^2}+\frac{\pi}{4}\right) - \frac{1}{2}\sin\left(f\sqrt{t^2-\lambda^2(y+l)^2}+\frac{\pi}{4}\right) - \frac{1}{2}\sin\left(f\sqrt{t^2-\lambda^2(y-l)^2}+\frac{\pi}{4}\right)\right]\right\}. \tag{5.7.6c}$$

By means of these two sets the response scenario can be described. Let us first consider u and v. Taking into account that the convolutions govern inertial waves in terms of the Bessel functions under the integral signs we find that immediatly after the onset of the forcing at $t = 0$ inertial waves propagate off shore starting from the coast, compare $C^{(2)}(y, t)$ and $C_t^{(2)}(y, t)$. Moreover, from the wind edge at $y = l$ inertial waves start to propagate as well on-shore as off-shore, described by the terms $C^{(2)}(|y - l|, t)$ and $C_t^{(2)}(|y - l|, t)$. The on-shore propagating wave with the phase $f(t^2 - \lambda^2(y - l)^2)^{1/2}$ reaches the coast $y = 0$ after the time $t = \lambda l$, and will be reflected there. Then the reflected wave propagates off-shore with the phase $f(t^2 - \lambda^2(y + l)^2)^{1/2}$. The corresponding terms in (5.7.5a, b) are $C^{(2)}(y + l, t)$ and $C_t^{(2)}(y + l, t)$.

The same happens for p and, since $w = -p_t$, for the vertical velocity, where, however, the corresponding terms are governed by $C^{(1)}$ and $C_t^{(1)}$.

Furthermore we observe the occurrence of inertial oscillations at the outset, which are confined to the forcing area $0 < y < l$. After sufficiently long time the asymptotic regime governed by the set (5.7.6) is established. First we note that the inertial oscillations are absorbed by the inertial waves. The inertial waves decay algebraically as $t^{-1/2}$. At this stage the importance of the phase conserving approximation mentioned in Sections 5.5. and 5.6. becomes evident because if we ignore the phase information in the asymptotic representations of $C^{(1)}$ and $C^{(2)}$ the inertial wave contributions in the set (5.7.6) would completely compensate each other.

The non-oscillatory part of the solution consists of a coastal jet as discussed in Section 5.4. and of a wind edge jet as discussed in Section 3.9. The coastal jet has twice the amplitude of the wind edge jet and is in downwind direction. The wind edge jet flows downwind within the forcing strip and forms a countercurrent outside, i.e., for $y > l$.

5.8. FORCING PERPENDICULAR TO THE COAST

In the present section we consider the case of a wind blowing perpendicular to the coast. In particular we first discuss the case of a wind confined to a coastal strip as indicated in Fig. 5.8.1

$$X = 0\,, \qquad Y = \frac{v_*^2}{H_{\text{mix}}}\,\theta(z + H_{\text{mix}})\,\theta(t)\,\theta(l - y)\,. \tag{5.8.1}$$

From the response of the semi-infinite ocean to the forcing (5.8.1) we may easily obtain the response to a large scale forcing of the form

$$X = 0\,, \qquad Y = \frac{v_*^2}{H_{\text{mix}}}\,\theta(z + H_{\text{mix}})\,\theta(t) \tag{5.8.2}$$

by considering $l \to \infty$.

Armed with the experience of the preceding sections the results are easy to obtain. First we expand Y in terms of the $F_n(z)$'s and Fourier transform Y with respect to t and x and we get

$$Y_n(\omega, ky) = \frac{v_*^2}{h_n}\frac{2\pi i}{\omega}\,\delta(k)\,\theta(l - y)\,. \tag{5.8.3}$$

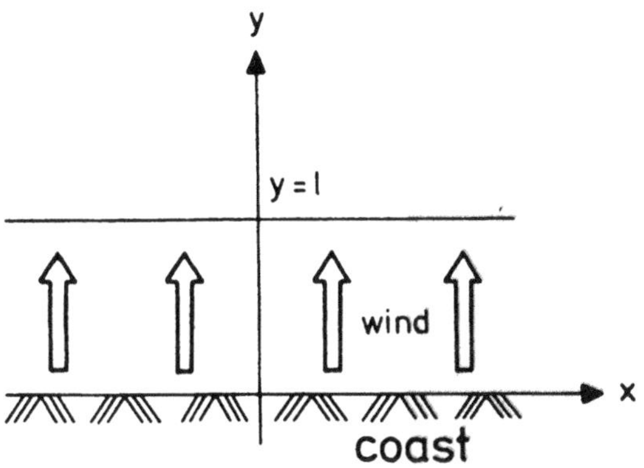

Fig. 5.8.1
Wind blowing perpendicular to the shore over a coastal strip

Again we suppress the index n and we find from the formal solution (5.2.6)

$$u(\omega, k, y) = \frac{v_*^2}{h}\frac{(-f\lambda^2)}{\omega} 2\pi i\, \delta(k)\, G * \theta(l - y') \tag{5.8.4a}$$

$$v(\omega, k, y) = \frac{v_*^2}{h}(-2\pi)\, \delta(k)\, \lambda^2 G * \theta(l - y') \tag{5.8.4b}$$

$$p(\omega, k, y) = \frac{v_*^2}{h}\frac{2\pi i}{\omega}\, \delta(k) \frac{\partial}{\partial y} G * \theta(l - y')\,. \tag{5.8.4c}$$

The integral $G * \theta(l - y')$ is already given by (5.4.5). Thus we have explicitly

$$u(\omega, k, y) = \frac{v_*^2}{h} 2\pi\, i\delta(k) \frac{(-f)}{\omega(\omega^2 - f^2)} \left\{\theta(l - y) - e^{i\sqrt{\omega^2 - f^2}\, \lambda y} + \frac{1}{2}\left[e^{i\sqrt{\omega^2 - f^2}\, \lambda(y + l)} + \operatorname{sgn}(y - l)\, e^{i\sqrt{\omega^2 - f^2}\, \lambda|y - l|}\right]\right\} \tag{5.8.5a}$$

$$v(\omega, k, y) = \frac{-v_*^2}{h}\frac{2\pi\delta(k)}{(\omega^2 - f^2)} \left\{\theta(l - y) - e^{i\sqrt{\omega^2 - f^2}\, \lambda y} + \frac{1}{2}\left[e^{i\sqrt{\omega^2 - f^2}\, \lambda(y + l)} + \operatorname{sgn}(y - l)\, e^{i\sqrt{\omega^2 - f^2}\, \lambda|y - l|}\right]\right\} \tag{5.8.5b}$$

$$p(\omega, k, y) = \frac{v_*^2}{h\lambda}\frac{2\pi\delta(k)}{\sqrt{\omega^2 - f^2}} \left\{e^{i\sqrt{\omega^2 - f^2}\, \lambda y} - \frac{1}{2}\left[e^{i\sqrt{\omega^2 - f^2}\, \lambda(y + l)} + e^{i\sqrt{\omega^2 - f^2}\, \lambda|y - l|}\right]\right\} \tag{5.8.5c}$$

After k-integration, which absorbs the factor $2\pi\, \delta(k)$, the set (5.8.5) represents the Eulerian frequency spectra of the modal components of the response to the forcing (5.8.1) at any location y. Comparing this with the corresponding expressions in case of an alongshore forcing, see (5.7.4), we observe in both cases the same peak locations, namely, at the inertial frequency, but for an alongshore forcing the spectral slope in the subinertial range appears to be steeper by one power of $1/\omega$.

In order to calculate the Fourier inversion of the set (5.8.5) we use again the convolutions $C^{(1)}$ and $C^{(2)}$ according to (5.5.1) and (5.6.1), respectively, and we find

$$u(y, t) = \frac{v_*^2}{hf}\left\{\theta(l - y)(1 - \cos ft) - C_t^{(2)}(y, t) + \frac{1}{2}\left[C_t^{(2)}(y + l, t) + C_t^{(2)}(|y - l|, t)\right]\right\} \tag{5.8.6a}$$

$$v(y, t) = \frac{v_*^2}{hf}\left\{\theta(l - y)\sin ft - C_{tt}^{(2)}(y, t) + \frac{1}{2}\left[C_{tt}^{(2)}(y + l, t) + C_{tt}^{(2)}(|y - l|, t)\right]\right\} \tag{5.8.6b}$$

$$p(y, t) = \frac{v_*^2}{h\lambda}\left\{-C_t^{(1)}(y, t) + \frac{1}{2}\left[C_t^{(1)}(y + l, t) + C_t^{(1)}(|y - l|, t)\right]\right\} \tag{5.8.6c}$$

The asymptotic behaviour of the response follows from (5.5.12), (5.5.16) and (5.6.17)

$$u(y, t) \approx \frac{v_*^2}{hf}\left\{\theta(l-y) - e^{-\frac{y}{R}} + \frac{1}{2}\left(e^{-\frac{(y+l)}{R}} + \operatorname{sgn}(y-l)\, e^{-\frac{|y-l|}{R}}\right)\right.$$
$$- \left(\frac{2}{\pi f t}\right)^{1/2}\left[\frac{y}{R}\cos\left(f\sqrt{t^2 - \lambda^2 y^2} + \frac{\pi}{4}\right)\right.$$
$$- \frac{y+l}{2R}\cos\left(f\sqrt{t^2 - \lambda^2 (y+l)^2} + \frac{\pi}{4}\right)$$
$$\left.\left. - \frac{y-l}{2R}\cos\left(f\sqrt{t^2 - \lambda^2 (y-l)^2} + \frac{\pi}{4}\right)\right]\right\} \tag{5.8.7a}$$

$$v(y, t) \approx \frac{v_*^2}{hf}\left(\frac{2}{\pi f t}\right)^{1/2}\left[\frac{y}{R}\sin\left(f\sqrt{t^2 - \lambda^2 y^2} + \frac{\pi}{4}\right) - \frac{y+l}{2R}\sin\left(f\sqrt{t^2 - \lambda^2 (y+l)^2} + \frac{\pi}{4}\right)\right.$$
$$\left. - \frac{y-l}{2R}\sin\left(f\sqrt{t^2 - \lambda^2 (y-l)^2} + \frac{\pi}{4}\right)\right] \tag{5.8.7b}$$

$$p(y, t) \approx \frac{v_*^2}{hf\lambda}\left\{-e^{-\frac{y}{R}} + \frac{1}{2}\left(e^{-\frac{(y+l)}{R}} + \operatorname{sgn}(y-l)\, e^{-\frac{|y-l|}{R}}\right) - \left(\frac{2}{\pi f t}\right)^{1/2}\right.$$
$$\times\left[\cos\left(f\sqrt{t^2 - \lambda^2 y^2} + \frac{\pi}{4}\right) - \frac{1}{2}\cos\left(f\sqrt{t^2 - \lambda^2 (y+l)^2} + \frac{\pi}{4}\right)\right.$$
$$\left.\left. - \frac{1}{2}\cos\left(f\sqrt{t^2 - \lambda^2 (y-l)^2} + \frac{\pi}{4}\right)\right]\right\} \tag{5.8.7c}$$

We summarize the response scenario: With the onset of the forcing inertial oscillations are excited within the coastal strip $0 < y < l$. Moreover, for each vertical mode there are three distinct inertial waves, one starts from the coast and propagates off-shore, and two start from the wind edge and propagate off-shore and on-shore. The on-shore propagating inertial wave reaches the coast after a time $t = \lambda l$, where it will be reflected and then the reflected wave also propagates off-shore.

For large times there are three off-shore radiating inertial waves with different phases, hence we have used the phase conserving asymptotic representations in the set (5.8.7). Obviously the pure inertial oscillations do not occur in the large times regime and the only oscillatory part of the response consists of inertial waves which decay algebraically. The non-oscillating part of the alongshore current consists of an Ekman transport which will be rectified near the coast and near the edge of the wind field. The terms which rectify the Ekman transport are in geostrophic balance with the pressure. Close to the coast and to the wind edge, i.e., $y \approx 0$ and $y \approx l$, respectively, the off-shore pressure gradient p_y balances the wind force.

The current perpendicular to the coast has no stationary part. Moreover, with $w = -p_t$ we see from (5.8.7c) that no non-oscillatory vertical velocity exists, hence the response pattern neither shows up- nor downwelling.

The response to the large scale forcing (5.8.2), where Y is no longer confined to a coastal strip of width l, may readily be obtained by considering the limit $l \to \infty$ in the above formulas. Then it follows

$$u(y, t) = \frac{v_*^2}{hf}[(1 - \cos ft) - C_t^{(2)}(y, t)] \tag{5.8.8a}$$

$$v(y, t) = \frac{v_*^2}{hf}[\sin ft - C_{tt}^{(2)}(y, t)] , \tag{5.8.8b}$$

$$p(y, t) = \frac{-v_*^2}{hf\lambda} C_t^{(1)}(y, t) . \tag{5.8.8c}$$

Finally we note the asymptotic behaviour of the response to the large scale forcing. In this case we may ignore the phase information and we have

$$u(y, t) \approx \frac{v_*^2}{hf}\left[1 - e^{-\frac{y}{R}} - \left(\frac{2}{\pi f t}\right)^{1/2} \frac{y}{R} \cos\left(ft + \frac{\pi}{4}\right)\right] \tag{5.8.9a}$$

$$v(y, t) \approx \frac{v_*^2}{hf}\left(\frac{2}{\pi f t}\right)^{1/2} \frac{y}{R} \sin\left(ft + \frac{\pi}{4}\right), \tag{5.8.9b}$$

$$p(y, t) \approx \frac{v_*^2}{hf\lambda}\left[e^{-\frac{y}{R}} - \left(\frac{2}{\pi f t}\right)^{1/2} \cos\left(ft + \frac{\pi}{4}\right)\right] \tag{5.8.9c}$$

5.9. ALONGSHORE FORCING OVER HALF PLANES

In the present and in the following sections of this chapter we consider wind fields which are no longer independent of the x-coordinate and, therefore, no longer proportional to $2\pi\,\delta(k)$ in the Fourier space. This implies that various terms of the formal solution (5.2.6) which have discarded by the relation $k\,\delta(k) = 0$ in the cases discussed above will now contribute to the response.

Let us consider again the time dependence of the forcing to be given by the step function $\theta(t)$ with the Fourier transform $\mathrm{i}/(\omega + \mathrm{i}\varepsilon)$, $\varepsilon \to +0$. Then from the general formal solution (5.2.6) it can be seen that the expressions of the longshore current and of the pressure have poles at $\omega = 0$ and $\omega = \pm k/\lambda$ in the ω-plane. The former pole corresponds to non-oscillatory, i.e., steady or accelerating motion, the latter ones to Kelvin waves. Moreover, there are singularities of the Green's function (5.2.4) in terms of the branch points $\omega = \pm(f^2 + k^2/\lambda^2)^{1/2}$. We have seen that these singularities correspond to inertial or Poincaré waves as discussed in the foregoing sections of this chapter and in Chapter 3. We may expect that those waves propagate off-shore and decay algebraically as $(ft)^{-1/2}$. Therefore, the most prominant signals for large times, i.e., $ft \gg 1$, can expected to be those which correspond to non-oscillatory motion and to Kelvin waves.

This implies that in leading order the response is governed by the contributions of the poles in the origin of the ω-plane, and by the Kelvin poles. The mathematical fact that the contributions of the poles closest to the origin yield the leading orders for large times can be illustrated by the examples $-1/\omega^2$ and $-1/(\omega^2 - f^2)$, with the Fourier transforms $t\theta(t)$ and $-\theta(t)\,(\sin ft)/f$, respectively.

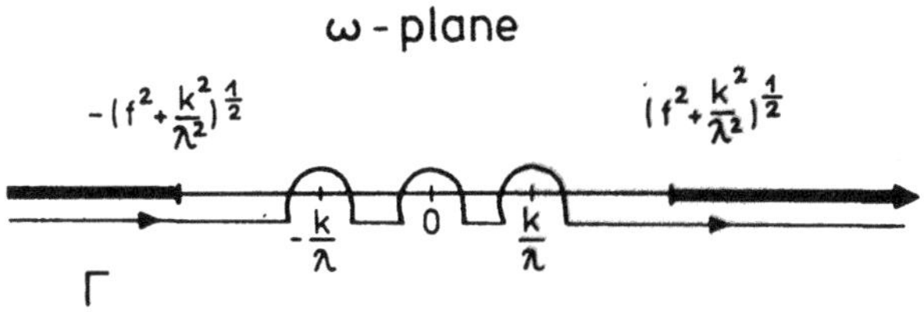

Fig. 5.9.1
The contour in the ω-plane with semicircular indentations at the origin and at the Kelvin poles
The branch cuts are indicated by fat lines.

For the Fourier inversion of (5.2.6a) and (5.2.6b) we, therefore, choose the cuts from $\omega = (f^2 + k^2/\lambda^2)^{1/2}$ along the positive real ω-axis to infinity, and from $\omega = -(f^2 + k^2/\lambda^2)^{1/2}$ along the negative real ω-axis to infinity. We integrate along the path Γ which is depicted in Fig. 5.9.1, i.e., we integrate along the contour just below the cuts including the poles at $\omega = 0$ and $\omega = \pm k/\lambda$ by means of semicircular indentations as indicated in Fig. 5.9.1. This amounts to filtering out the inertial waves. Then using Jordan's Lemma we can complete the contour by an infinite semicircle in the lower half plane and use the residues theorem. Thus the response governed by the general set (5.2.6) can be calculated for large times by means of the residues of the poles in the origin of the ω-plane and of the Kelvin poles.

We note that this reasoning is equivalent to that of CREPON and RICHEZ (1982) based on the Laplace transformation with respect to t. These authors used Sutton's algorithm (SUTTON 1934) to evaluate the asymptotic expansion of the response whose Laplace transform is known.

Let us consider the cases of alongshore winds blowing over the half planes $x > 0$, see Fig. 5.9.2, and $x < 0$, see Fig. 5.9.3, i.e.,

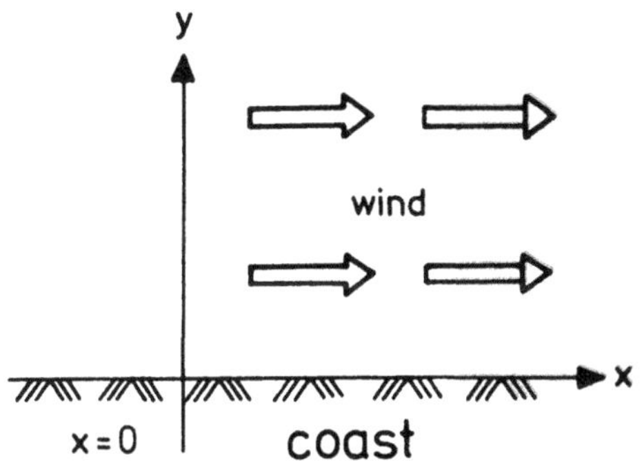

Fig. 5.9.2
Wind blowing alongshore over the half plane $x > 0$ according to (5.9.1), upper case

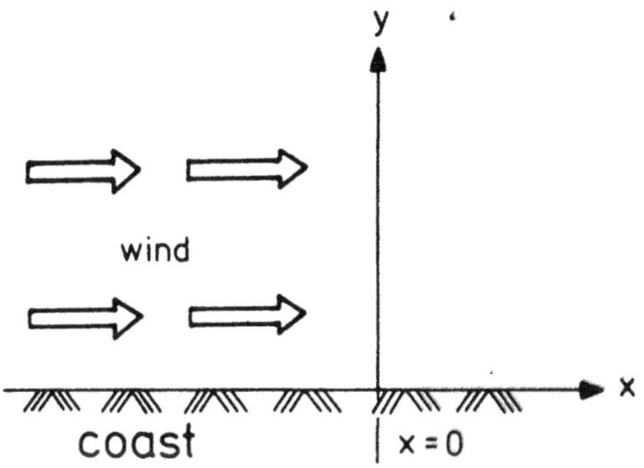

Fig. 5.9.3
Wind blowing alongshore over the half plane $x < 0$ according to (5.9.1), lower case

$$X(x, y, z, t) = \frac{u_*^2}{H_{mix}} \theta(z + H_{mix})\, \theta(t)\, \theta(\pm x)\,, \qquad Y = 0\,. \tag{5.9.1}$$

Again we expand X in terms of vertical eigenfunctions and Fourier transform X with respect to t and x to obtain

$$X_n(\omega, k, y) = \frac{u_*^2}{h_n} \frac{\mathrm{i}}{(\omega + \mathrm{i}\varepsilon)} \frac{\mp \mathrm{i}}{(k \mp \mathrm{i}\varepsilon)}\,, \tag{5.9.2}$$

where the upper sign correspond to $\theta(x)$ and the lower to $\theta(-x)$. Inserting (5.9.2) into the set (5.2.6) and dropping again the index n we find

$$u(\omega, k, y) = \frac{u_*^2}{h} \frac{\pm \mathrm{i}}{k \mp \mathrm{i}\varepsilon} \frac{1}{\omega(\omega^2\lambda^2 - k^2)} [\omega\lambda^2 + f^2\omega\lambda^4 G * 1 - k\lambda^2 f G_y * 1] \tag{5.9.3a}$$

$$v(\omega, k, y) = \frac{u_*^2}{h} \frac{\pm 1}{k \mp \mathrm{i}\varepsilon} \frac{\lambda^2 f}{\omega} G * 1\,, \tag{5.9.3b}$$

$$p(\omega, k, y) = \frac{u_*^2}{h} \frac{\pm 1}{k \mp \mathrm{i}\varepsilon} \frac{\mathrm{i}}{\omega(\omega^2\lambda^2 - k^2)} [k - \omega\lambda^2 f G_y * 1 + k\lambda^2 f^2 G * 1]\,. \tag{5.9.3c}$$

Next we expand these expressions around the poles at $\omega = 0$ and $\omega = \pm k/\lambda$. Using the identity (5.3.1c) we find

$$u(\omega, k, y) = \frac{u_*^2}{h} \frac{\pm \mathrm{i}}{k \mp \mathrm{i}\varepsilon} \left[\frac{1}{\omega^2 - \frac{k^2}{\lambda^2}} \left(1 + \frac{G * 1}{R^2} - \frac{\operatorname{sgn}(\omega)}{R} G_y * 1\right)\bigg|_{\omega^2 = \frac{k^2}{\lambda^2}} + \frac{\lambda^2 f}{k\omega} G_y * 1 \bigg|_{\omega = 0} \right] \tag{5.9.4a}$$

$$v(\omega, k, y) = \frac{u_*^2}{h} \frac{\pm 1}{k \mp \mathrm{i}\varepsilon} \frac{\lambda^2 f}{\omega} G * 1 \bigg|_{\omega = 0} \tag{5.9.4b}$$

$$p(\omega, k, y) = \frac{u_*^2}{h} \frac{\pm \mathrm{i}}{k \mp \mathrm{i}\varepsilon} \left\{ \frac{1}{\lambda} \left(\frac{1}{\omega^2 - \frac{k^2}{\lambda^2}} \right) \left[\operatorname{sgn}(\omega) \left(1 + \frac{G * 1}{R^2}\right) - \frac{G_y * 1}{R} \right]\bigg|_{\omega^2 = \frac{k^2}{\lambda^2}} - \frac{1}{k\omega} \left(1 + \frac{G * 1}{R^2}\right)\bigg|_{\omega = 0} \right\}. \tag{5.9.4c}$$

From (5.4.6) we have for $\omega^2 = k^2/\lambda^2$ with $\alpha^2 = -\lambda^2 f^2 = -1/R^2$,

$$1 + \frac{G * 1}{R^2} = \mathrm{e}^{-\frac{y}{R}}\,, \tag{5.9.5a}$$

$$G_y * 1 = -R\, \mathrm{e}^{-\frac{y}{R}}\,, \tag{5.9.5b}$$

and for $\omega = 0$ we find with

$$\alpha^2 = -\lambda^2 \left(f^2 + \frac{k^2}{\lambda^2} \right) = -\frac{(1 + k^2 R^2)}{R^2}$$

$$1+\frac{G*1}{R^2}=1+\frac{e^{-\sqrt{1+k^2R^2}\frac{y}{R}}-1}{1+k^2R^2}, \tag{5.9.5c}$$

$$G_y*1=\frac{-R\,e^{-\sqrt{1+k^2R^2}\frac{y}{R}}}{\sqrt{1+k^2R^2}}. \tag{5.9.5d}$$

Inserting these expressions into the set (5.9.4) we arrive at

$$u(\omega,k,y)=\frac{u_*^2}{h}\frac{\pm i}{(k\mp i\varepsilon)}\frac{\lambda}{k}\left[\frac{e^{-\frac{y}{R}}}{\omega-\frac{k}{\lambda}}-\frac{1}{\omega}\frac{e^{-\sqrt{1+k^2R^2}\frac{y}{R}}}{\sqrt{1+k^2R^2}}\right]. \tag{5.9.6a}$$

$$v(\omega,k,y)=\frac{u_*^2}{h}\frac{\mp 1}{(k\mp i\varepsilon)f\omega(1+k^2R^2)}\left(1-e^{-\sqrt{1+k^2R^2}\frac{y}{R}}\right) \tag{5.9.6b}$$

$$p(\omega,k,y)=\frac{u_*^2}{h}\frac{\pm i}{(k\mp i\varepsilon)}\frac{1}{k}\left[\frac{e^{-\frac{y}{R}}}{\omega-\frac{k}{\lambda}}-\frac{1}{\omega(1+k^2R^2)}\left(k^2R^2+e^{-\sqrt{1+k^2R^2}\frac{y}{R}}\right)\right]. \tag{5.9.6c}$$

We observe that only one of the Kelvin poles, namely $\omega=k/\lambda$, remains in accordance with our general discussion of Section 5.3.

We perform the inverse Fourier transformation with respect to ω in the usual manner, i.e., we replace ω by $\omega+i\varepsilon$ and close the contour in the lower half plane, and we obtain

$$u(k,y,t)=\frac{u_*^2}{h}\frac{\mp i}{(k\mp i\varepsilon)}\frac{i\lambda}{k}\left[\left(e^{-it\frac{k}{\lambda}}-1\right)e^{-\frac{y}{R}}-\left(\frac{e^{-\sqrt{1+k^2R^2}\frac{y}{R}}}{\sqrt{1+k^2R^2}}-e^{-\frac{y}{R}}\right)\right] \tag{5.9.7a}$$

$$v(k,y,t)=\frac{u_*^2}{h}\frac{\mp i}{(k\mp i\varepsilon)}\frac{1}{f}\left[\left(e^{-\frac{y}{R}}-1\right)+\frac{k^2R^2}{1+k^2R^2}+\left(\frac{e^{-\sqrt{1+k^2R^2}\frac{y}{R}}}{1+k^2R^2}-e^{-\frac{y}{R}}\right)\right] \tag{5.9.7b}$$

$$p(k,y,t)=\frac{u_*^2}{h}\frac{\mp i}{(k\mp i\varepsilon)}\frac{i}{k}\left[\left(e^{-it\frac{k}{\lambda}}-1\right)e^{-\frac{y}{R}}-\frac{k^2R^2}{1+k^2R^2}-\left(\frac{e^{-\sqrt{1+k^2R^2}\frac{y}{R}}}{1+k^2R^2}-e^{-\frac{y}{R}}\right)\right]. \tag{5.9.7c}$$

In these expressions we have added and substracted the term $\exp(-y/R)$ in order to separate the leading contributions which arise from the poles in the origin of the k-plane. By means of the Taylor expansion of $\exp(-\sqrt{1+k^2R^2}\,y/R)$ around $k\approx 0$, which yields

$$e^{-\sqrt{1+k^2R^2}\frac{y}{R}}=e^{-\frac{y}{R}}-yR\,e^{-\frac{y}{R}}\frac{k^2}{2}+\dots$$

we may readily show that the rightmost terms in the set (5.9.7) do not contribute to the residues of the $k = 0$ pole.

Taking only the $k = 0$ contributions into account we find the approximate set

$$u(x, y, t) = \frac{u_*^2}{h} \lambda \left[\pm \left(\frac{t}{\lambda} - x \right) \theta \left(\pm \left(x - \frac{t}{\lambda} \right) \right) \pm x\theta(\pm x) \right] e^{-\frac{y}{R}} \tag{5.9.8 a}$$

$$v(x, y, t) = \frac{u_*^2}{hf} \left(e^{-\frac{y}{R}} - 1 \right) \theta(\pm x) \tag{5.9.8 b}$$

$$p(x, y, t) = \frac{u_*^2}{h} \left[\pm \left(\frac{t}{\lambda} - x \right) \theta \left(\pm \left(x - \frac{t}{\lambda} \right) \right) \pm x\theta(\pm x) \right] e^{-\frac{y}{R}} \tag{5.9.8 c}$$

where the integral (B 19) was used.

The approximate solutions (5.9.8 a, b) describe the propagation of Kelvin waves which start from the wind edge at $x = 0$ and move alongshore with the coast to the right, and the development of coastal jets in the associated forcing areas.

There is a remarkable asymmetry in the response of the ocean to the two different forcing patterns described by (5.9.1).

In the case of $X \sim \theta(x)$ which corresponds to the upper signs in (5.9.8) we observe the usual coastal jet in the region $x > t/\lambda$, see Fig. 5.9.4, that is before the Kelvin wave reaches the point $x = t/\lambda$. For $x \geqq t/\lambda$, however, the jet appears to be arrested by the Kelvin wave. Clearly this happens for every vertical mode n with the associated eigenfunction $F_n(z)$ and the corresponding eigenvalue λ_n.

In the case $\theta(-x)$ we find a jet along the coast with $x < 0$. At $x = 0$ there is a transition of the coastal jet to a Kelvin wave which now radiates outside the forcing area $x < 0$, as indicated in Fig. 5.9.5.

We will discuss the physical meaning in more detail in the following section. We add the complete result of the Fourier inversion of (5.9.7). With the aid of the integrals (B 19), (B 22), and (B 29) to (B 32) we find in the x-domain

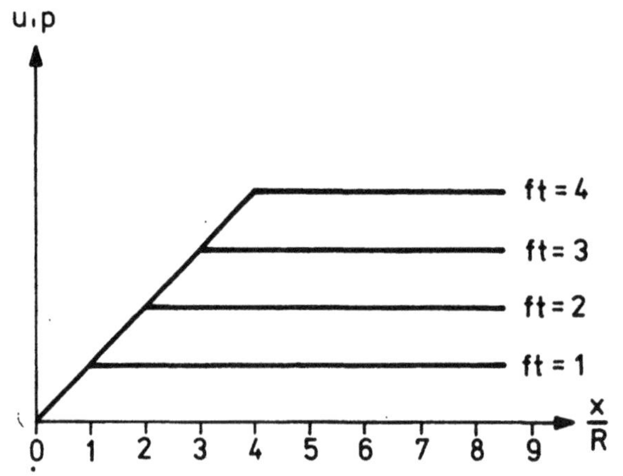

Fig. 5.9.4
The arrest of the coastal jet through the Kelvin wave according to (5.9.8 a, c), upper case

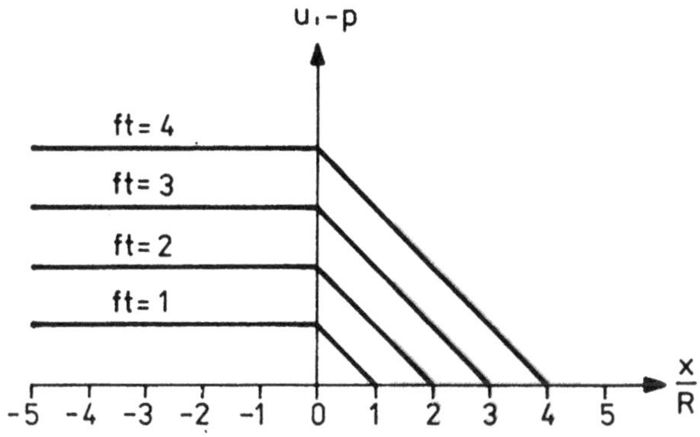

Fig. 5.9.5
Development of a coastal jet and the propagation of the Kelvin wave outside the forcing region

$$u(x, y, t) = \frac{u_*^2}{h}\left[\pm\lambda\left(\frac{t}{\lambda} - x\right)\theta\left(\pm\left(x - \frac{t}{\lambda}\right)\right)e^{-\frac{y}{R}} \pm \frac{\lambda}{\pi R}\int_0^\infty dx'\, x' K_0\left(\frac{\sqrt{(x \mp x')^2 + y^2)}}{R}\right)\right] \tag{5.9.9 a}$$

$$v(x, y, t) = \frac{u_*^2}{hf}\left[\frac{1}{\pi R^2}\int_y^\infty dy'\, K_0\left(\frac{\sqrt{(x \mp x')^2 + y'^2}}{R}\right) - \theta(\pm x) \pm \frac{\operatorname{sgn}(x)}{2}\, e^{-\frac{|x|}{R}}\right] \tag{5.9.9 b}$$

$$p(x, y, t) = \frac{u_*^2}{h}\left[\pm\left(\frac{t}{\lambda} - x\right)\theta\left(\pm\left(x - \frac{t}{\lambda}\right)\right)e^{-\frac{y}{R}} \pm \frac{1}{\pi R^2}\int_y^\infty dy' \int_0^\infty dx'\, x' K_0\left(\frac{\sqrt{(x \mp x')^2 + y'^2}}{R}\right) \pm \frac{R}{2}\, e^{-\frac{|x|}{R}}\right] \tag{5.9.9 c}$$

where K_0 is the modified Bessel function.

5.10. EXPORT OF UPWELLING AND THE GENERATION OF COASTAL UNDERCURRENTS

In the present section we write the vertical mode index n explicitly. With $w_n = -p_{nt}$ we find from the approximate solution (5.9.8c)

$$w_n(x, y, t) = \mp \frac{u_*^2}{h_n \lambda_n}\,\theta\left(\pm\left(x - \frac{t}{\lambda_n}\right)\right)e^{-\frac{y}{R_n}} \tag{5.10.1}$$

where the upper sign corresponds to the case $X \sim \theta(x)$ and the lower one to $X \sim \theta(-x)$, respectively. This formula implies that in the first case, i.e., upper sign, the upwelling

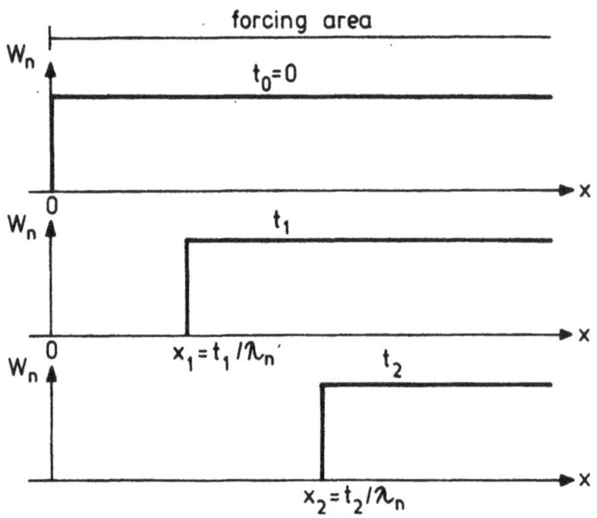

Fig. 5.10.1
Vertical modal velocity in response to a wind over the half plane $x > 0$, for three different times
Note that the vertical motion has been terminated after the Kelvin front has passed through.

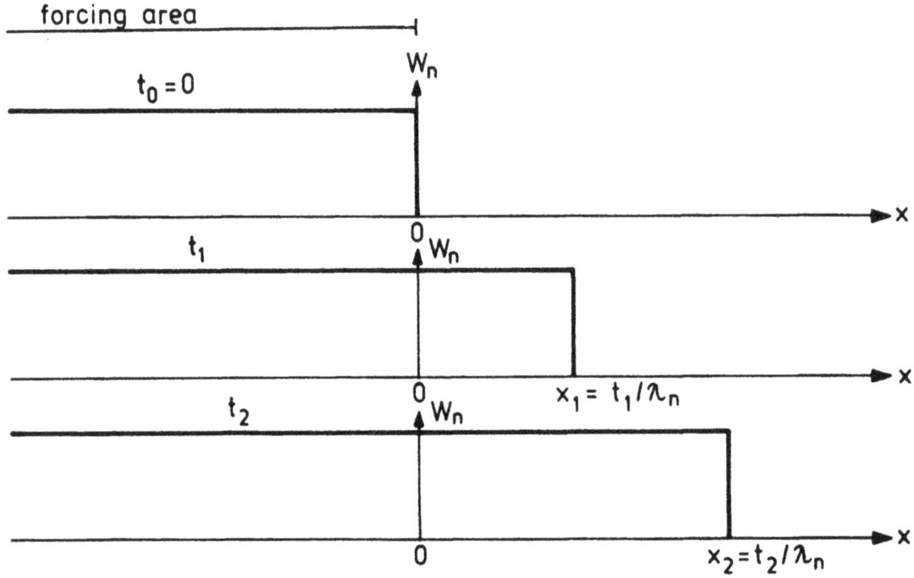

Fig. 5.10.2
Vertical modal velocity in response to a wind over the half plane $x < 0$ for three different times
In this case the vertical motion has been exported outside the forcing area by means of the Kelvin wave.

stops after the passage of the Kelvin wave front, as indicated in Fig. 5.10.1. In the second case, i.e., lower sign, however, the upwelling will be exported even outside the forcing area through the Kelvin wave as depicted in Fig. 5.10.2. Clearly, it depends on the sign of the wind stress whether we deal with up- or downwelling. The above reasoning holds true for both kinds of vertical motion. The important result is that, e.g., upwelling may occur at places where the local wind would not generate it.

A further important effect described by (5.9.8) is the generation of coastal undercurrents in the case of $X \sim \theta(x)$. Let us consider the alongshore current according to (5.9.8 a) taking the upper sign

$$u_n(x, y, t) = \frac{u_*^2}{h_n f} \left[\left(tf - \frac{x}{R_n} \right) \theta \left(\frac{x}{R_n} - tf \right) + \frac{x}{R_n} \theta(x) \right] e^{-\frac{y}{R}} \tag{5.10.2}$$

The full solution follows after summation over all vertical modes

$$u(x, y, z, t) = \sum_n u_n(x, y, t) F_n(z) . \tag{5.10.3}$$

For distances far from the coast, say y being of the order of the barotropic Rossby radius, we have with $R_0 \gg R_1$, and, therefore, $\exp(-y/R_0) \gg \exp(-y/R_1)$, a practically barotropic current

$$u(x, y, z, t) = \frac{u_*^2}{h_0 f} F_0(z) e^{-\frac{y}{R_0}} \left[\left(tf - \frac{x}{R_0} \right) \theta \left(\frac{x}{R_0} - tf \right) + \frac{x}{R_0} \theta(x) \right] . \tag{5.10.4}$$

However, when y approaches the coast the baroclinic modes become more and more important. Let y be in the interval $R_1 \gtrsim y > 0$, so that because of $\exp(-y/R_1) \gg \exp(-y/R_{n+1})$ the first baroclinic mode dominates the baroclinic portion of the response.

Now we consider a snapshot at a fixed time t. At this time the barotropic, and the first baroclinic front have reached the points $x_0 = t/\lambda_0$ and $x_1 = t/\lambda_1$, respectively, where x_0 is considerably larger than x_1 because the barotropic front moves faster than the baroclinic front. In the region $x > x_0$, see Fig. 5.10.3, we have

$$u(x, y, t) = u_*^2 \left(\frac{F_0}{h_0} e^{-\frac{y}{R_0}} + \frac{F_1}{h_1} e^{-\frac{y}{R_1}} \right) t . \tag{5.10.5 a}$$

This is the usually known coastal jet.

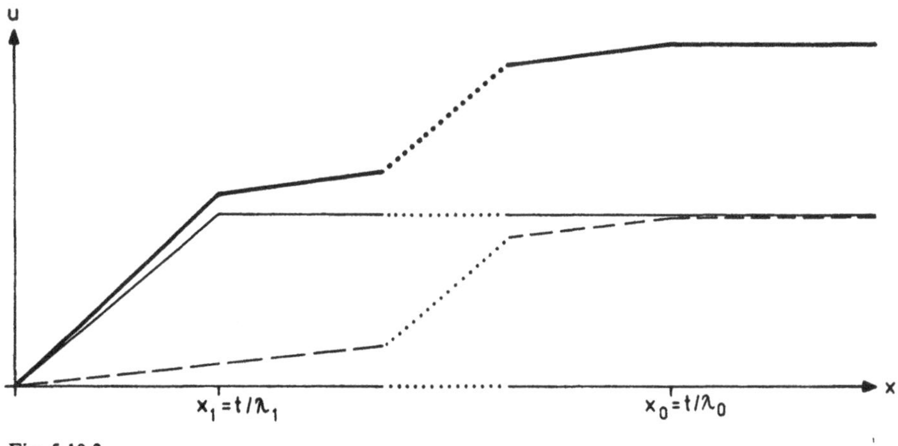

Fig. 5.10.3
A scheme of the development of the undercurrent behind the Kelvin fronts according to (5.10.5d) The dashed line is the barotropic signal, the solid line is the baroclinic one. The fat line is the sum of both contributions. In the region $x < x_1$ the signal is dominated by the baroclinic mode. Note that the x-axis as well the curves are broken as indicated by the dots.

Now let us consider an interval in the wake of the first barotropic front, i.e., $x_1 < x < x_0$, where

$$u(x, y, z, t) = \frac{u_*^2}{f}\left(\frac{F_0}{h_0} e^{-\frac{y}{R_0}} \frac{x}{R_0} + \frac{F_1}{h_1} e^{-\frac{y}{R_1}} tf\right) \tag{5.10.5b}$$

with

$$\frac{x}{R_0} = \frac{1}{R_0}[x_0 - (x_0 - x)] = tf - \frac{x_0 - x}{R_0}$$

we can put this into the form

$$u(x, y, z, t) = u_*^2 \left\{t\left[\frac{F_0}{h_0} e^{-\frac{y}{R_0}} + \frac{F_1(z)}{h_1} e^{-\frac{y}{R_1}}\right] - \frac{x_0 - x}{fR_0} \frac{F_0}{h_0} e^{-\frac{y}{R}}\right\}. \tag{5.10.5c}$$

This expression describes again the coastal jet which is confined to the upper half of the water column, however, this current is somewhat weakened by the rightmost term which does not vary with depth. Hence, after the passage of the barotropic front a marginal undercurrent may develop.

Finally we consider the region $0 < x < x_1$, that is the area where the first baroclinic front has passed through. It follows

$$u(x, y, z, t) = \frac{u_*^2}{f}\left[\frac{F_0}{h_0 R_0} e^{-\frac{y}{R_0}} + \frac{F_1}{h_1 R_1} e^{-\frac{y}{R_1}}\right] x\, \theta(x) \tag{5.10.5d}$$

Here the modes F_0 and F_1 are weighted by their inverse Rossby radii. Since $R_0 \gg R_1$, and consequently $1/R_0 \ll 1/R_1$, it appears that the baroclinic mode dominates this part of the response. Because $F_1(z)$ varies in depth and changes sign, compare, for instance, $F_1(z) = \sqrt{2} \cos(z\pi/H)$ for constant BVF, it follows a well developed undercurrent, which establishes in the wake of the baroclinic front. Clearly, this current is coastally trapped with the trapping scale being the first baroclinic Rossby radius. Thus the equation set (5.10.5) describes the scenario of the generation of a coastal undercurrent in the wake of those Kelvin waves, which arrest the coastal jet.

Let us assume that after sufficient long time all Kelvin modes have passed through a certain point x. Then we get

$$u(x, y, z) = \frac{u_*^2}{f} x\, \theta(x) \sum_{n=1}^{\infty} \frac{F_n(z)}{h_n R_n} e^{-\frac{y}{R_n}}. \tag{5.10.6}$$

With the aid of (1.9.25) and (1.9.34) we can perform the mode summation and we find explicitly

$$u(x, y, z) = \frac{u_*^2 x\, \theta(x)}{2fH_{mix}R_1}\left[\frac{\sin\frac{\pi}{H}(H_{mix} + z)}{\cosh\frac{y}{R_1} - \cos\frac{\pi}{H}(H_{mix} + z)} + \frac{\sin\frac{\pi}{H}(H_{mix} - z)}{\cosh\frac{y}{R_1} - \cos\frac{\pi}{H}(H_{mix} - z)}\right]. \tag{5.10.7}$$

With the numerical values of the parameters $u_*^2 = 1\ \mathrm{cm}^2\ \mathrm{s}^{-2}$, $x = 20$ km, $H = 250$ m, $H_{\mathrm{mix}} = 25$ m, $R_1 = 8$ km, $f = 10^{-4}\ \mathrm{s}^{-1}$, and $N = 10^{-2}\ \mathrm{s}^{-1}$ we obtain the contours of the alongshore velocity as shown in Fig. 5.10.4. Beneath the downwind surface current an undercurrent has established.

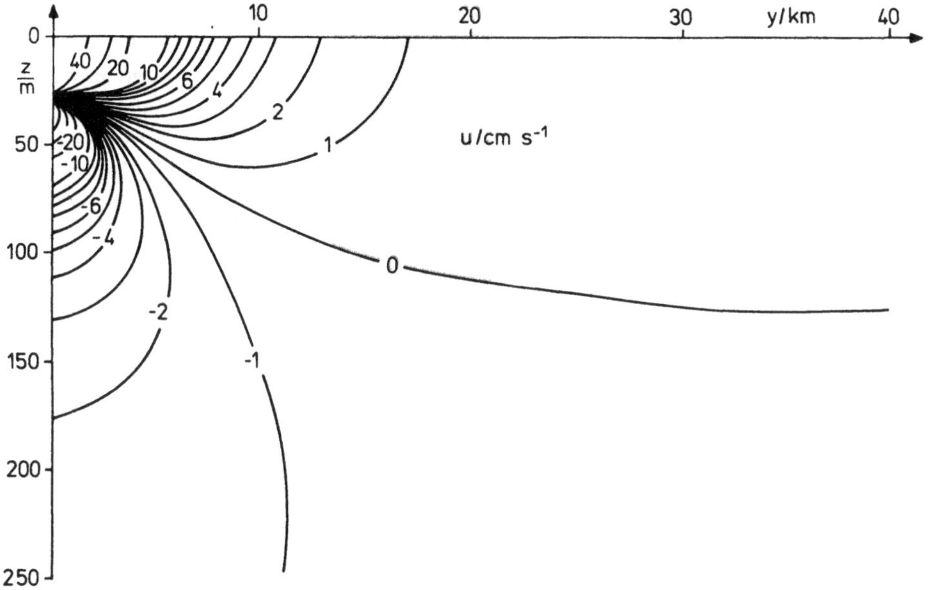

Fig. 5.10.4
Contours of velocity in cm/s of the alongshore current according to (5.10.7)
The adopted numerical values are listed in the text. Note that beneath the downwind coastal jet an undercurrent flowing in the opposite direction has established.

5.11. FORCING PERPENDICULAR TO THE COAST OVER HALF PLANES

Next we study the response to a forcing close to (5.9.1) where, however, X and Y are exchanged, i.e.,

$$X = 0\,, \qquad Y = \frac{v_*^2}{H_{\mathrm{mix}}}\,\theta(z + H_{\mathrm{mix}})\,\theta(t)\,\theta(\pm x)\,. \tag{5.11.1}$$

After Fourier transformation with respect to t and x and after expansion of Y in terms of vertical eigenfunctions we get

$$Y(\omega, k, y) = \frac{v_*^2}{h}\,\frac{\mathrm{i}}{\omega}\left(\frac{\mp\mathrm{i}}{k \mp \mathrm{i}\varepsilon}\right) \tag{5.11.2}$$

where the vertical mode index n was suppressed. Then from our general formal solution (5.2.6) we find

$$u(\omega, k, y) = \frac{v_*^2}{h}\,\frac{\mathrm{i}}{\omega}\left(\frac{\mp\mathrm{i}}{k \mp \mathrm{i}\varepsilon}\right)\left[-f\lambda^2 G * 1 + \frac{k}{\omega}\,G_y * 1\right] \tag{5.11.3a}$$

$$v(\omega, k, y) = \frac{v_*^2}{h}\frac{\mathrm{i}}{\omega}\left(\frac{\mp \mathrm{i}}{k \mp \mathrm{i}\varepsilon}\right)\frac{\mathrm{i}(\omega^2\lambda^2 - k^2)}{\omega} G * 1 \tag{5.11.3b}$$

$$p(\omega, k, y) = \frac{v_*^2}{h}\frac{\mathrm{i}}{\omega}\left(\frac{\mp \mathrm{i}}{k \mp \mathrm{i}\varepsilon}\right)\left[-\frac{fk}{\omega} G * 1 + G_y * 1\right], \tag{5.11.3c}$$

The general discussion in Section 5.3. has revealed that a forcing perpendicular to the coast cannot excite Kelvin waves, and so the Kelvin poles $\omega = \pm k/\lambda$ are absent in (5.11.3). Thus, ignoring inertial waves, only the residues of the poles in the origin of the ω-plane contribute to the response. The zero frequency values of the expressions $G * 1$ and $G_y * 1$ are already given by (5.9.5c) and (5.9.5d), respectively, and we find in the t-domain

$$u(y, k, t) = \frac{v_*^2}{h}\left(\frac{\mp \mathrm{i}}{k \mp \mathrm{i}\varepsilon}\right)\left[\frac{\left(1 - \mathrm{e}^{-\sqrt{1+k^2R^2}\frac{y}{R}}\right)}{f(1 + k^2R^2)} + \frac{\mathrm{i}kRt\,\mathrm{e}^{-\sqrt{1+k^2R^2}\frac{y}{R}}}{\sqrt{1 + k^2R^2}}\right] \tag{5.11.4a}$$

$$v(y, k, t) = \frac{v_*^2}{h}\left(\frac{\mp \mathrm{i}}{k \mp \mathrm{i}\varepsilon}\right)\frac{k^2R^2t}{1 + k^2R^2}\left(1 - \mathrm{e}^{-\sqrt{1+k^2R^2}\frac{y}{R}}\right) \tag{5.11.4b}$$

$$p(y, k, t) = \frac{v_*^2}{h}\left(\frac{\mp \mathrm{i}}{k \mp \mathrm{i}\varepsilon}\right)\left[-ft\frac{\mathrm{i}kR^2}{1 + k^2R^2}\left(1 - \mathrm{e}^{-\sqrt{1+k^2R^2}\frac{y}{R}}\right) - \frac{R\,\mathrm{e}^{-\sqrt{1+k^2R^2}\frac{y}{R}}}{\sqrt{1 + k^2R^2}}\right]. \tag{5.11.4c}$$

Using the integrals (B 22), (B 28), and (B 32), see Appendix B, we obtain after inverse Fourier transformation

$$u(x, y, t) = \frac{v_*^2}{hf}\left[\theta(\pm x) \mp \operatorname{sgn}(x)\cdot\frac{1}{2}\mathrm{e}^{-\frac{|x|}{R}} - \int_y^\infty \frac{\mathrm{d}y'}{R}\int_0^\infty \frac{\mathrm{d}x'}{R} K_0\left(\frac{\sqrt{(x \mp x')^2 + y'^2}}{R}\right)\right.$$
$$\left. \pm \frac{ft}{\pi} K_0\left(\frac{\sqrt{x^2 + y^2}}{R}\right)\right] \tag{5.11.5a}$$

$$v(x, y, t) = \frac{v_*^2}{h} t\left[\pm \operatorname{sgn}(x)\cdot\frac{1}{2}\mathrm{e}^{-\frac{|x|}{R}} \pm \frac{1}{\pi}\frac{\partial}{\partial x}\int_y^\infty \mathrm{d}y'\, K_0\left(\frac{\sqrt{x^2 + y'^2}}{R}\right)\right] \tag{5.11.5b}$$

$$p(x, y, t) = \frac{v_*^2}{h}\left[\pm Rft\cdot\frac{1}{2}\mathrm{e}^{-\frac{|x|}{R}} \pm \frac{ft}{\pi}\int_y^\infty \mathrm{d}y'\, K_0\left(\frac{\sqrt{x^2 + y'^2}}{R}\right)\right.$$
$$\left. + \frac{1}{\pi}\int_0^\infty \mathrm{d}x'\, K_0\left(\frac{\sqrt{(x \mp x')^2 + y^2}}{R}\right)\right]. \tag{5.11.5c}$$

Far from the coast, i.e., $y \gg R$, the first terms in the set (5.11.5) dominates the response. We observe the usual Ekman flux which has been rectified near the discontinuity line of the wind at $x = 0$, where a geostrophically adjusted wind edge jet has developed. Thus, far off-shore we recognize the pattern (3.9.5) which describes the response of an infinite ocean on the f-plane to a rotational wind.

When approaching the coast the jet will be rectified by the remaining terms involving the modified Bessel function K_0. In the vicinity of the corner $x = y = 0$ an alongshore jet appears which is balanced by the alongshore pressure gradient, compare the rightmost terms in (5.11.5a) and (5.11.5c). This jet diminishes for distances larger than the associated Rossby radius, i.e., $x \gg R$.

Close to the corner we observe a logarithmic divergence in the u-field which follows from the small argument behaviour of the modified Bessel function, i.e.,

$$K_0(x) \sim -\ln\frac{x}{2} - C \quad \text{for} \quad x \to 0\,.$$

We may attribute this divergence to the high degree of idealization of the model geometry and the solution (5.11.5) can be expected to provide a reasonable description of the response outside a small semicircle around the corner point $x = y = 0$.

Finally we show that if we add the solutions corresponding to $Y \sim \theta(x)$ and $Y \sim \theta(-x)$, which are both governed by (5.11.5) taking the upper and the lower sign, respectively, we obtain (5.8.9) with the inertial waves being ignored. This is the response to a large scale forcing perpendicular to the coast. In fact, it follows from (5.11.5)

$$u(y,t) = \frac{v_*^2}{hf}\left\{\theta(x) + \theta(-x) - \frac{1}{\pi R^2}\int_y^\infty dy' \int_0^\infty dx' \left[K_0\left(\frac{\sqrt{(x-x')^2+y'^2}}{R}\right) + K_0\left(\frac{\sqrt{(x+x')^2+y'^2}}{R}\right)\right]\right\} \tag{5.11.6a}$$

$$v = 0 \tag{5.11.6b}$$

$$p(y,t) = \frac{v_*^2}{h}\frac{1}{\pi}\int_0^\infty dx' \left[K_0\left(\frac{\sqrt{(x-x')^2+y^2}}{R}\right) + K_0\left(\frac{\sqrt{(x+x')^2+y^2}}{R}\right)\right]. \tag{5.11.6c}$$

Using the integral (e.g. GRADSTEIN, RYSHIK 1971, p. 719)

$$\int_{-\infty}^{+\infty} dx'\, K_0\left(\frac{\sqrt{x'^2+y^2}}{R}\right) = 2Ry\Gamma\left(\frac{1}{2}\right)K_{-1/2}\left(\frac{y}{R}\right) = \pi R\, e^{-\frac{y}{R}} \tag{5.11.7}$$

with

$$\Gamma\left(\frac{1}{2}\right) = \sqrt{\pi} \quad \text{and} \quad K_{\pm 1/2}(x) = \sqrt{\frac{\pi}{2x}}\, e^{-x}\,,$$

and, moreover,

$$\theta(x) + \theta(-x) = 1$$

we arrive at (5.8.9).

5.12. FORCING OVER A RECTANGULAR AREA CLOSE TO THE COAST, ALONGSHORE WIND

In this and in the following section we consider a wind confined to a rectangular patch close to the coast. We start with a longshore wind and study in the next section a wind blowing perpendicular to the coast, see Fig. 5.12.1. Thus we deal first with the case

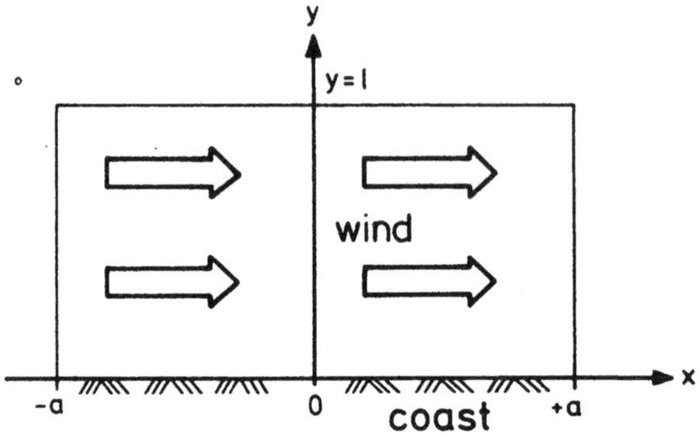

Fig. 5.12.1
Wind blowing alongshore over a rectangular coastal area according to (5.12.1)

$$X(x, y, z, t) = \frac{u_*^2}{H_{mix}} \theta(z + H_{mix})\, \theta(t)\, \theta(a - |x|)\, \theta(l - y)\,, \qquad Y = 0\,. \tag{5.12.1}$$

After Fourier transformation and normal mode expansion we have

$$X(\omega, k, y) = \frac{u_*^2}{h} \frac{\mathrm{i}}{\omega} \frac{2 \sin ka}{k} \theta(l - y)\,. \tag{5.12.2}$$

Where the vertical mode index n was dropped, and we obtain from our general formal solution (5.2.6)

$$u(\omega, k, y) = \frac{u_*^2}{h} \frac{2 \sin ka}{k} \frac{(-1)}{\omega(\omega^2\lambda^2 - k^2)} \left[\omega\lambda^2 \theta(l - y) + \frac{\omega\lambda^2}{R^2} G * \theta - \frac{\lambda k}{R} G(y, l) \right.$$
$$\left. - \frac{\lambda k}{R} G_y * \theta + \frac{k^2}{\omega} G_y(y, l) \right] \tag{5.12.3a}$$

$$v(\omega, k, y) = \frac{u_*^2}{h} \frac{2 \sin ka}{k} \mathrm{i} \left[\frac{\lambda}{\omega R} G * \theta - \frac{k}{\omega^2} G(y, l) \right] \tag{5.12.3b}$$

$$p(\omega, k, y) = \frac{u_*^2}{h} \frac{2 \sin ka}{k} \frac{(-1)}{\omega(\omega^2\lambda^2 - k^2)} \left[k\theta(l - y) - \frac{\omega\lambda}{R} G_y + k G_y(y, l) \right.$$
$$\left. + \frac{k^2}{R^2} G * \theta - \frac{k^2 f}{\omega} G(l, y) \right]. \tag{5.12.3c}$$

The terms $G(y, l)$ and $G_y(y, l)$ result from $G * \theta_{y'}$ and $G_y * \theta_{y'}$, respectively, with the aid of $\theta_{y'}(l - y') = -\delta(l - y')$. In accordance with the introductionary remarks of Section 5.9. we have to expand the expressions of the set (5.12.3) around the poles at $\omega = 0$, and $\omega = \pm k/\lambda$ in order to find the leading contributions for large times, i.e., $tf \gg 1$. This amounts to filter out the inertial waves. Thus, utilizing the identity (5.3.1 c) we have

$$u(\omega, k, y) = \frac{u_*^2}{h}\frac{2\sin ka}{k}(-1)\left\{\frac{1}{\omega^2 - \frac{k^2}{\lambda^2}}\left[\theta(l-y) + \frac{G*\theta}{R^2} + G_y(y, l) - \frac{\operatorname{sgn}(\omega)}{R}(G(y, l) + G_y*\theta)\right]_{\omega^2=\frac{k^2}{\lambda^2}} + \frac{\lambda^2 f}{k\omega}(G(y, l) + G_y*\theta)\bigg|_{\omega=0} - \frac{1}{\omega^2}G_y(y, l)\bigg|_{\omega=0}\right\} \quad (5.12.4\text{a})$$

$$v(\omega, k, y) = \frac{u_*^2}{h}\frac{2\sin ka}{k}\,\mathrm{i}\left[\frac{f\lambda^2}{\omega}G*\theta\bigg|_{\omega=0} - \frac{k}{\omega^2}G(y, l)\bigg|_{\omega=0}\right] \quad (5.12.4\text{b})$$

$$p(\omega, k, y) = \frac{u_*^2}{h}\frac{2\sin ka}{k}\left\{\frac{1}{\omega^2 - \frac{k^2}{\lambda^2}}\left[fG(y, l) + fG_y*\theta - \frac{\operatorname{sgn}(\omega)}{\lambda}\times\left(\theta(l-y) + G_y(y, l) + \frac{G*\theta}{R^2}\right)\right]_{\omega^2=\frac{k^2}{\lambda^2}} - \frac{f}{\omega^2}G(y, l)\bigg|_{\omega=0} + \frac{1}{\omega k}\left(\theta(l-y) + G_y(y, l) + \frac{G*\theta}{R^2}\right)\bigg|_{\omega=0}\right\}. \quad (5.12.4\text{c})$$

From (5.2.4) and (5.4.5) we obtain for $\omega^2 = k^2/\lambda^2$ with $\alpha^2 = -\lambda^2 f^2 = -1/R^2$

$$\theta(l-y) + \frac{G*\theta}{R^2}\bigg|_{\omega^2=\frac{k^2}{\lambda^2}} = \mathrm{e}^{-\frac{y}{R}} - \frac{1}{2}\mathrm{e}^{-\frac{y+l}{R}} - \operatorname{sgn}(y-l)\frac{1}{2}\mathrm{e}^{-\frac{|y-l|}{R}} \quad (5.12.5\text{a})$$

$$G_y*\theta\bigg|_{\omega^2=\frac{k^2}{\lambda^2}} = R\left[-\mathrm{e}^{-\frac{y}{R}} + \frac{1}{2}\mathrm{e}^{-\frac{(y+l)}{R}} + \frac{1}{2}\mathrm{e}^{-\frac{|y-l|}{R}}\right] \quad (5.12.5\text{b})$$

$$G(y, l)\bigg|_{\omega^2=\frac{k^2}{\lambda^2}} = \frac{R}{2}\left[\mathrm{e}^{-\frac{(y+l)}{R}} - \mathrm{e}^{-\frac{|y-l|}{R}}\right] \quad (5.12.5\text{c})$$

$$G_y(y, l)\bigg|_{\omega^2=\frac{k^2}{\lambda^2}} = \frac{1}{2}\left[\operatorname{sgn}(y-l)\,\mathrm{e}^{-\frac{|y-l|}{R}} - \mathrm{e}^{-\frac{(y+l)}{R}}\right], \quad (5.12.5\text{d})$$

and for $\omega = 0$ with

$$\alpha^2 = -(\lambda^2 f^2 + k^2) = -\frac{(1 + k^2R^2)}{R^2}$$

$$\theta(l-y)+\left.\frac{G*\theta}{R^2}\right|_{\omega=0}=\frac{k^2R^2\theta(l-y)}{1+k^2R^2}+\frac{1}{1+k^2R^2}$$
$$\times\left[e^{-\sqrt{1+k^2R^2}\frac{y}{R}}-\frac{1}{2}e^{-\sqrt{1+k^2R^2}\frac{(y+l)}{R}}\right.$$
$$\left.-\operatorname{sgn}(y-l)\cdot\frac{1}{2}e^{-\sqrt{1+k^2R^2}\frac{|y-l|}{R}}\right] \quad (5.12.5e)$$

$$G_y*\theta|_{\omega=0}=\frac{R}{\sqrt{1+k^2R^2}}\left[-e^{-\sqrt{1+k^2R^2}\frac{y}{R}}+\frac{1}{2}e^{-\sqrt{1+k^2R^2}\frac{(y+l)}{R}}+\frac{1}{2}e^{-\sqrt{1+k^2R^2}\frac{|y-l|}{R}}\right] \quad (5.12.5f)$$

$$G(y,l)=\frac{R}{2\sqrt{1+k^2R^2}}\left[e^{-\sqrt{1+k^2R^2}\frac{y+l}{R}}-e^{-\sqrt{1+k^2R^2}\frac{|y-l|}{R}}\right] \quad (5.12.5g)$$

$$G_y(y,l)=\frac{1}{2}\left[\operatorname{sgn}(y-l)\,e^{-\sqrt{1+k^2R^2}\frac{|y-l|}{R}}-e^{-\sqrt{1+k^2R^2}\frac{(y+l)}{R}}\right] \quad (5.12.5h)$$

We note that for $l \to \infty$ the expressions (5.12.5c, d, g, h) vanish and (5.12.5a, b, e, f) equal (5.9.5a, b, c, d), respectively. Inserting these expressions into the equation set (5.12.4) and performing the elementary inverse Fourier transformation with respect to ω we arrive at

$$u(y,k,t)=\frac{u_*^2}{h}\frac{2\sin ka}{k}\left\{\frac{i}{k}e^{-i\frac{kt}{\lambda}}\left[e^{-\frac{y}{R}}-e^{-\frac{(y+l)}{R}}\right]-\frac{i\lambda}{k}\frac{e^{-\sqrt{1+k^2R^2}\frac{y}{R}}-e^{-\sqrt{1+k^2R^2}\frac{(y+l)}{R}}}{\sqrt{1+k^2R^2}}\right.$$
$$\left.-\frac{t}{2}\left[\operatorname{sgn}(y-l)\,e^{-\sqrt{1+k^2R^2}\frac{|y-l|}{R}}-e^{-\sqrt{1+k^2R^2}\frac{(y+l)}{R}}\right]\right\}, \quad (5.12.6a)$$

$$v(y,k,t)=\frac{u_*^2}{h}\frac{2\sin ka}{k}\left\{\frac{1}{f(1+k^2R^2)}\left[-\theta(l-y)+e^{-\sqrt{1+k^2R^2}\frac{y}{R}}-\frac{1}{2}e^{-\sqrt{1+k^2R^2}\frac{(y+l)}{R}}\right.\right.$$
$$\left.-\operatorname{sgn}(y-l)\cdot\frac{1}{2}e^{-\sqrt{1+k^2R^2}\frac{|y-l|}{R}}\right]$$
$$\left.+\frac{itkR}{2\sqrt{1+k^2R^2}}\left[e^{-\sqrt{1+k^2R^2}\frac{(y+l)}{R}}-e^{-\sqrt{1+k^2R^2}\frac{|y-l|}{R}}\right]\right\}, \quad (5.12.6b)$$

$$p(y,k,t)=\frac{u_*^2}{h}\frac{2\sin ka}{k}\left\{\frac{i}{k}e^{-i\frac{kt}{\lambda}}\left(e^{-\frac{y}{R}}-e^{-\sqrt{1+k^2R^2}\frac{y+l}{R}}\right)\right.$$
$$-\frac{i}{k(1+k^2R^2)}\left(e^{-\sqrt{1+k^2R^2}\frac{y}{R}}-e^{-\sqrt{1+k^2R^2}\frac{(y+l)}{R}}\right)$$

$$- \frac{ikR^2}{1+k^2R^2}\left[\theta(l-y) + \operatorname{sgn}(y-l)\cdot\frac{1}{2}\,e^{-\sqrt{1+k^2R^2}\frac{|y-l|}{R}} - \frac{1}{2}\,e^{-\sqrt{1+k^2R^2}\frac{(y+l)}{R}}\right]$$

$$- \frac{t}{2\lambda\sqrt{1+k^2R^2}}\left[e^{-\sqrt{1+k^2R^2}\frac{|y-l|}{R}} - e^{-\sqrt{1+k^2R^2}\frac{(y+l)}{R}}\right]\Bigg\}. \quad (5.12.6c)$$

Since for $a \to \infty$ it follows $(2 \sin ka)/k = 2\pi\,\delta(k)$ we find that for large a the set (5.12.6) yields the non-oscillatory part of the set (5.7.6). That is the response to a longshore forcing over an infinite long coastal strip.

Furthermore, for large l, i.e., $l \to \infty$, from (5.12.6) we obtain the response to a longshore forcing within an infinite strip orientated perpendicular to the coast as indicated in Fig. 5.12.2. Let us discuss this example in some detail. First we note that all exponentials in (5.12.6) involving l vasnish for large l and we have

$$u(y,k,t) = \frac{u_*^2}{h}\frac{2\sin ka}{k}\left[\frac{i\lambda}{k}\left(e^{-i\frac{kt}{\lambda}} - 1\right)e^{-\frac{y}{R}} - \frac{i\lambda}{k}\left(\frac{e^{-\sqrt{1+k^2R^2}\frac{y}{R}}}{\sqrt{1+k^2R^2}} - e^{-\frac{y}{R}}\right)\right] \quad (5.12.7a)$$

$$v(y,k,t) = \frac{u_*^2}{h}\frac{2\sin ka}{k}\left[-\frac{1-e^{-\frac{y}{R}}}{f} + \frac{k^2R^2}{f(1+k^2R^2)} + \frac{e^{-\sqrt{1+k^2R^2}\frac{y}{R}}}{f(1+k^2R^2)} - \frac{1}{f}\,e^{-\frac{y}{R}}\right] \quad (5.12.7b)$$

$$p(y,k,t) = \frac{u_*^2}{h}\frac{2\sin ka}{k}\left[\frac{i}{k}\left(e^{-i\frac{kt}{\lambda}} - 1\right)e^{-\frac{y}{R}} - \frac{ikR}{1+k^2R^2} - \frac{i}{k}\left(\frac{e^{-\sqrt{1+k^2R^2}\frac{y}{R}}}{1+k^2R^2} - e^{-\frac{y}{R}}\right)\right]$$

(5.12.7c)

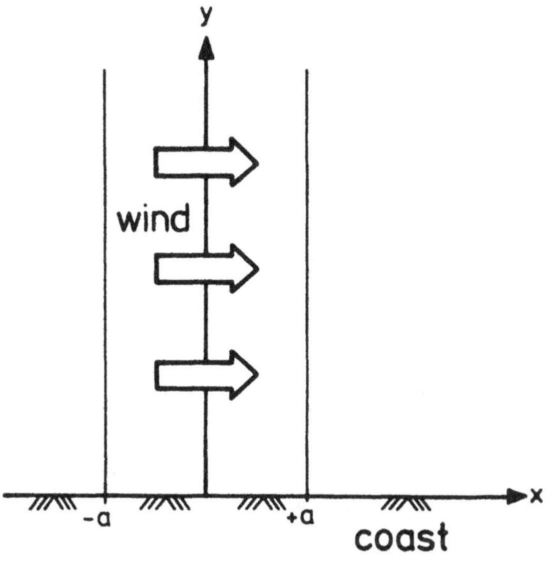

Fig. 5.12.2
Wind blowing alongshore over an infinite strip orientated perpendicular to the coast

Here we have added and substracted the exponential in order to separate the leading contributions for small k. Then we may find in the manner outlined in Section 5.9. the approximate solution set

$$u(x, y, t) = \frac{u_*^2}{2h}\left(\left|x - \frac{t}{\lambda} - a\right| - \left|x - \frac{t}{\lambda} + a\right| - |x - a| + |x + a|\right) e^{-\frac{y}{R}} \tag{5.12.8 a}$$

$$v(x, y, t) = \frac{u_*^2}{hf}\,\theta(a - |x|)\left(e^{-\frac{y}{R}} - 1\right) \tag{5.12.8 b}$$

$$p(x, y, t) = \frac{u_*^2}{2h}\left(\left|x - \frac{t}{\lambda} - a\right| - \left|x - \frac{t}{\lambda} + a\right| - |x - a| + |x + a|\right) e^{-\frac{y}{R}} \tag{5.12.8 c}$$

where the Fourier integral

$$\int_{-\infty}^{+\infty} \frac{dk}{2\pi}\,\frac{2\sin ka}{k}\,\frac{1}{k}\,e^{ik\left(x-\frac{t}{\lambda}\right)} = -\frac{i}{2}\left(\left|x - \frac{t}{\lambda} - a\right| - \left|x - \frac{t}{\lambda} + a\right|\right) \tag{5.12.9}$$

was used, see (B 25) in the appendix. The set (5.12.8) describe Kelvin waves which are generated at the wind discontinuity lines $x = \pm a$ and which propagate alongshore with the coast to the right, having the modal phase speed $1/\lambda$, see Fig. 5.12.3. This motion is coastally trapped with the trapping scale R.

We may distinguish two different phases. Initially, when the Kelvin wave generated at $x = -a$ has not reached a point x within the forcing area, say e.g., $x = 0$, there develops a coastal jet according to

$$u(0, y, t) = \frac{u_*^2}{h}\, t\, e^{-\frac{y}{R}} \quad \text{for} \quad t < a\lambda\,.$$

After the passage of the Kelvin front, however, the coastal jet has been arrested. At $t = 2a\lambda$ the Kelvin wave reaches the point $x = a$ and an alongshore pressure gradient has established in the forcing area. Outside the forcing area, i.e., $x > a$, both Kelvin waves add and leave a steady state in their wake.

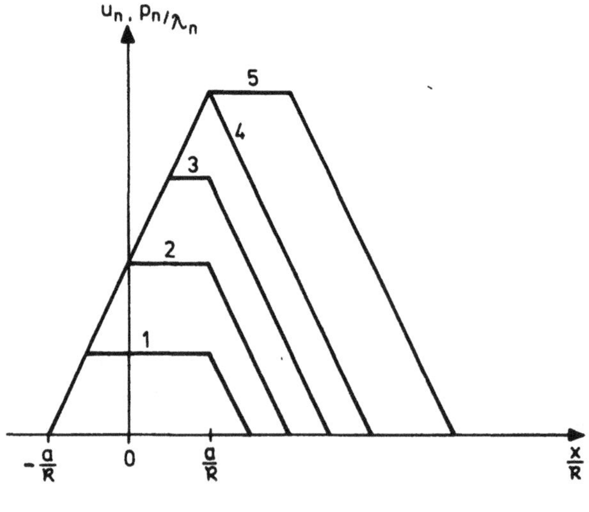

Fig. 5.12.3
The development of the Kelvin wave excited at the wind edges shown in Fig. 5.12.2 according to (5.12.8a, c)

Clearly, in the region $x < -a$ the response vanishes.
Thus we have for $t > 2a\lambda$ within the strip $-a \leqq x \leqq a$

$$u(x, y, t) = \frac{u_*^2}{h} \lambda(x + a) \, e^{-\frac{y}{R}} \tag{5.12.10a}$$

$$v(x, y, t) = \frac{u_*^2}{h} \frac{e^{-\frac{y}{R}} - 1}{f} \tag{5.12.10b}$$

$$p(x, y, t) = \frac{u_*^2}{h} (x + a) \, e^{-\frac{y}{R}} \tag{5.12.10c}$$

and for $x > a$ in the wake of the Kelvin waves

$$u(x, y, t) = \frac{u_*^2}{h} \lambda \cdot 2a \, e^{-\frac{y}{R}} \tag{5.12.11a}$$

$$v(x, y, t) = 0 \tag{5.12.11b}$$

$$p(x, y, t) = \frac{u_*^2}{h} \cdot 2a \, e^{-\frac{y}{R}} . \tag{5.12.11c}$$

Obviously, for $y \gg R$ the off-shore Ekman transport, i.e.,

$$\int_{-a}^{+a} \mathrm{d}x \, v(x, y, t) = \frac{-u_*^2 \cdot 2a}{hf}$$

equals the mass flux in the wake of the Kelvin waves as given by

$$\int_0^\infty \mathrm{d}y \, u(x, y, t) = \frac{u_*^2}{h} \cdot 2a\lambda \int_0^\infty \mathrm{d}y \, e^{-\frac{y}{R}} = \frac{u_*^2 \cdot 2a}{hf} .$$

Thus the mass flux associated with the Kelvin waves is fed by the off-shore Ekman transport. Near the coast, i.e., $y \ll R$, the Ekman transport vanishes and it appears that the wind force is balanced by the alongshore pressure gradient p_x, compare (5.12.10c). Clearly the process described happens for every vertical mode. Since the longshore current is according to (5.12.10a), (5.12.11a) weighted by the vertical eigenvalue λ_n and because of $\lambda_0 \ll \lambda_1$ the baroclinic signals dominate the response. Let y be of the order of the first baroclinic Rossby radius R_1. Then the first baroclinic mode is the gravest and we have $u(x, y, t) \approx u_1(x, y, t) \, F_1(z)$, i.e., the longshore current varies in the vertical like $F_1(z)$. In other words, the longshore current consists of a downwind flow in the upper part of the water columne and an opposite undercurrent below.

The modal vertical velocity according to $w_n = -p_{nt}$ follows from (5.12.8c) to be

$$w(x, y, t) = -\frac{u_*^2}{h\lambda} \theta \left(a - \left| x - \frac{t}{\lambda} \right| \right) .$$

That means that an upwelling (or downwelling) region of length $2a$ moves alongshore. The upwelling occurs behind the Kelvin wave front which was excited at $x = a$ and terminates after the arrival of the second wave front generated at $x = -a$.

The described response scenario turns out to be of importance in the theory of eastern boundary currents, see PHILLANDER and YOON (1982) and YOON and PHILLANDER (1982).

Finally we add the Fourier inversion of the set (5.12.6). With the integrals (B 25), (B 27), (B 33), (B 34), (B 35) we arrive at

$$
\begin{aligned}
u(x, y, t) = \frac{u_*^2}{h} \Bigg\{ & \frac{\lambda}{2} \left(\left| x - \frac{t}{\lambda} - a \right| - \left| x - \frac{t}{\lambda} + a \right| \right) \left(e^{-\frac{y}{R}} - e^{-\frac{y+l}{R}} \right) \\
& - \frac{\lambda}{2\pi R} \int_{-\infty}^{+\infty} dx' (|x - x' - a| - |x - x' + a|) \\
& \times \left[K_0 \left(\frac{\sqrt{x'^2 + y^2}}{R} \right) - K_0 \left(\frac{\sqrt{x'^2 + (y + l)^2}}{R} \right) \right] \\
& - \frac{t}{2\pi R} \int_{-a}^{+a} dx' \left[\frac{|y - l| \, K_1 \left(\frac{\sqrt{(y - l)^2 + (x - x')^2}}{R} \right)}{\sqrt{(y - l)^2 + (x - x')^2}} \right. \\
& \left. - \frac{(y + l) \, K_1 \left(\frac{\sqrt{(y + l)^2 + (x - x')^2}}{R} \right)}{\sqrt{(y + l)^2 + (x - x')^2}} \right] \Bigg\}
\end{aligned}
\tag{5.12.12a}
$$

$$
\begin{aligned}
v(x, y, t) = \frac{u_*^2}{h} \Bigg\{ & - \frac{\theta(l - y)}{f} \left[\theta(a - |x|) + \operatorname{sgn}(x - a) \cdot \frac{1}{2} e^{-\frac{|x - a|}{R}} - \operatorname{sgn}(x + a) \cdot \frac{1}{2} e^{-\frac{|x + a|}{R}} \right] \\
& + \frac{t}{2\pi R} \left[K_0 \left(\frac{1}{R} \sqrt{(y - l)^2 + (a + x)^2} \right) - K_0 \left(\frac{1}{R} \sqrt{(y - l)^2 + (a - x)^2} \right) \right. \\
& \left. - K_0 \left(\frac{1}{R} \sqrt{(y + l)^2 + (a + x)^2} \right) + K_0 \left(\frac{1}{R} \sqrt{(y + l)^2 + (a - x)^2} \right) \right] \\
& + \frac{\lambda}{\pi R} \int_{y}^{\infty} dy' \int_{-a}^{a} dx' \left[K_0 \left(\frac{1}{R} \sqrt{y'^2 + (x - x')^2} \right) \right. \\
& - \frac{1}{2} K_0 \left(\frac{1}{R} \sqrt{(y' + l)^2 + (x - x')^2} \right) \\
& \left. - \frac{1}{2} K_0 \left(\frac{1}{R} \sqrt{(y' - l)^2 + (x - x')} \right) \right] \Bigg\}
\end{aligned}
\tag{5.12.12b}
$$

$$p(x, y, t) = \frac{u_*^2}{h}\left\{\frac{1}{2}\left(\left|x - \frac{t}{\lambda} - a\right| - \left|x - \frac{t}{\lambda} + a\right|\right)\left(e^{-\frac{y}{R}} - e^{-\frac{(y+l)}{R}}\right)\right.$$

$$- \frac{R}{2}\,\theta(l - y)\left(e^{-\frac{|x+a|}{R}} - e^{-\frac{|x-a|}{R}}\right)$$

$$- \int_y^\infty \frac{dy'}{2\pi R^2}\int_{-\infty}^{+\infty} dx'(|x - x' - a| - |x - x' + a|)$$

$$\times\left[K_0\left(\frac{1}{R}\sqrt{x'^2 + y'^2}\right) - K_0\left(\frac{1}{R}\sqrt{x'^2 + (y - l)^2}\right)\right]$$

$$+ \frac{1}{2\pi}\int_y^\infty dy'\left[K_0\left(\frac{1}{R}\sqrt{(a - x)^2 + (y' - l)^2}\right)\right.$$

$$\left. - K_0\left(\frac{1}{R}\sqrt{(a + x)^2 + (y' + l)^2}\right)\right]$$

$$+ \frac{tf}{2\pi}\int_{-a}^{+a} dx'\left[K_0\left(\frac{1}{R}\sqrt{(y - l)^2 + (x - x')^2}\right)\right.$$

$$\left.\left. - K_0\left(\frac{1}{R}\sqrt{(y + l)^2 + (x - x')^2}\right)\right]\right\}. \qquad (5.12.12c)$$

We discuss some properties of this solution: For large arguments the modified Bessel function $K_n(x)$ behaves like $\sqrt{\pi/(2z)}\,e^{-z}$. Therefore, we find the case of large l, i.e., $l \to \infty$, by neglecting of all exponentials and all K_0's, and K_1's which involve l in the set (5.12.12). Then it follows the response to a longshore wind blowing over an infinite strip orientated perpendicular to the coast

$$u(x, y, t) = \frac{u_*^2}{h}\left[\frac{\lambda}{2}\left(\left|x - \frac{t}{\lambda} - a\right| - \left|x - \frac{t}{\lambda} + a\right|\right)\right.$$

$$\left.\times\, e^{-\frac{y}{R}}\frac{\lambda}{2\pi R}\int_{-\infty}^{+\infty} dx'(|x - x' - a| - |x - x' + a|)\,K_0\left(\frac{\sqrt{x'^2 + y^2}}{R}\right)\right]$$

(5.12.13a)

$$v(x, y, t) = \frac{u_*^2}{h}\left\{\frac{1}{f}\left[\theta(a - |x|) + \operatorname{sgn}(x - a)\cdot\frac{1}{2}e^{-\frac{|x-a|}{R}} - \operatorname{sgn}(x + a)\cdot\frac{1}{2}e^{-\frac{|x+a|}{R}}\right]\right.$$

$$\left. + \frac{\lambda}{\pi R}\int_y^\infty dy'\int_{-a}^{+a} dx'\,K_0\left(\frac{\sqrt{y'^2 + (x - x')^2}}{R}\right)\right\} \qquad (5.12.13b)$$

$$p(x, y, t) = \frac{u_*^2}{h}\left[\frac{1}{2}\left(\left|x - \frac{t}{\lambda} - a\right| - \left|x - \frac{t}{\lambda} + a\right|\right) e^{-\frac{y}{R}} - \frac{R}{2}\left(e^{-\frac{|x+a|}{R}} - e^{-\frac{|x-a|}{R}}\right)\right.$$

$$\left. - \frac{1}{2\pi R^2}\int_y^\infty dy' \int_{-\infty}^\infty dx'(|x - x' - a| - |x - x' + a|)\, K_0\left(\frac{\sqrt{x'^2 + y'^2}}{R}\right)\right]. \tag{5.12.13c}$$

The approximate solution corresponding to this set is given by (5.12.8).

Moreover, for $a \to \infty$ from (5.12.12) we obtain the non-oscillatory part of (5.7.6), i.e., the response to an alongshore forcing over an infinite coastal strip. This follows from the relation

$$\left|x - \frac{t}{\lambda} - a\right| - \left|x - \frac{t}{\lambda} + a\right| - |x - a| + |x + a| = -2\left(x - \frac{t}{\lambda}\right) \quad \text{for} \quad a \to \infty\,,$$

and using the integral (5.11.12).

Moreover, we need the integral

$$\int_{-\infty}^{\infty} dx' \frac{K_1\left(\frac{1}{R}\sqrt{x'^2 + y^2}\right)}{\sqrt{x'^2 + y^2}} = \frac{\pi R}{y} e^{-\frac{y}{R}} \tag{5.12.14}$$

which follows if we differentiate (5.11.12) by y.

Finally we mention the occurrence of slight logarithmic divergences at the corner points of the wind patch due to the small argument behaviour of the modified Bessel function K_0, see e.g. the accelerating term in (5.12.12b). These divergences can be attributed to the high degree of idealization of the forcing function involving sharp corner points. Clearly, the validity of the solution (5.12.12) is confined to the region outside small circles around the corners.

5.13. FORCING OVER A RECTANGULAR AREA CLOSE TO THE COAST, CROSS-SHORE WIND

In this section we add the calculation of the oceanic response to the forcing

$$X = 0$$

$$Y = \frac{v_*^2}{H_{mix}} \theta(z + H_{mix})\, \theta(t)\, \theta(l - y)\, \theta(a - |x|)\,. \tag{5.13.1}$$

The mathematical structure of (5.13.1) is close to that of (5.12.1) discussed in the foregoing section, however, X and Y are exchanged.

After expansion of (5.13.1) in vertical eigenfunctions and suppressing the index n we get in the ω—k domain

$$Y(\omega, k, y) = \frac{v_*^2}{h} \frac{i}{\omega} \frac{2 \sin ka}{k} \theta(l - y)\,. \tag{5.13.2}$$

Equations (5.13.2) and (5.2.6) give

$$u(\omega, k, y) = \frac{v_*^2}{h}\frac{2\sin ka}{k}\frac{\mathrm{i}}{\omega}\left(-\frac{1}{fR^2}G*\theta + \frac{k}{\omega}\frac{\partial G}{\partial y}*\theta\right) \tag{5.13.3a}$$

$$v(\omega, k, y) = \frac{v_*^2}{h}\frac{2\sin ka}{k}\frac{-1}{\omega^2}(\omega^2\lambda^2 - k^2)\,G*\theta \tag{5.13.3b}$$

$$p(\omega, k, y) = \frac{v_*^2}{h}\frac{\mathrm{i}}{\omega}\frac{2\sin ka}{k}\left(-\frac{fk}{\omega}G*\theta + G_y*\theta\right). \tag{5.13.3c}$$

Filtering out the inertial waves the remaining singularities of the set (5.13.3) are the poles at $\omega = 0$. With the aid of (5.12.5e) and (5.12.5f) the corresponding residues can readily be obtained. The Fourier inversion of the resulting formulas can be performed with the help of the integrals (B 25), (B 27), (B 33), and (B 35).

Thus, in the $x - t$ domain we get

$$\begin{aligned} u(x, y, t) = \frac{v_*^2}{hf}\Bigg\{&\left[\theta(a-|x|) + \operatorname{sgn}(x-a)\cdot\frac{1}{2}\mathrm{e}^{-\frac{|x-a|}{R}} - \operatorname{sgn}(x+a)\cdot\frac{1}{2}\mathrm{e}^{-\frac{|x+a|}{R}}\right]\theta(l-y) \\ &- \frac{1}{\pi R^2}\int_y^\infty \mathrm{d}y' \int_{-a}^{+a}\mathrm{d}x'\left[K_0\left(\frac{1}{R}\sqrt{(x-x')^2+y'^2}\right)\right. \\ &\qquad - \frac{1}{2}K_0\left(\frac{1}{R}\sqrt{(x-x')^2+(l+y')^2}\right) \\ &\qquad \left. - \frac{1}{2}K_0\left(\frac{1}{R}\sqrt{(x-x')^2+(l-y')^2}\right)\right] \\ &- \frac{ft}{\pi}\left[K_0\left(\frac{1}{R}\sqrt{y^2+(a-x)^2}\right) - K_0\left(\frac{1}{R}\sqrt{y^2+(a+x)^2}\right)\right. \\ &- \frac{1}{2}K_0\left(\frac{1}{R}\sqrt{(y+l)^2+(a-x)^2}\right) + \frac{1}{2}K_0\left(\frac{1}{R}\sqrt{(y+l)^2+(a+x)^2}\right) \\ &\left.\left. - \frac{1}{2}K_0\left(\frac{1}{R}\sqrt{(y-l)^2+(a-x)^2}\right) + \frac{1}{2}K_0\left(\frac{1}{R}\sqrt{(y-l)^2+(a+x)^2}\right)\right]\right\} \end{aligned} \tag{5.13.4a}$$

$$\begin{aligned} v(x,y,t) = \frac{v_*^2}{h}t\Bigg\{&\frac{\theta(l-y)}{2}\left[\operatorname{sgn}(x-a)\,\mathrm{e}^{-\frac{|x-a|}{R}} - \operatorname{sgn}(x+a)\,\mathrm{e}^{-\frac{|x+a|}{R}}\right] \\ &+ \frac{\partial}{\partial x}\frac{1}{\pi}\int_y^\infty \mathrm{d}y'\left[K_0\left(\frac{1}{R}\sqrt{y'^2+(a-x)^2}\right) - K_0\left(\frac{1}{R}\sqrt{y'^2+(a+y)^2}\right)\right. \\ &\qquad - \frac{1}{2}K_0\left(\frac{1}{R}\sqrt{(y'+l)^2+(a-x)^2}\right) \end{aligned}$$

$$+\frac{1}{2}K_0\left(\frac{1}{R}\sqrt{(y'+l)^2+(a+x)^2}\right)$$

$$-\frac{1}{2}K_0\left(\frac{1}{R}\sqrt{(y'-l)^2+(a-x)^2}\right)$$

$$\left.\left.+\frac{1}{2}K_0\left(\frac{1}{R}\sqrt{(y-l)^2+(a+x)^2}\right)\right]\right\} \tag{5.13.4b}$$

$$p(x,y,t)=\frac{v_*^2}{h}\left\{\int_{-a}^{+a}\frac{\mathrm{d}x'}{\pi}\left[K_0\left(\frac{1}{R}\sqrt{(x-x')^2+y^2}\right)-\frac{1}{2}K_0\left(\frac{1}{R}\sqrt{(x-x')^2+(y+l)^2}\right)\right.\right.$$

$$\left.-\frac{1}{2}\operatorname{sgn}(y-l)\,K_0\left(\frac{1}{R}\sqrt{(x-x')^2+(y-l)^2}\right)\right]$$

$$+ft\left[\frac{R}{2}\,\theta(l-y)\left(\mathrm{e}^{-\frac{|x+a|}{R}}-\mathrm{e}^{-\frac{|x-a|}{R}}\right)\right.$$

$$+\int_{y}^{\infty}\frac{\mathrm{d}y'}{\pi}\left(K_0\left(\frac{1}{R}\sqrt{y'^2+(a-x)^2}\right)-K_0\left(\frac{1}{R}\sqrt{y'^2+(a+x)^2}\right)\right.$$

$$-\frac{1}{2}K_0\left(\frac{1}{R}\sqrt{(y'+l)^2+(a-x)^2}\right)$$

$$+\frac{1}{2}K_0\left(\frac{1}{R}\sqrt{(y'+l)^2+(a+x)^2}\right)$$

$$-\frac{1}{2}K_0\left(\frac{1}{R}\sqrt{(y'-l)^2+(a-x)^2}\right)$$

$$\left.\left.\left.+\frac{1}{2}K_0\left(\frac{1}{R}\sqrt{(y'-l)^2+(a+x)^2}\right)\right)\right]\right\}. \tag{5.13.4c}$$

For $a \to \infty$ all exponentials and K_0-functions involving the 'a' vanish and we obtain with the aid of (5.11.7) the non-oscillating part of the set (5.8.7) which describes the response to a cross-shore wind blowing over the strip $y < l$.

For $l \to \infty$ the set (5.13.4) describes the response to a cross-shore wind blowing over an infinite strip of width $2a$ orientated perpendicular to the coast. Then the set (5.13.4) reduces to

$$u(x,y,t)=\frac{v_*^2}{hf}\left\{\theta(a-|x|)+\operatorname{sgn}(x-a)\cdot\frac{1}{2}\,\mathrm{e}^{-\frac{|x-a|}{R}}-\operatorname{sgn}(x+a)\cdot\frac{1}{2}\,\mathrm{e}^{-\frac{|x+a|}{R}}\right.$$

$$-\frac{1}{R^2}\int_{y}^{\infty}\mathrm{d}y'\int_{-a}^{+a}\frac{\mathrm{d}x'}{\pi}K_0\left(\frac{1}{R}\sqrt{(x-x')^2+y'^2}\right)$$

$$\left.-tf\left[K_0\left(\frac{1}{R}\sqrt{y^2+(a-x)^2}\right)-K_0\left(\frac{1}{R}\sqrt{y^2+(a+x)^2}\right)\right]\right\} \tag{5.13.5a}$$

$$v(x, y, t) = \frac{v_*^2}{h} t \left\{ \operatorname{sgn}(x-a) \cdot \frac{1}{2} e^{-\frac{|x-a|}{R}} - \operatorname{sgn}(x+a) \cdot \frac{1}{2} e^{-\frac{|x+a|}{R}} \right.$$

$$\left. + \frac{\partial}{\partial x} \int_y^\infty \frac{dy'}{\pi} \left[K_0\left(\frac{1}{R}\sqrt{y'^2 + (a-x)^2}\right) - K_0\left(\frac{1}{R}\sqrt{y'^2 + (a+x)^2}\right) \right] \right\}$$

(5.13.5 b)

$$p(x, y, t) = -\frac{v_*^2}{h} \left\{ \int_{-a}^{a} \frac{dx'}{\pi} K_0\left(\frac{1}{R}\sqrt{(x-x')^2 + y^2}\right) \right.$$

$$+ ft \left[\frac{R}{2}\left(e^{-\frac{|x+a|}{R}} - e^{-\frac{|x-a|}{R}} \right) + \int_y^\infty \frac{dy'}{\pi} \left(K_0\left(\frac{1}{R}\sqrt{y'^2 + (a-x)^2}\right) \right. \right.$$

$$\left. \left. \left. - K_0\left(\frac{1}{R}\sqrt{y'^2 + (a+x)^2}\right) \right) \right] \right\}.$$

(5.13.5 c)

Far off-shore, i.e., $y \gg R$, it remains the response to a rotational non-divergent wind blowing over a strip. This is the case considered in Section 3.10., compare in particular (3.10.5).

When approaching the coast the terms involving the K_0-function which describe the coastal rectification become important. Let $a \gg R$ and consider $x \ll a$, i.e., a location far from the wind edges. Then we have approximately

$$u(x, y, t) \approx \frac{v_*^2}{h} \frac{1}{f} \left(1 - e^{-\frac{y}{R}}\right)$$

$$v \approx 0$$

$$p \approx -\frac{v_*^2}{h} R\, e^{-\frac{y}{R}}.$$

Thus, close to the coast the Ekman current diminishes and the wind is balanced by the cross-shore pressure gradient p_y.

Finally, we note that in the full solution (5.13.4) there occur some logarithmic divergences at the corner points of the wind edge. Mathematically these divergences are due to the small argument behaviour of the Bessel functions K_0. From the physical point of view these slight inconsistencies are the consequence of the high degree of idealization of the assumed rectangular wind patch. Thus, the solution (5.13.4) is valid everywhere except within small circles around the artifical corner points.

CHAPTER 6

Forcing of a channel

6.1. FORMULATION OF THE PROBLEM

In this chapter we consider the response patterns of a fluid bounded by two coasts. Such a channel model may be valuable in order to gain some insight into the dynamical behaviour of marginal seas, e.g. the Baltic, and of elongated lakes. The forcing problem has been considered by various authors, notably CHARNEY (1955), CREPON (1969), see also the surveys given by CSANADY (1977), (1982). These authors considered some examples of external forcing function and determined the barotropic response and the response of a two layer fluid. In this chapter we will again elaborate a systematic theory based on the Green's functions technique. We consider a continuously stratified system. The theory outlined below is close to FENNEL (1986).

We consider a flat bottomed infinite long canal where the coasts have assumed to be straight walls. The x-coordinate is alongshore and y being the cross-coordinate, z is

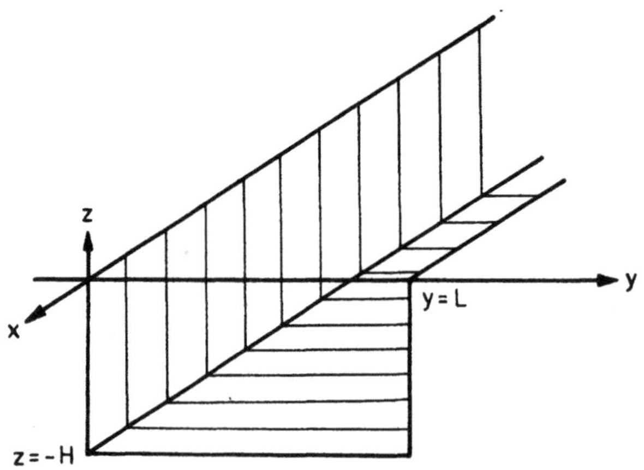

Fig. 6.1.1
Geometry the channel model

positive upward. The shores coincide with the lines $y = 0$ and $y = L$, i.e., L is the width of the canal. The model geometry is drawn in Fig. 6.1.1.

Again we start with the Boussinesq equations

$$u_{nt} - fv_n + p_{nx} = X_n \tag{6.1.1 a}$$

$$v_{nt} + fu_n + p_{ny} = Y_n \tag{6.1.1 b}$$

$$u_{nx} + v_{ny} + \lambda_n^2 p_{nt} = 0 \tag{6.1.1 c}$$

where

$$(u, v, p) = \sum_n (u_n, v_n, p_n)\, F_n(z) \tag{6.1.2 a}$$

and

$$(X, Y) = \sum_n (X_n, Y_n)\, F_n(z) \tag{6.1.2 b}$$

are the expansions of the dynamical fields and of the external forcing, respectively, in terms of the vertical eigenfunctions $F_n(z)$. Let the forcing start to act at $t = 0$ on a fluid at rest. This implies the initial conditions

$$u_n = v_n = p_n = 0 \quad \text{for} \quad t \leqq 0\,. \tag{6.1.3}$$

All these equations are identical to those considered in Section 5.1. However, an important difference results from the boundary conditions imposed by the two coasts, i.e.,

$$v_n(x, 0, t) = v_n(x, L, t) = 0\,. \tag{6.1.4}$$

The boundary conditions have the simplest form for the cross-current. Hence, we will consider the Green's function which correspond to this velocity component. Thus we may derive an equation for v alone and calculate the corresponding Green's function $G(y, y')$. If G is known we have the formal solution of v_n and then we may express the remaining dynamical fields in terms of v_n and the external forcing functions X_n and Y_n as outlined in Section 5.1. Obviously, the derivation of those equations is independent of the boundary conditions. Thus, we get according to (5.1.7) to (5.1.12) with the index n being dropped

$$v_{yy} + \alpha^2 v = \mathrm{i}\,\frac{(\omega^2\lambda^2 - k^2)}{\omega}\, Y + \left(\lambda^2 f + \frac{k}{\omega}\frac{\partial}{\partial y}\right) X \tag{6.1.5 a}$$

$$u = \frac{\mathrm{i}\omega}{\omega^2\lambda^2 - k^2}\left[\lambda^2 X + \left(\lambda^2 f - \frac{k}{\omega}\frac{\partial}{\partial y}\right) v\right] \tag{6.1.5 b}$$

$$p = \frac{\mathrm{i}k}{\omega^2\lambda^2 - k^2}\left[X + \left(f - \frac{\omega}{k}\frac{\partial}{\partial y}\right) v\right] \tag{6.1.5 c}$$

where $\alpha^2 = \lambda^2(\omega^2 - f^2) - k^2$. This set forms together with the boundary conditions (6.1.4) and the initial conditions (6.1.3) the starting point of our theory. In the following sections we first estimate the Green's function in order to have a general formal solution, and then we discuss the response of the canal model to several examples of wind forcing patterns.

6.2. THE GREEN'S FUNCTION OF THE CHANNEL MODEL, AND THE FORMAL SOLUTION

The Green's function corresponding to the cross current v, is governed by the differential equation

$$\left(\frac{d^2}{dy^2} + \alpha^2\right) G(y, y') = \delta(y - y') \tag{6.2.1}$$

with the boundary conditions

$$G(0, y') = G(L, y') = 0 , \tag{6.2.2}$$

where $\alpha^2 = \lambda^2(\omega^2 - f^2) - k^2$. The two linear independent solutions of (6.2.1) for $y \neq y'$ are $\sin(\alpha y)$ and $\sin(y - L)$ which have the Wronskian $\alpha \sin(\alpha L)$. Thus the Green's function may readily be constructed

$$G(y, y') = \theta(y - y')\, G^{>}(y, y') + \theta(y' - y)\, G^{<}(y, y') \tag{6.2.3a}$$

and

$$G^{>}(y, y') = \frac{\sin \alpha y' \sin \alpha(y - L)}{\alpha \sin \alpha L} = G^{<}(y', y) . \tag{6.2.3b}$$

Using standard identities of the trigonometric functions we may rewrite the Green's function in a more compact form

$$G(y, y') = \frac{\cos(\alpha(|y - y'| - L)) - \cos(\alpha(y + y' - L))}{2\alpha \sin \alpha L} . \tag{6.2.4}$$

Next we discuss the analytic properties of the Green's function. First we note that the G is regular for $\alpha = 0$, i.e.,

$$G(y, y')|_{\alpha=0} = \frac{yy'}{L} - \frac{1}{2}(y + y') + \frac{1}{2}|y - y'| . \tag{6.2.5}$$

Moreover, the Green's function has an countable infinite number of poles at $\alpha L = m\pi$ $(m = 1, 2, 3, \ldots)$ or if we introduce the discrete cross-wavenumber $\varkappa_m = m\pi/L$, at $\alpha = \varkappa_m$. These poles determine a dispersion relation, namely

$$\omega^2 = \omega_m^2(k) = f^2 + \frac{k^2 + \varkappa_m^2}{\lambda^2} . \tag{6.2.6}$$

We recognize that (6.2.6) is the well-known dispersion relation of Poincaré waves in a channel. We note that the frequencies of the Poincaré waves are superinertial, i.e., $\omega > f$, although for small wavenumbers the frequency of the gravest mode is rather close to f.

The expansion of the Green's function in terms of cross-channel modes yields

$$G(y, y') = \frac{2}{L} \sum_{m=1}^{\infty} \frac{\sin \varkappa_m y \sin \varkappa_m y'}{\lambda^2(\omega^2 - \omega_m^2(k))} \tag{6.2.7}$$

with the normalized cross-channel eigenfunctions $(2/L)^{1/2} \cdot \sin \varkappa_m y$. For frequencies larger than f the Green's function varies sinusoidally with y.

In the subinertial range $\omega < f$, however, α becomes imaginary, $\alpha = \mathrm{i}\,|\alpha|$ with $|\alpha| = [\lambda^2(f^2 - \omega^2) + k^2]^{1/2}$, and the trigonometric functions in (6.2.3) or (6.2.4) transform into hyperbolic ones. Hence, in the subinertial domain G is a localized function of y. We note that there are no poles at $\omega = \pm f$. This is due to the existence of two coasts which prevent the water masses from oscillating as a whole with the inertial frequency.

Finally we remark that the formal solution of the problem is completely given by the set (5.2.6). The important difference, however, is that we have to use the Green's function of the channel model in one of the representations (6.2.3), (6.2.4) or (6.2.7).

Moreover, the convolution integrals $G * X$ and $G * Y$ appearing in (5.2.6) must now be replaced by

$$G * X = \int_0^L G(y, y')\, X(y')\, \mathrm{d}y' , \qquad G * Y = \int_0^L G(y, y')\, Y(y')\, \mathrm{d}y' . \tag{6.2.8}$$

Thus, apart from the analytical behaviour of the forcing functions, the analytical properties of the response of a channel model are determined by the poles in the origin of the ω-plane, which correspond to non-oscillatory motion, by the Poincaré poles, governed by the Green's function, and by the Kelvin poles which occur only in the expressions of the longshore current u and the pressure p. In Fig. 6.2.1 the locations of the various poles have been indicated. We note that for $L \to \infty$ the infinite number of Poincaré poles degenerate to a continuum as shown in Fig. 5.9.1.

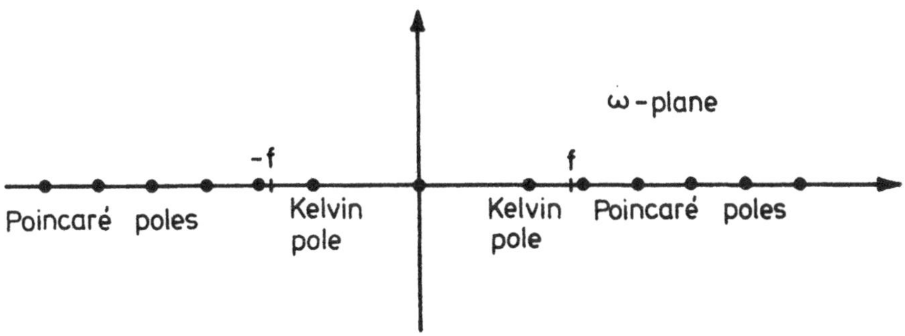

Fig. 6.2.1
Location of the Poincaré poles and Kelvin poles on the real frequency axis in the ω-plane
Note that for $L \to \infty$ the Poincaré poles degenerate to a continuum as indicated in Fig. 5.9.1.

6.3. KELVIN WAVES IN A CHANNEL

Next we discuss the generation of Kelvin waves in a channel in a general manner. The analysis is close to that of Section 5.3. and amounts to the investigation of the residues of the Kelvin poles. We may put $Y = 0$, because the cross wind is not associated with any Kelvin pole. First we rewrite the solutions of u and p according to (5.3.3) and (5.3.8), i.e.,

$$u(\omega, k, y) = \frac{i}{\omega} G_y * X_{y'}$$
$$+ \frac{i}{2} \frac{X*}{\left(\omega - \frac{k}{\lambda}\right)} \left[\delta(y - y') + \frac{G}{R^2} + G_{yy'} - \frac{1}{R}(G_y + G_{y'})\right]$$
$$+ \frac{i}{2} \frac{X*}{\left(\omega + \frac{k}{\lambda}\right)} \left[\delta(y - y') + \frac{G}{R^2} + G_{yy'} + \frac{1}{R}(G_y + G_{y'})\right] \qquad (6.3.1\,a)$$

and

$$p(\omega, k, y) = -\frac{if}{\omega} G * X_{y'} + \frac{i}{2\lambda} \frac{X*}{\left(\omega - \frac{k}{\lambda}\right)} \left[\delta(y - y') + \frac{G}{R^2} + G_{yy'} - \frac{1}{R}(G_y + G_{y'})\right]$$
$$+ \frac{i}{2\lambda} \frac{X*}{\left(\omega + \frac{k}{\lambda}\right)} \left[\delta(y - y') + \frac{G}{R^2} + G_{yy'} + \frac{1}{R}(G_y + G_{y'})\right]. \qquad (6.3.1\,b)$$

Next we estimate the contributions of the brackets at $\omega = \pm k/\lambda$. For $\omega^2 = k^2/\lambda^2$ we have

$$\alpha^2 = -\frac{1}{R^2}, \qquad G_{y'y'} - \frac{G}{R^2} = \delta(y - y').$$

Moreover, from (6.2.4) we get

$$G_y + G_{y'} = \frac{\sin \alpha(y + y' - L)}{2 \sin \alpha L}$$

and differentiating this by y

$$G_{yy} + G_{yy'} = \frac{\alpha \cos \alpha(y' + y - L)}{2 \sin \alpha L}.$$

Thus we obtain

$$\left[G_{yy} + G_{yy'} - \frac{1}{R}(G_y + G_{y'})\right]_{\omega = \frac{k}{\lambda}} = \frac{1}{2R} \frac{e^{-\frac{y'}{R}} e^{-\frac{y-L}{R}}}{\sinh (L/R)} \qquad (6.3.2\,a)$$

$$\left[G_{yy} + G_{yy'} + \frac{1}{R}(G_y + G_{y'})\right]_{\omega = -\frac{k}{\lambda}} = \frac{1}{2R} \frac{e^{-\frac{y'}{R}} e^{-\frac{L-y}{R}}}{\sinh (L/R)}. \qquad (6.3.2\,b)$$

Suppose that the y-scale of X is large compared to L so that X varies only insignificantly with y. Then we get

$$X * e^{\frac{y'}{R}} = X \int_0^L dy' \, e^{\frac{y'}{R}} = RX\left(e^{\frac{L}{R}} - 1\right)$$

and

$$X * e^{-\frac{y'}{R}} = RX\left(1 - e^{-\frac{L}{R}}\right).$$

Moreover, let us assume $L \gg R$ which, as a rule, is valid for the baroclinic modes. Thus, with $\sinh L/R = (1/2) \exp (L/R)$ we arrive at

$$u_{\text{Kelvin}}(\omega, k, y) = \frac{i}{2}\left[X\left(\frac{k}{\lambda}, k\right)\frac{e^{-\frac{y}{R}}}{\omega - \frac{k}{\lambda}} + X\left(-\frac{k}{\lambda}, k\right)\frac{e^{-\frac{L-y}{R}}}{\omega + \frac{k}{\lambda}}\right] \tag{6.3.3 a}$$

and

$$p_{\text{Kelvin}}(\omega, k, y) = \frac{i}{2\lambda}\left[X\left(\frac{k}{\lambda}, k\right)\frac{e^{-\frac{y}{R}}}{\omega - \frac{k}{\lambda}} - X\left(-\frac{k}{\lambda}, k\right)\frac{e^{-\frac{L-y}{R}}}{\omega + \frac{k}{\lambda}}\right] \tag{6.3.3 b}$$

where $X(\omega, k)$ has been replaced by $X(\pm k/\lambda, k)$, respectively. These relations show that at both sides of the canal Kelvin waves may be generated which propagates with the coast to the right. The cross-channel pressure gradient balances geostrophically the longshore current. Without the simplification $L \gg R$ and X being independent of y, we obtain the general result

$$u_{\text{Kelvin}} = \frac{i}{4R \sinh \frac{L}{R}}\left\{\int_0^L dy' \frac{X\left(\frac{k}{\lambda}, k, y'\right)}{\omega - \frac{k}{\lambda}} e^{-\frac{y'}{R}} e^{\frac{L-y}{R}} + \int_0^L dy' \frac{X\left(-\frac{k}{\lambda}, k, y'\right)}{\omega + \frac{k}{\lambda}} e^{\frac{y'}{R}} e^{-\frac{L-y}{R}}\right\} \tag{6.3.4 a}$$

$$p_{\text{Kelvin}} = \frac{if}{4 \sinh \frac{L}{R}}\left\{\int_0^L dy' \frac{X\left(\frac{k}{\lambda}, k, y'\right)}{\omega - \frac{k}{\lambda}} e^{-\frac{y'}{R}} e^{\frac{(L-y)}{R}} - \int_0^L dy' \frac{X\left(-\frac{k}{\lambda}, k, y'\right)}{\omega + \frac{k}{\lambda}} e^{\frac{y'}{R}} e^{-\frac{(L-y)}{R}}\right\}. \tag{6.3.4 b}$$

Similar to the case of the semi-infinite ocean we have derived the formulas (6.3.4) using only properties of the Green's function without any further specification of the alongshore forcing X. Obviously, Kelvin waves are forced through inhomogenities of the longshore wind in alongshore direction. Otherwise, i.e., if X being independent of x in the Fourier space $X \sim \delta(k)$ would follow and then no Kelvin waves can be excited.

6.4. COASTAL JETS IN A CHANNEL

In this and in the following section we will consider several examples of external forcing functions. We start with a longshore wind which is independent of the longshore coordinate x

$$X(x, z, t) = \frac{u_*^2}{H_{\text{mix}}} \theta(z + H_{\text{mix}})\, \theta(t)\, Q(y), \qquad Y = 0. \tag{6.4.1}$$

The wind may vary with y. This is modelled by the function $Q(y)$. Fourier transformation of X regarding x and t, and expansion in terms of vertical eigenfunction yields

$$X_n(\omega, k, y) = \frac{u_*^2}{h_n} \frac{2\pi i}{\omega} \delta(k)\, Q(y). \tag{6.4.2}$$

Then, from (5.2.6) we get, utilizing $\delta(k)\, \alpha(\omega, k) = \delta(k)\, \alpha(\omega, 0)$

$$u(\omega, k, y) = -\frac{u_*^2}{h} \frac{2\pi\delta(k)}{\omega^2} \left(Q + \frac{1}{R^2} G * Q\right) \tag{6.4.3 a}$$

$$v(\omega, k, y) = \frac{u_*^2}{h} \frac{2\pi i\, \delta(k)}{\omega f R^2} G * Q \tag{6.4.3 b}$$

$$p(\omega, k, y) = \frac{u_*^2}{h} \cdot 2\pi\, \delta(k) \frac{f}{\omega^2} \frac{\partial}{\partial y} G * Q. \tag{6.4.3 c}$$

Here the index n has been suppressed, and $G * Q$ is according to (6.2.8)

$$G * Q = \int_0^L \mathrm{d}y'\, G(y, y')\, Q(y'). \tag{6.4.4}$$

In order to calculate the convolution (6.4.4) it appears to be convenient to express G under the integral sign by $(1/\alpha^2)\, [\delta(y - y') - G_{y'y'}]$, compare (6.2.1). Then after twice integration by parts we find

$$G * Q = \frac{1}{\alpha^2} \left[Q(y) - Q(L)\, G_{y'}(y, L) + Q(0)\, G_{y'}(y, 0) - \int_0^L \mathrm{d}y'\, G(y, y')\, Q_{y'y'}(y') \right]. \tag{6.4.5}$$

In particular, for constant or linearly varying $Q(y)$ the rightmost term vanishes. Thus the convolution $G * Q$ consists of a term proportional to the wind pattern (first term) plus coastal rectifications (second and third term).

Let us consider three explicit examples of $Q(y)$, as plotted in Fig. 6.4.1

$$Q(y) = 1 \tag{6.4.6 a}$$

$$Q(y) = \frac{y}{L} \tag{6.4.6 b}$$

$$Q(y) = \frac{y}{L} - \frac{1}{2}. \tag{6.4.6 c}$$

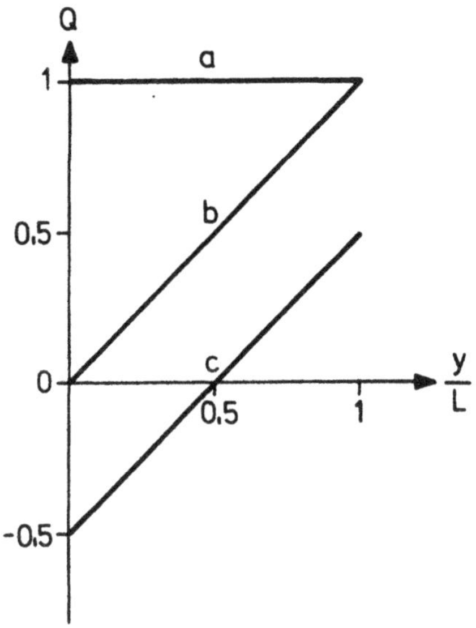

Fig. 6.4.1
Sketch of the function $Q(y)$ according to (6.4.6a, b, c), which models the cross-channel structure of the along-shore wind

In the first case, $Q = 1$, we get using both representations of G, i.e., (6.2.3) and (6.2.7),

$$G * Q = \frac{1}{\alpha^2}\left(1 - \frac{\cos\alpha\left(\frac{L}{2} - y\right)}{\cos\frac{\alpha L}{2}}\right) = \frac{4}{L}\sum_{n=0}^{\infty} \frac{\sin \varkappa_{n'} y}{\lambda^2 \varkappa_{n'}(\omega^2 - \omega_{n'}^2(0))} \tag{6.4.7}$$

with $n' = 2n + 1$. Then (6.4.3a) to (6.4.3c) becomes

$$u(\omega, k, y) = -\frac{u_*^2}{h}\frac{2\pi\delta(k)}{\omega^2}\left[1 + \frac{\lambda^2 f^2}{\alpha^2}\left(1 - \frac{\cos\frac{\alpha L}{2}\left(1 - \frac{2y}{L}\right)}{\cos\frac{\alpha L}{2}}\right)\right] \tag{6.4.8a}$$

$$v(\omega, k, y) = \frac{u_*^2}{h}\frac{2\pi\,\mathrm{i}\delta(k)}{\omega}\cdot\frac{\lambda^2 f}{\alpha^2}\left(1 - \frac{\cos\frac{\alpha L}{2}\left(1 - \frac{2y}{L}\right)}{\cos\frac{\alpha L}{2}}\right) \tag{6.4.8b}$$

$$p(\omega, k, y) = -\frac{u_*^2}{h}\frac{2\pi\delta(k)\, f}{\omega^2}\cdot\frac{\sin\frac{\alpha L}{2}\left(1 - \frac{2y}{L}\right)}{\alpha\cos\frac{\alpha L}{2}}. \tag{6.4.8c}$$

For $f < \omega$ the fields (6.4.8a) to (6.4.8c) possess peaks at the odd Poincaré frequencies $\omega_{2n+1}^2(0) = f^2 + \varkappa_{2n+1}^2/\lambda^2$.

In the subinertial range $f > \omega$, particularly in the low frequency limit $\omega \to 0$, the well known coastal jet follows, see CSANADY (1977). Note that for $\omega = 0$, $\alpha = \mathrm{i}/R$, and so the trigonometric function being transformed into hyperbolic ones. Taking into acount that the vertical component w is associated with the pressure by means of

$$w = -\frac{p_{zt}}{N^2},$$

the expression (6.4.8c) implies the occurrence of upwelling at one coast and downwelling at the other coast.

In the second case $Q = y/L$, we obtain

$$G * Q = \frac{1}{\alpha^2}\left(\frac{y}{L} - \frac{\sin \alpha y}{\sin \alpha L}\right) = -\frac{2}{L}\sum_{n=1}^{\infty} \frac{(-1)^n \sin \varkappa_n y}{\lambda^2 \varkappa_n(\omega^2 - \omega_n^2(0))} \tag{6.4.9}$$

and, hence,

$$u(\omega, k, y) = -\frac{u_*^2}{h}\frac{2\pi\delta(k)}{\omega^2}\left[\frac{y}{L}\left(1 + \frac{f^2\lambda^2}{\alpha^2}\right) - \frac{f^2\lambda^2}{\alpha^2}\frac{\sin \alpha L}{\sin \alpha L}\right] \tag{6.4.10a}$$

$$v(\omega, k, y) = \frac{u_*^2}{h}\frac{2\pi\, i\delta(k)}{\omega}\frac{\lambda^2 f}{\alpha^2}\left[\frac{y}{L} - \frac{\sin \alpha y}{\sin \alpha L}\right] \tag{6.4.10b}$$

$$p(\omega, k, y) = \frac{u_*^2}{h}\frac{2\pi\delta(k)}{\omega^2} f\left[\frac{1}{\alpha^2 L} - \frac{\cos \alpha y}{\alpha \sin \alpha L}\right]. \tag{6.4.10c}$$

For $f < \omega$ the response possesses peaks at all Poincaré frequencies $\omega_n(0)$. In the sub-inertial range coastal trapped motion occurs only at one of the coasts, i.e., $y = L$. From (6.4.10c), it follows that in addition to the coastal upwelling the windstress curl in terms of $(\partial/\partial y)\, Q$ yields a further source of vertical motion.

Third case: $Q = \left(\frac{y}{L} - \frac{1}{2}\right)$

$$G * Q = \frac{1}{\alpha^2}\left(Q - \frac{1}{2}\frac{\sin \alpha\left(\frac{y-L}{2}\right)}{\sin \frac{\alpha L}{2}}\right) = -\frac{2}{L}\sum_{n=1}^{\infty} \frac{\sin \varkappa_{2n} y}{\lambda^2 \varkappa_{2n}(\omega^2 - \omega_{2n}^2(0))}. \tag{6.4.11}$$

In this case the response is described by

$$u(\omega, k, y) = -\frac{u_*^2}{h}\frac{2\pi\delta(k)}{\omega^2}\left[Q\left(1 + \frac{f^2\lambda^2}{\alpha^2}\right) - \frac{\sin \frac{\alpha L}{2}\left(\frac{2y}{L} - 1\right)}{\sin \frac{\alpha L}{2}}\right] \tag{6.4.12a}$$

$$v(\omega, k, y) = \frac{u_*^2}{h}\frac{2\pi\, i\delta(k)}{\omega}\frac{\lambda^2 f}{\alpha^2}\left[Q - \frac{\sin \frac{\alpha L}{2}\left(\frac{2y}{L} - 1\right)}{\sin \frac{\alpha L}{2}}\right] \tag{6.4.12b}$$

$$p(\omega, k, y) = \frac{u_*^2}{h}\frac{2\pi\delta(k)}{\omega^2} f \left[\frac{1}{\alpha^2 L} - \frac{\cos\frac{\alpha L}{2}\left(\frac{2y}{L} - 1\right)}{2\alpha \sin\frac{\alpha L}{2}} \right]. \tag{6.4.12c}$$

Thus an alongshore forcing that changes its direction at $y = L/2$ excites only the even Poincaré oscillations $\omega = \omega_{2n}(0)$. In the subinertial regime coastal jets with opposite directions appear at both coasts. In (6.4.12c) we recognize again the wind stress curl as further source of up- or downwelling. Note that in all three examples in the zero frequency limit u and p are in geostrophic balance.

The general solution in terms of (5.2.6) represents the full amount of possible response pattern. The above three examples illustrate how slightly different forcing functions yield different realizations of various possible patterns. In the following two sections we will discuss the solution of the first example (6.4.6a) in some detail.

6.5. LARGE SCALE ALONGSHORE FORCING AND THE ASSOCIATED COASTAL JET

Generally the spatial scales of the meteorological forces exceed by far the scales of elongated lakes and straits which can be modelled by canals. Hence, for practical discussions the most relevant solution is given by the set (6.4.8).

We consider now the solution (6.4.8) in some detail. We start with the subinertial portion of the response which is determined by the residues of the poles in the origin of the ω-plane. The inverse Fourier transformation of the subinertial part of (6.4.8) yields

$$u_n(y, t) = \frac{u_*^2}{h_n} t\theta(t) \frac{\cosh\frac{L}{2R_n}\left(1 - \frac{2y}{L}\right)}{\cosh\frac{L}{2R_n}} \tag{6.5.1a}$$

$$v_n(y, t) = -\frac{u_*^2}{h_n f}\theta(t)\left(1 - \frac{\cosh\frac{L}{2R_n}\left(1 - \frac{2y}{L}\right)}{\cosh\frac{L}{2R_n}}\right) \tag{6.5.1b}$$

$$p_n(y, t) = \frac{u_*^2}{h_n} t\theta(t) f R_n \frac{\sinh\frac{L}{2R_n}\left(1 - \frac{2y}{L}\right)}{\cosh\frac{L}{2R_n}} \tag{6.5.1c}$$

where the vertical mode index n has been written explicitly. Clearly, the full response is given after summation over all vertical modes according to (6.1.2a). Because u and p are linearly increasing with time, some limiting dissipation becomes important since otherwise these fields would unreasonably grow up for large time. We choose as dissipation mechanism Rayleigh damping and Newtonian cooling with the same relaxation parameter r. This implies that all ω's appearing in the response have to be replaced by $\bar{\omega} = \omega + ir$ with the exception of those ω which are explicite variables of the forcing. We shall, however, neglect the r occurring in α, i.e., $\alpha(\bar{\omega}, 0) \approx \alpha(\omega, 0)$. Then with

$$\frac{1}{\omega^2} \to \frac{1}{(\omega + i\varepsilon)} \frac{1}{(\omega + ir)},$$

it follows

$$-\int \frac{d\omega}{2\pi} \frac{e^{-i\omega t}}{(\omega + i\varepsilon)(\omega + ir)} = \frac{\theta(t)}{r}(1 - e^{-rt}) \approx \begin{cases} t & \text{for} \quad rt \ll 1 \\ \dfrac{1}{r} & \text{for} \quad rt \gg 1 \,. \end{cases} \tag{6.5.2}$$

For large times, i.e., $rt \gg 1$ this simplifies to $1/r$ and, hence, we have only to replace the factor t, which occurs in (6.5.1 a, c) by $1/r$. Thus for large times the full solution of u and v reads

$$u(y, z, t) = \frac{u_*^2}{r} \sum_{n=0} C_n(y) \frac{F_n(z)}{h_n} \tag{6.5.3 a}$$

$$v(y, z, t) = -\frac{u_*^2 \theta(z + H_{\text{mix}})}{H_{\text{mix}} f} + \frac{u_*^2}{f} \sum_{n=0}^{\infty} C_n(y) \frac{F_n(z)}{h_n} \tag{6.5.3 b}$$

where the abbreviation

$$C_n(y) = \frac{\cosh \dfrac{L}{2R_n}\left(1 - \dfrac{2y}{L}\right)}{\cosh \dfrac{L}{2R_n}} \tag{6.5.4}$$

has been introduced. Let us consider a constant BVF, implying according to Subsection 1.9.2.

$$F_0(z) = 1 \quad \text{and} \quad F_n(z) = (2)^{1/2} \cos \frac{n\pi}{H} z$$

and let us adopt the following numerical values:

$$N = 2\pi \cdot 10^{-2}\,\text{s}^{-1}, \qquad f = 10^{-4}\,\text{s}^{-1}, \qquad L = 40\,\text{km}, \qquad H = 20\,\text{m},$$

$$H_{\text{mix}} = \frac{1}{2} H, \qquad u_* = 1\,\text{cm s}^{-1}, \quad \text{and} \quad r = 10^{-5}\,\text{s}^{-1}.$$

This implies a barotropic Rossby radius $R_0 = 140$ km, i.e., $R_0 \gg L$, and the first baroclinic Rossby radius being $R_1 = 4$ km. Then the alongshore current plotted in Fig. 6.5.1 results. The figure shows a coastal jet with a typical scale of the first baroclinic Rossby radius R_1.

This downwind jet is confined to the upper layer. Outside the jet, i.e., $y > R_1$, a barotropic downwind current remains, with $u \approx u_*^2/(rH) = 50\,\text{cm s}^{-1}$.

Since the expression (6.5.4) has the same sign at both coasts the structure of the current appears to be the same in the vicinity of both coasts.

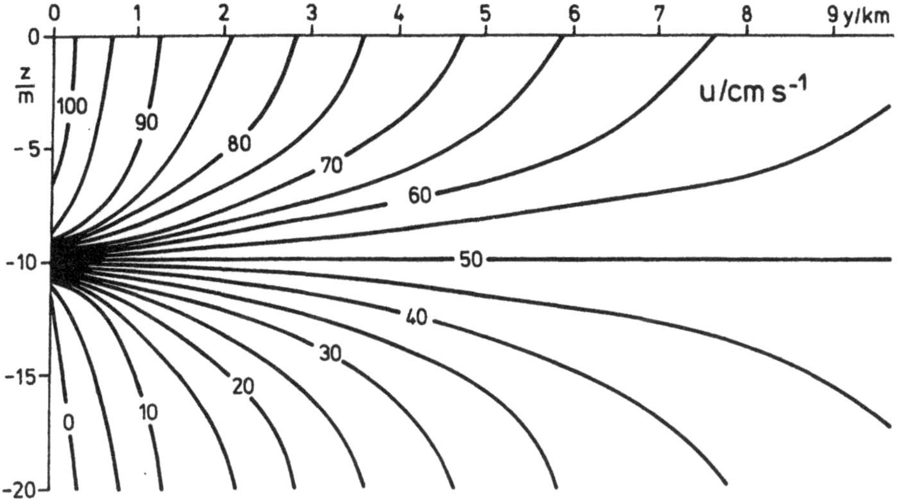

Fig. 6.5.1
The velocity structure of a coastal jet according to (6.5.3a)
The numerical values of the involved parameters are listed in the text. Shown is the alongshore current which is close to the coast confined to the upper layer. For distances larger than the first baroclinic Rossby radius, $R_1 = 4$ km, a barotropic current has established.

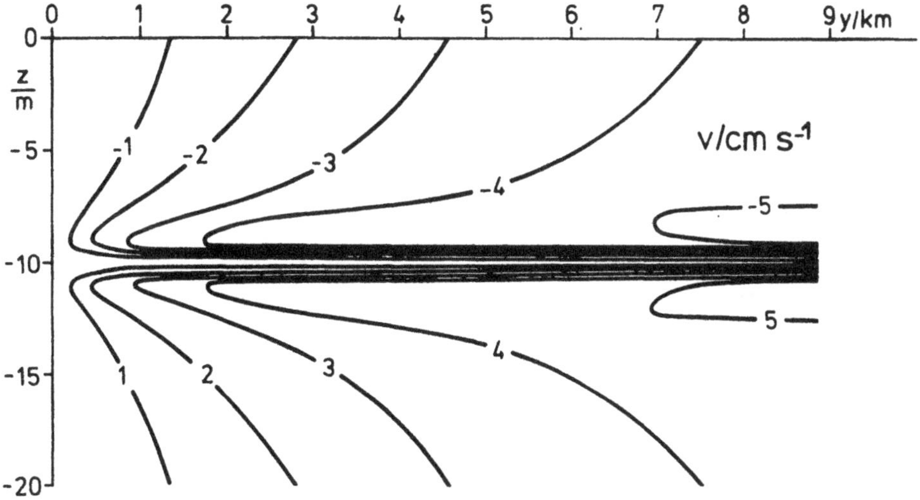

Fig. 6.5.2
The contours of the velocity of the cross-circulation associated with the coastal jet

The corresponding cross circulation has been depicted in Fig. 6.5.2. We may separate the barotropic mode in (6.5.3b) and find

$$v(y, z, t) = -\frac{u_*^2}{H_{\text{mix}} f}\left[\theta(z + H_{\text{mix}}) - \frac{H_{\text{mix}}}{H}\right] + \frac{u_*^2}{f}\sum_{n=1}^{\infty} C_n(y)\,\frac{F_n(z)}{h_n}\,. \tag{6.5.5}$$

For $y > R_1$, the rightmost term is negligible and it remains a reduced Ekman flux in the upper layer and a countercurrent below. When approaching the coast the baroclinic modes come into play which reduce the current to zero close to the coast.

The sea level behaviour is governed by the barotropic mode of the pressure divided by g, i.e., $\eta = p_0/g$. Thus from (6.5.1b, c), with t being replaced by r^{-1} in (6.5.1c), it may be inferred that through the Ekman transport water piles up at the coast $y = 0$. It appears that the ratio of the channel width and the barotropic Rossby radius affects the amplitude of the sea level change. For the parameters given above the sea level increases by 10 cm. However, for a channel width $L = R_0$ the same wind would pile up the water by 50 cm.

Finally we consider the vertical current w which follows from

$$w = \frac{(-1)}{N^2} p_{zt} ,$$

hence from (6.5.1c) we get

$$w(y, z, t) = -\theta(t)\, u_*^2 \frac{f}{N^2} \sum_{n=0}^{\infty} \frac{R_n}{h_n} \cdot \frac{\sinh \frac{L}{2R_n}\left(1 - \frac{2y}{L}\right)}{\cosh \frac{L}{2R_n}} \frac{d}{dz} F_n(z) . \qquad (6.5.6)$$

This expression appears to be exact as well with as without friction. The resulting picture is drawn in Fig. 6.5.3. Note that at the coast $y = 0$, which is shown in the Fig. 6.5.3 the vertical is negative, that means downwelling. At the opposite coast, however, there occurs upwelling. This follows because the hyperbolic sine is an odd function. The structure of w is the same at both coasts, only the sign changes.

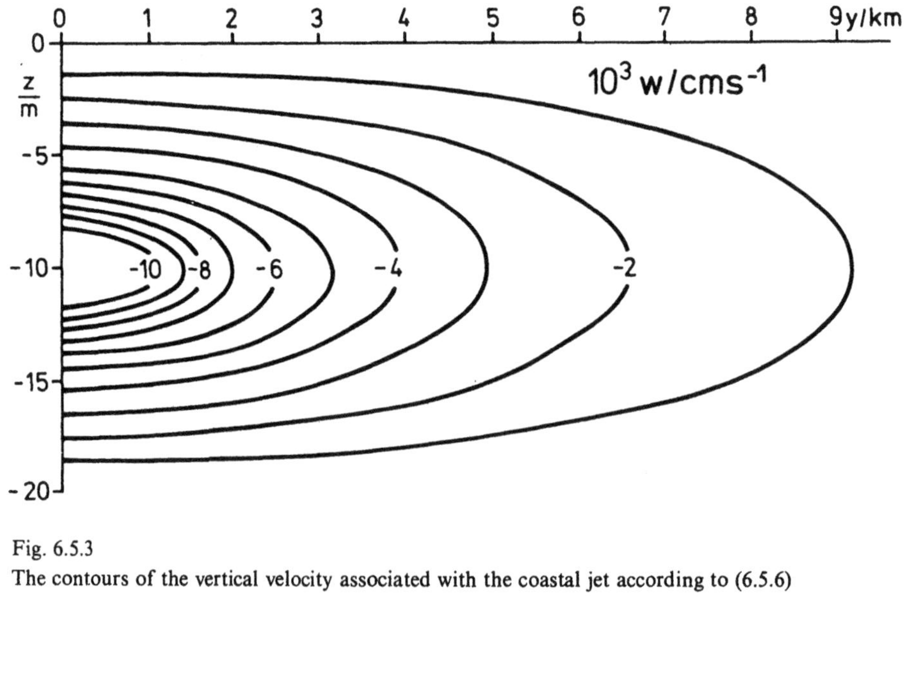

Fig. 6.5.3
The contours of the vertical velocity associated with the coastal jet according to (6.5.6)

6.6. FREQUENCY SPECTRA OF THE HORIZONTAL CURRENT COMPONENTS

In the foregoing section we have studied the subinertial part of the solution set (6.4.8), i.e., the response to a large scale forcing $X \sim \theta(t)$, $Y = 0$. Now we consider the complete response. First we estimate the Fourier inversion of (6.4.8) which consists of the expression of the set (6.5.1) plus the sum of the residues of the Poincaré poles. We get

$$u(y, t) = \frac{u_*^2}{h}\,\theta(t)\left\{t\,\frac{\cosh\dfrac{L}{2R}\left(1-\dfrac{2y}{L}\right)}{\cosh\dfrac{L}{2R}} + \frac{4f^2}{L}\sum_{m=0}^{\infty}\frac{\sin\varkappa_{m'}y\,\sin\omega_{m'}(0)\,t}{\varkappa_{m'}\omega_{m'}^3(0)}\right\} \tag{6.6.1 a}$$

$$v(y, t) = \frac{u_*^2}{hf}\,\theta(t)\left\{-1 + \frac{\cosh\dfrac{L}{2R}\left(1-\dfrac{2y}{L}\right)}{\cosh\dfrac{L}{2R}} + \frac{4f^2}{L}\sum_{m=0}^{\infty}\frac{\sin\varkappa_{m'}y\,\cos\omega_{m'}(0)\,t}{\varkappa_{m'}\omega_{m'}^2(0)}\right\} \tag{6.6.1 b}$$

$$p(y, t) = \frac{u_*^2}{h}\,\theta(t)\left\{\frac{t}{\lambda}\,\frac{\sinh\dfrac{L}{2R}\left(1-\dfrac{2y}{L}\right)}{\cosh\left(\dfrac{L}{2R}\right)} - \frac{4f}{L\lambda^2}\sum_{m=0}^{\infty}\frac{\cos\varkappa_{m'}y\,\sin\omega_{m'}(0)\,t}{\omega_{m'}^3(0)}\right\} \tag{6.6.1 c}$$

with $m' = 2m + 1$. The sums in (6.6.1) describe cross channel modes which oscillate with the odd Poincaré frequencies $\omega_{2m+1}(0)$. This is due to the presence of the two opposite coasts. Contrary to an infinite and a semi-infinite ocean the inertial waves cannot propagate in the cross channel direction because of the coasts where they are reflected and hence standing modes will be established with the discrete cross wave numbers $\varkappa_m = m\pi/L$.

The inertial circles mentioned in Section 3.1. degenerate in a channel to elliptically polarized or, close to the coast, to approximately linearly polarized motion. In Fig. 6.6.1 the trajectories of water particles are plotted for the simplest case of Poincaré oscillations

$$\frac{dx}{dt} = u_1(y, t) = a - b\sin\frac{\pi y}{L}\cos ft \tag{6.6.2 a}$$

$$\frac{dy}{dt} = v_1(y, t) = b\sin\frac{\pi y}{L}\sin ft \tag{6.6.2 b}$$

where a and b are constants.

The resulting curves are Lagrangian trajectories. Note that these equations are non-linear because of the occurrence of y in the argument of the sine. The resulting trajectories drawn in Fig. 6.6.1 were estimated numerically.

Respecting also the vertical mode index n the Poincaré frequencies occurring in (6.6.1) reads

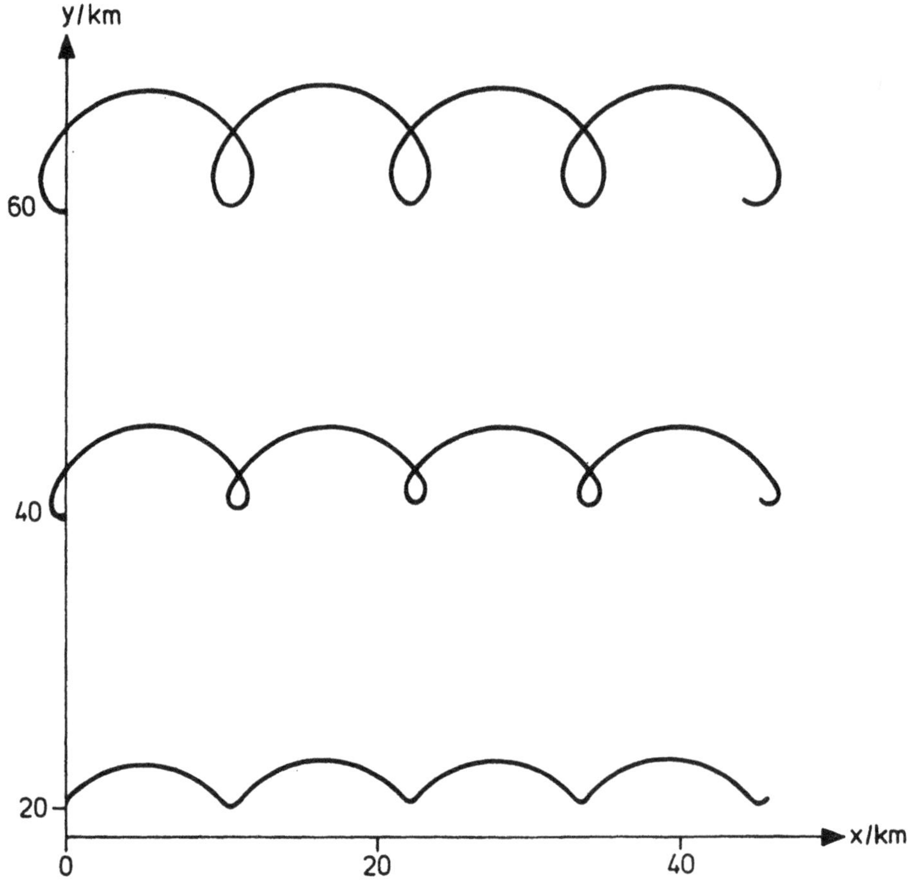

Fig. 6.6.1
Sketch of Lagrangian trajectories of water particles in different distances from the coast for a simple example of Poincaré oscillations
The curves, which are evaluated numerically, are not identical with the progressive vector diagrams.

$$\omega_{mn}(0) = \left(f^2 + \frac{\varkappa_m^2}{\lambda_n^2}\right)^{1/2}, \qquad m = 1, 3, 5, \ldots, \qquad n = 0, 1, 2, \ldots$$

The gravest baroclinic modes are associated with the frequencies

$$\omega_{1n}(0) = \left(f^2 + \frac{\pi^2}{L^2\lambda_n^2}\right)^{1/2}, \qquad n \geqq 1,$$

which are close to f. The corresponding blue shift, i.e., $(\omega_{1n}(0) - f)/f$ is approximately $(1/2)\,(H^2/L^2)\,(N^2/f^2)$. Adopting the numerical values used in the foregoing section a blue shift of few percent follows. It appears that $\omega_{mn}(0)$ are the frequencies of zero group velocity. This can easily by verified from (6.2.6), i.e., $C_{gr} = \dfrac{d\omega}{dk} = k/(\lambda^2\omega(k))$ which vanishes for $k = 0$.

Due to the smallness of λ_0 the barotropic Poincaré frequencies $\omega_{m0}(0)$ are much higher than the baroclinic ones. Thus for a channel with parameters adopted in the foregoing section the period of the gravest barotropic Poincaré mode is about one hour.

We may calculate the Eulerian spectra from (6.4.8). We introduce again the friction parameter r, i.e., replace all ω's by $\bar{\omega} = \omega + ir$ except those ω which are the variables of the forcing function. Then after the trivial k-integration from (6.4.8a, b) we obtain

$$u(\omega, y) = -\frac{u_*^2}{h\omega\bar{\omega}}\left[1 + \frac{(\lambda f)^2}{\alpha^2(\bar{\omega})}\left(1 - \frac{\cos\alpha(\bar{\omega})\left(\frac{L}{2} - y\right)}{\cos\frac{\alpha(\bar{\omega})\,L}{2}}\right)\right] \tag{6.6.3}$$

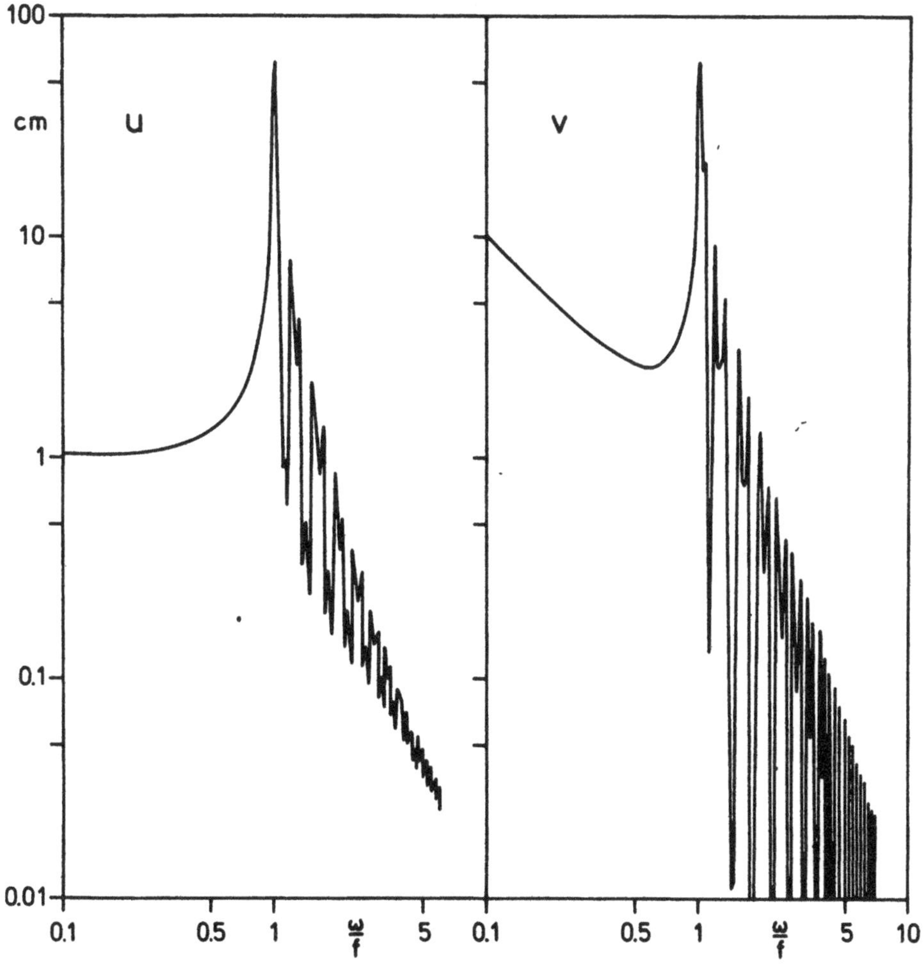

Fig. 6.6.2
The theoretically estimated Eulerian spectra of the alongshore and cross-shore current, u and v, in the middle of the channel, $y = L/2$, according to (6.6.3) and (6.6.4), respectively

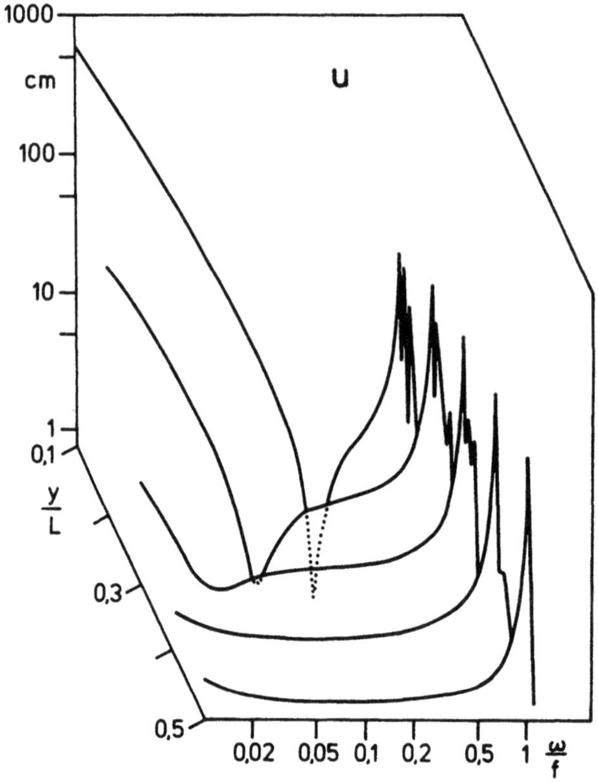

Fig. 6.6.3
The u-spectrum for different distances from the coast

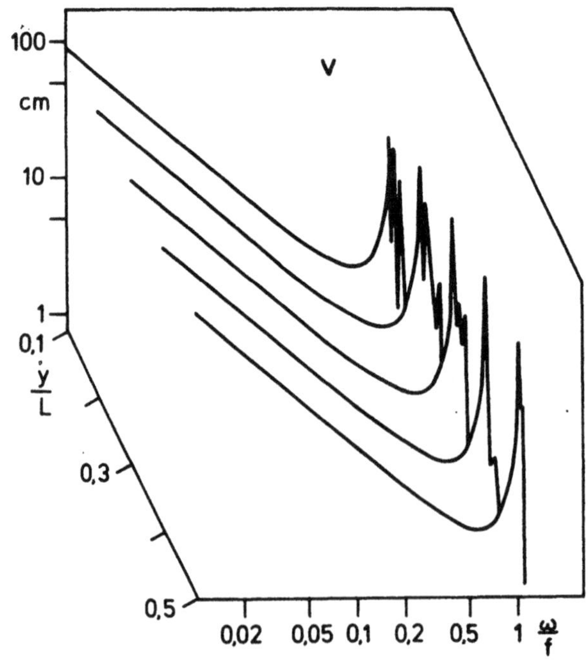

Fig. 6.6.4
Same as Fig. 6.6.3 for the v-spectrum

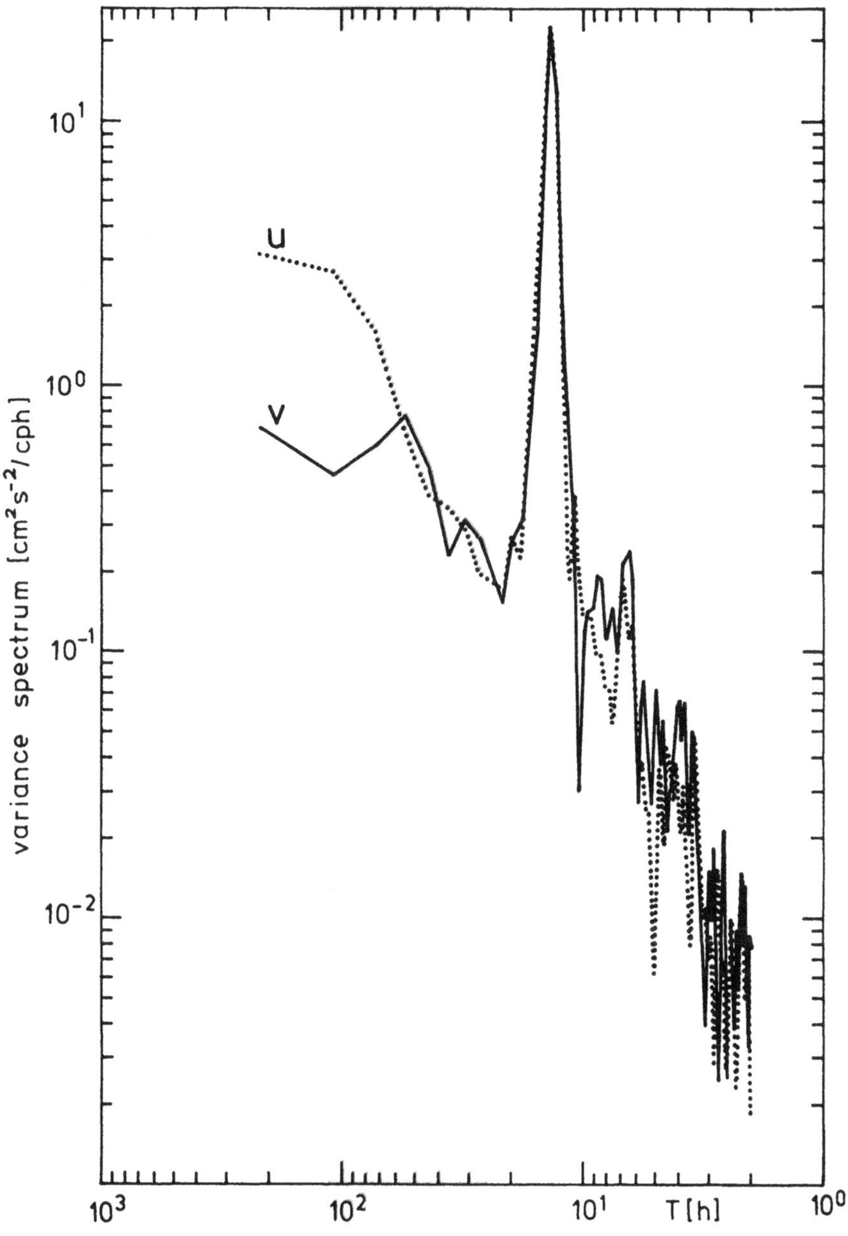

Fig. 6.6.5
A typical spectrum observed in the Bothnian Sea (redrawn from ALENIUS (1982))

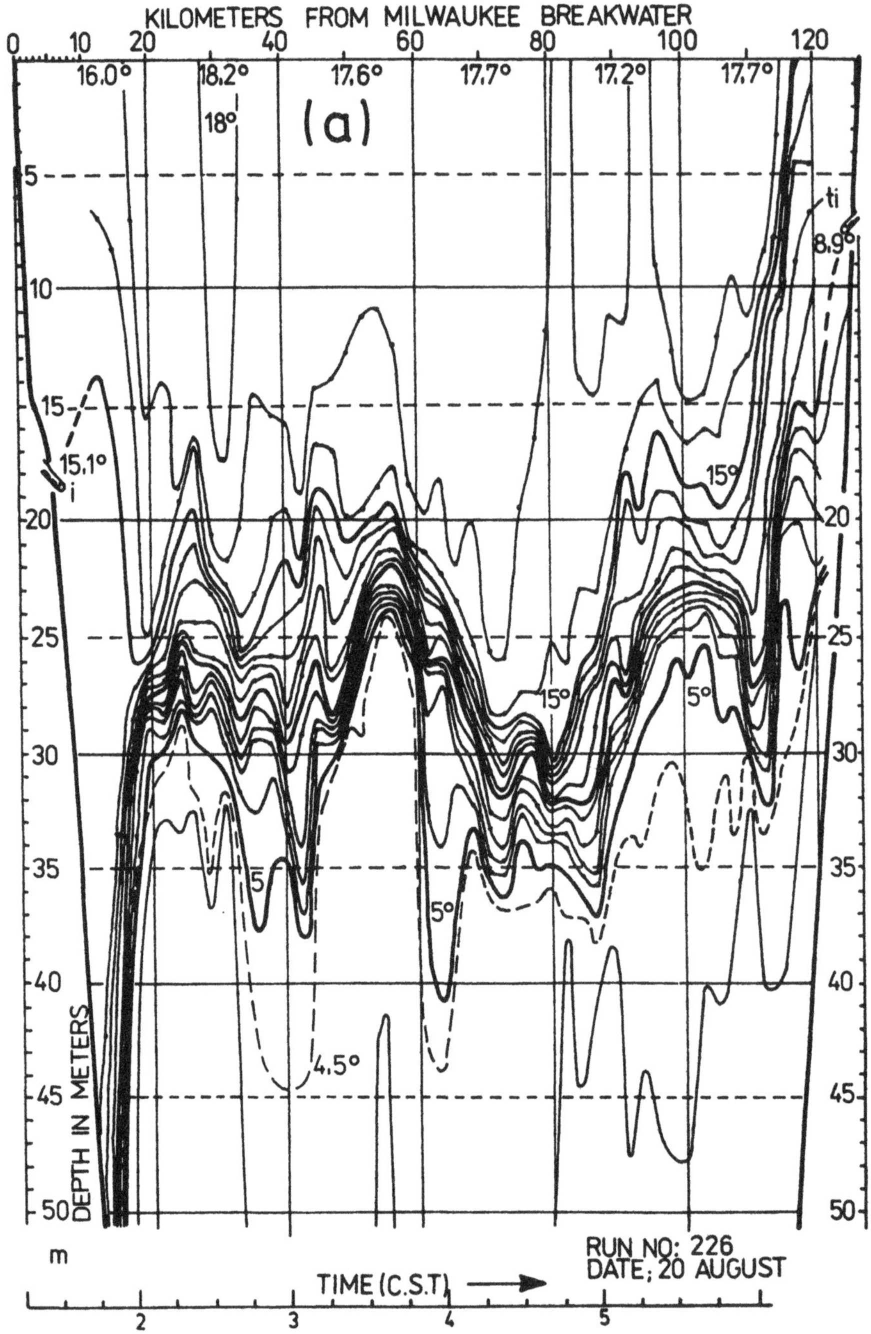

Fig. 6.6.6
Distribution of isotherms (°C) in the Milwaukee-Muskegon section of Lake Michigan, as measured during a ferry crossing on 20 August 1963 (redrawn from Mortimer (1968))

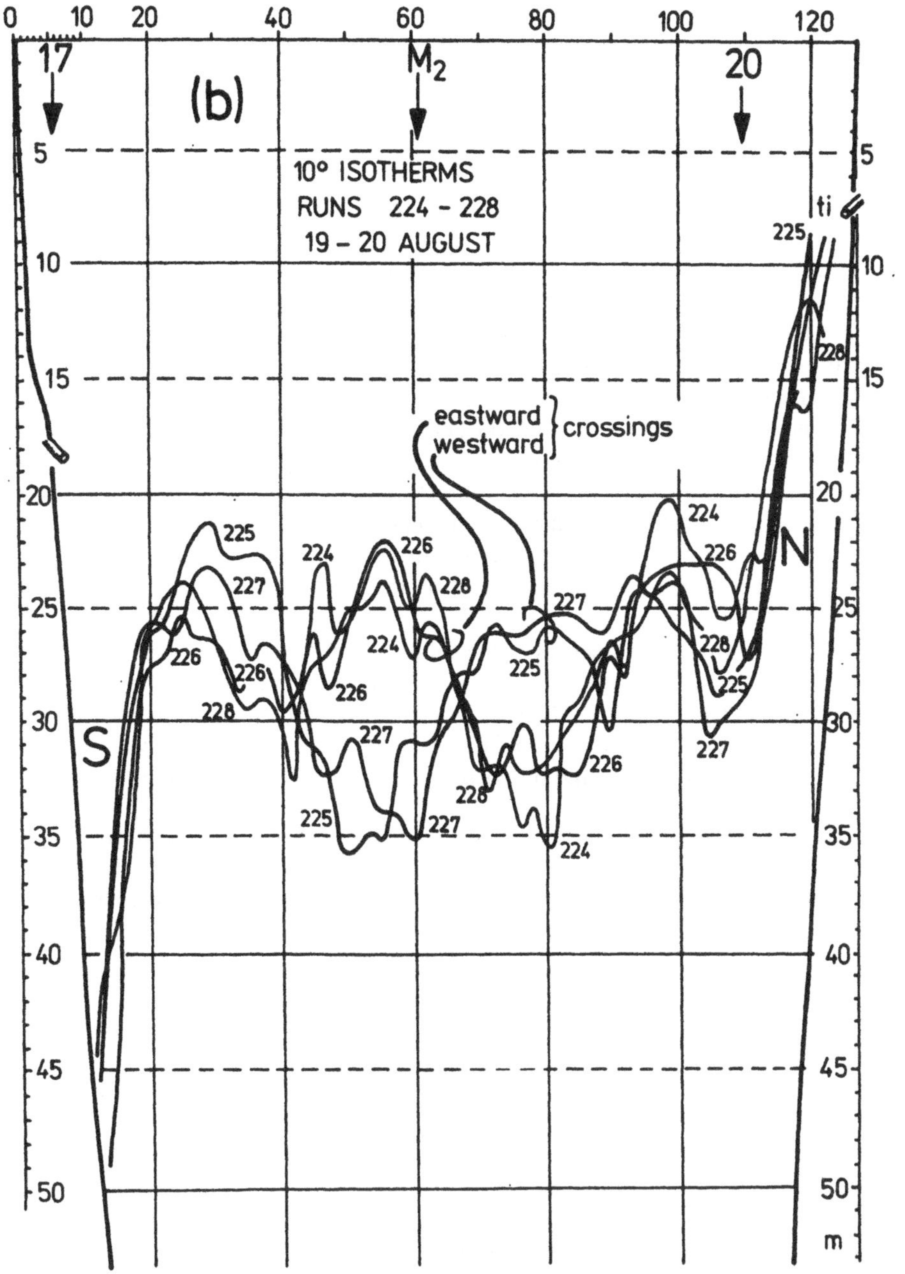

Fig. 6.6.7
Distribution of the 10 °C isotherms in the same section as in Fig. 6.6.6 during five consecutive ferry crossings, being about half a pendulum day apart in the center of the lake (redrawn from MORTIMER (1968))

and

$$v(\omega, y) = \frac{u_*^2 \, i\lambda^2 f}{h\omega\alpha^2(\bar{\omega})} \left[1 - \frac{\cos \alpha(\bar{\omega}) \left(\frac{L}{2} - y \right)}{\cos \frac{\alpha(\bar{\omega}) L}{2}} \right] \tag{6.6.4}$$

with $\bar{\omega} = \omega + ir$.

Let us define the modulus of u and v to be the Eulerian spectra. Then we obtain for $y = L/2$, i.e., in the middle of the channel, the spectra shown in Fig. 6.6.2. In Fig. 6.6.3 and Fig. 6.6.4 the low frequency part including the gravest inertial peak of the u- and v-spectrum, respectively, are displayed for different distances from the coast.

The general shape of the spectra is as expected from measurements in the Baltic Sea, see e.g. KIELMANN et al. (1973), MÄLKKI (1975), KRAUSS (1981), ALENIUS (1982). In Fig. 6.6.5 a typical spectrum is shown. Generally, there exist a well developed peak close to but a little bit larger than the local inertial frequency f. For frequencies higher than f the spectra oscillate wildly due to the higher Poincaré oscillations. Thus all peaks occur at frequencies of zero group velocity. In the subinertial range the spectra are smooth. Note the influence of the coastal jet on the u-spectrum in the proximity of the coast, Fig. 6.6.3.

Finally we note that the properties of the solution (6.6.1 c) compare well with the fascinating observations of isotherms in Lake Michigan which were reported by MORTIMER (1968), see Figs. 6.6.6 and 6.6.7. The lake was in response to strong winds which have blown two days in southward direction, i.e., out of the page. Strong downwelling occurs at the western (left-hand) shore and the corresponding upwelling is found at the opposite coast.

Moreover, several odd-numbered Poincaré modes appear to have been excited. The standing-wave character is clearly revealed by the unchanging position of the mid-lake nodal point of the 10 °C isotherms which were observed during five consecutive vessel crossings, see Fig. 6.6.7.

6.7. RESPONSE TO FORCING PERPENDICULAR TO THE COAST

In this section we add the solution which describes the response to a large scale forcing perpendicular to the coast, i.e.,

$$X = 0\,, \qquad Y(t, z) = \frac{v_*^2}{H_{\text{mix}}}\,\theta(z + H_{\text{mix}})\,\theta(t)\,. \tag{6.7.1}$$

After Fourier transformation and vertical mode expansion of (6.7.1) we get

$$Y_n(\omega, k) = \frac{v_*^2}{h_n}\frac{2\pi i}{\omega}\,\delta(k)\,. \tag{6.7.2}$$

Inserting this into the set (5.2.6) and dropping the index n we find

$$u(\omega, k, y) = -\frac{v_*^2}{h}\frac{f\lambda^2}{\omega} \cdot 2\pi i\delta(k)\, G * 1 \tag{6.7.3a}$$

$$v(\omega, k, y) = -\frac{v_*^2}{h}\lambda^2 \cdot 2\pi\,\delta(k)\,G * 1 \tag{6.7.3b}$$

$$p(\omega, k, y) = \frac{v_*^2}{h}\frac{2\pi i\delta(k)}{\omega}\frac{\partial}{\partial y}G * 1\,. \tag{6.7.3c}$$

The convolution $G * 1$ is already given by (6.4.7) and so the solution in the Fourier domain is complety known. The Fourier inversion of (6.7.3) yields

$$u(y, t) = +\frac{v_*^2}{h}\theta(t)\left[\frac{1}{f} - \frac{\cosh\left(\frac{L}{2R} - \frac{y}{R}\right)}{f\cosh\frac{L}{2R}} - \frac{4f}{L}\sum_{m=0}^{\infty}\frac{\sin \varkappa_{m'} y \cos \omega_{m'}(0)\, t}{\varkappa_{m'}\omega_{m'}^2(0)}\right] \tag{6.7.4a}$$

$$v(y, t) = \frac{v_*^2}{h}\theta(t)\frac{4}{L}\sum_{m=0}^{\infty}\frac{\sin \varkappa_{m'}\, y \sin \omega_{m'}(0)\, t}{\varkappa_{m'}\omega_{m'}(0)} \tag{6.7.4b}$$

$$p(y, t) = -\frac{v_*^2}{h}\theta(t)\left[R\frac{\sinh\left(\frac{L}{2R} - \frac{y}{R}\right)}{\cosh\frac{L}{2R}} - \frac{4}{L\lambda^2}\sum_{m=0}^{\infty}\frac{\cos \varkappa_{m'} y \cos \omega_{m'}(0)\, t}{\omega_{m'}^2(0)}\right] \tag{6.7.4c}$$

with $m' = 2m + 1$.

In the inertial domain, i.e., for frequencies which exceed the inertial, this solution shows a similar behaviour as the response to the longshore forcing case (6.6.1). In the subinertial domain, however, there exist striking differences. First, the response according to (6.7.4) has no accelerating jet. The dominating stationary signal is an Ekman flux parallel to the coast, which decreases to zero close to the coast, see (6.7.4a). This coastal rectification is geostrophically balanced by the pressure gradient p_y, see first term in (6.7.4c). Moreover, close to the coasts the cross-shore gradient p_y balances the wind force. Since w is proportional to p_t as well the vertical velocity as the cross current v, given by (6.7.4) are completely governed by Poincaré oscillations. Again we observe that these vibrations oscillate with those frequencies that correspond to zero group velocities.

6.8. RESPONSE TO A MOVING FRONT

In the present section we consider a vasnishing alongshore wind and a cross-shore wind with a moving frontal edge

$$Y(x, y, t) = \frac{v_*^2}{H_{\text{mix}}}\theta(z + H_{\text{mix}})\,\theta\left(t - \frac{x}{c}\right). \tag{6.8.1}$$

This wind pattern moves with the phase speed c parallel to the channel axis. For $c \to \infty$ (6.8.1) converges to the large scale cross wind switched-on at $t = 0$ which was considered in the previous section. After Fourier transformation and normal mode expansion we get

$$Y = \frac{v_*^2}{h}\frac{2\pi i}{\omega}\delta\left(k - \frac{\omega}{c}\right). \tag{6.8.2}$$

Inserting (6.8.2) into the general solution set (5.2.6) it follows

$$u(\omega, k, y) = \frac{v_*^2}{h}\frac{2\pi i}{\omega}\delta\left(k - \frac{\omega}{c}\right)\left(-\lambda^2 f + \frac{1}{c}\frac{\partial}{\partial y}\right) G * 1 \tag{6.8.3a}$$

$$v(\omega, k, y) = -\frac{v_*^2}{h}\cdot 2\pi\delta\left(k - \frac{\omega}{c}\right)\lambda^2\left(1 - \frac{1}{c^2\lambda^2}\right) G * 1 \tag{6.8.3b}$$

$$p(\omega, k, y) = \frac{v_*^2}{h}\frac{2\pi i}{\omega}\delta\left(k - \frac{\omega}{c}\right)\left(-\frac{f}{c} + \frac{\partial}{\partial y}\right) G * 1 . \tag{6.8.3c}$$

Here the relation

$$F(k)\,\delta\left(k - \frac{\omega}{c}\right) = F\left(\frac{\omega}{c}\right)\delta\left(k - \frac{\omega}{c}\right)$$

was used. The convolution $G * 1$ is given by (6.4.7), i.e.,

$$G * 1 = \frac{1}{\alpha^2}\left(1 - \frac{\cos\frac{\alpha L}{2}\left(1 - \frac{2y}{L}\right)}{\cos\frac{\alpha L}{2}}\right) = \frac{4}{L}\sum_{m=0}^{\infty}\frac{\sin \varkappa_{m'} y}{\lambda^2\varkappa_{m'}(\omega^2\gamma^2 - \omega_{m'}^2(0))}, \qquad m' = 2m + 1 , \tag{6.8.4}$$

where

$$\gamma^2 = 1 - \frac{1}{c^2\lambda^2}, \qquad \text{and} \qquad \alpha^2 = \lambda^2(\omega^2\gamma^2 - f^2) = \alpha^2\left(\omega, \frac{\omega}{c}\right).$$

Note that the only difference of (6.4.7) and (6.8.4) is the appearance of the factor γ^2 that compares the phase velocity c of the wind pattern with the phase speed λ^{-1}. For $\gamma^2 < 0$, i.e., $c^2 < \lambda^{-2}$, $G * 1$ is coastally trapped for all frequencies ω since α is then imaginary. For $\gamma^2 > 0$ or $c^2 > \lambda^{-2}$, $G * 1$ is coastally trapped in the subinertial frequency range, i.e., $\omega < \frac{f}{\gamma}$, whereas für $\omega > \frac{f}{\gamma}$ the convolution $G * 1$ varies over the whole cross section and possess poles at the enhanced odd Poincaré frequencies $\frac{\omega_{2m+1}(0)}{\gamma}$. Thus the response pattern changes drastically if γ^2 changes its sign.

The Eulerian spectra of (6.8.3a) to (6.8.3c) can easily be obtained by k-integration. Only for positive γ^2 peaks appear at the odd Poincaré frequencies $\omega_{2m+1}(0)$ somewhat enhanced by the factor $1/\gamma$. In the subinertial range and, moreover, for negative γ^2 the frequency spectra are smooth.

Next we transform the solution set (6.8.3) with (6.8.4) into the $y - t$ space. Using for $G * 1$ the sum representation according to (6.8.4) we have for $\gamma^2 > 0$ the ω's to replace by $\omega + i\varepsilon$ ($\varepsilon \to +0$) in order to be consistent with causality. Then we find from the residue theorem

$$u(x, y, t) = \frac{v_*^2}{h} \frac{\theta\left(t - \frac{x}{c}\right)}{f} \left[1 - \frac{\cosh \frac{L}{2R}\left(1 - \frac{2y}{L}\right) + \frac{1}{c\lambda} \sinh \frac{L}{2R}\left(1 - \frac{2y}{L}\right)}{\cosh \frac{L}{2R}} \right.$$

$$\left. + \frac{4f}{L} \sum_{m=0}^{\infty} \left(\frac{\cos \varkappa_{m'} y}{c\lambda^2} - \frac{f \sin \varkappa_{m'} y}{\varkappa_{m'}} \right) \frac{1}{\omega_{m'}^2(0)} \cos \frac{\omega_{m'}(0)}{\gamma} \left(t - \frac{x}{c}\right) \right] \tag{6.8.5 a}$$

$$v(x, y, t) = \frac{v_*^2}{h} \theta\left(t - \frac{x}{c}\right) \frac{4}{L} \sum_{m=0}^{\infty} \frac{\gamma \sin \varkappa_{m'} y}{\varkappa_{m'} \omega_{m'}(0)} \sin \frac{\omega_{m'}(0)}{\gamma} \left(t - \frac{x}{c}\right) \tag{6.8.5 b}$$

$$p(x, y, t) = \frac{v_*^2}{h} \theta\left(t - \frac{x}{c}\right) \left\{ \frac{R}{c\lambda} \left[1 - \frac{\cosh \frac{L}{2R}\left(1 - \frac{2y}{L}\right)}{\cosh \frac{L}{2R}} \right] + R \frac{\sinh \frac{L}{2R}\left(1 - \frac{2y}{L}\right)}{\cosh \frac{L}{2R}} \right.$$

$$\left. + \frac{4}{L\lambda^2} \sum_{m=0}^{\infty} \left[\frac{\cos \varkappa_{m'} y}{\omega_{m'}^2(0)} - \frac{f \sin \varkappa_{m'} y}{c \varkappa_{m'} \omega_{m'}^2(0)} \right] \cos \frac{\omega_{m'}(0)}{\gamma} \left(t - \frac{x}{c}\right) \right\} \tag{6.8.5 c}$$

with $m' = 2m + 1$.

The response governed by (6.8.5a) to (6.8.5c) propagates with the phase speed c of the wind pattern and vanishes ahead of the frontal edge at $x = ct$. Behind the front, the response consists of an Ekman current with a geostrophically balanced coastal rectification, see (6.8.5a), (6.8.5c). Moreover, Poincaré oscillations with the odd frequencies $\omega_{2m+1}(0)/\gamma$ occur. For $c \to \infty$ the response to a large scale Y-forcing follows, which has been calculated in the foregoing section, compare (6.7.4).

Next we study the case $\gamma^2 < 0$, where the phase of the response moves faster than the phase of the forcing. In this case the integrand of (6.8.4) has two complex poles at

$$\omega = \pm i \frac{\omega_{2m+1}(0)}{\gamma}.$$

Thus the inverse Fourier transformation of (6.8.3a) to (6.8.3c) with (6.8.4) yields finite contributions as well for $x > ct$ as for $x < ct$ since poles in both the upper and lower half plane of the complex ω-plane exist. The results are

$$u(x, y, t) = \frac{v_*^2}{h} \left\{ \frac{\theta\left(t - \frac{x}{c}\right)}{f} \left[1 - \frac{\cos \frac{L}{2R}\left(1 - \frac{2y}{L}\right) + \frac{1}{c\lambda} \sinh \frac{L}{2R}\left(1 - \frac{2y}{L}\right)}{\cosh \frac{L}{2R}} \right] \right.$$

$$+ \operatorname{sgn}\left(t - \frac{x}{c}\right) \frac{2}{L} \sum_{m=0}^{\infty} \left(- \frac{f \sin \varkappa_{m'} y}{\varkappa_{m'}} + \frac{\cos \varkappa_{m'} y}{c\lambda^2}\right) \frac{e^{-\frac{\omega_{m'}(0)}{|\gamma|}\left|t - \frac{x}{c}\right|}}{\omega_{m'}^2(0)} \Bigg\} \tag{6.8.6 a}$$

$$v(x, y, t) = -\frac{v_*^2}{h} \frac{2}{L} \sum_{m=0}^{\infty} \frac{|\gamma| \sin \varkappa_{m'} y}{\varkappa_{m'} \omega_{m'}^2(0)} e^{-\frac{\omega_{m'}(0)}{|\gamma|}\left|t - \frac{x}{c}\right|} \tag{6.8.6b}$$

$$p(x, y, t) = \frac{v_*^2}{h} \left\{ \theta\left(t - \frac{x}{c}\right) \left[\frac{R}{c\lambda} \left(1 - \frac{\cosh \frac{L}{2R}\left(1 - \frac{2y}{L}\right)}{\cosh \frac{L}{2R}} \right) + \frac{R \sinh \frac{L}{2R}\left(1 - \frac{2y}{L}\right)}{\cosh \frac{L}{2R}} \right] \right.$$

$$\left. - \operatorname{sgn}\left(t - \frac{x}{c}\right) \frac{2}{L} \sum_{m=0}^{\infty} \left(\frac{f}{c\lambda^2} \frac{\sin \varkappa_{m'} y}{\varkappa_{m'}} - \frac{1}{\lambda^2} \cos \varkappa_{m'} y \right) \frac{1}{\omega_{m'}^2(0)} e^{-\frac{\omega_{m'}(0)}{|\gamma|}\left|t - \frac{x}{c}\right|} \right\}$$

(6.8.6c)

Thus in the case of a slowly moving front instead of Poincaré oscillations there occurs a signal trapped at the frontal edge which decays exponentially with an e-folding scale $c|\gamma|/\omega_{2m1}(0)$, and which propagates as well ahead as behind the front.

It appears that the stationary portion of the response in the wake of the front is in both cases the same. The adaption of the response to the geostrophic regime is, however, quite different. In particular, only in the case of a fast moving front inertial motion in terms of the Poincaré oscillations occur.

If we assume that c lies between the barotropic and the baroclinic phase speed, i.e., $\lambda_0^{-1} > c > \lambda_1^{-1}$, as we did in Section 3.11., it follows that the barotropic portion of the response is governed by the set (6.8.6) while the baroclinic part is given by (6.8.5).

6.9. ALONGSHORE FORCING OVER A HALF PLANE

In this section we wish to analyze the response of the channel model to an alongshore wind blowing over the half plane $x > 0$, i.e., we consider

$$X = \frac{u_*^2}{H_{\text{mix}}} \theta(z + H_{\text{mix}})\, \theta(t)\, \theta(x)\,, \qquad Y = 0\,, \tag{6.9.1}$$

see Fig. 6.9.1. Fourier transformation concerning t and x and expansion in vertical eigenfunctions yield

$$X_m(\omega, k) = \frac{v_*^2}{h_m} \frac{\mathrm{i}}{\omega + \mathrm{i}\varepsilon} \frac{(-\mathrm{i})}{k - \mathrm{i}\varepsilon}\,. \tag{6.9.2}$$

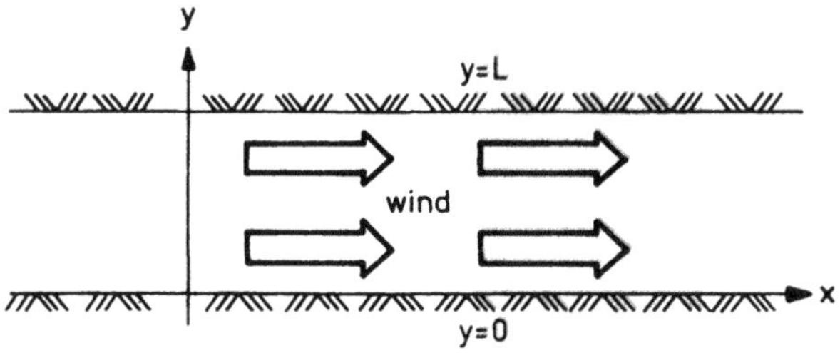

Fig. 6.9.1
Sketch of the wind pattern according to (6.9.1)

The general solution of this problem is formally given by the set (5.9.3) where, however, in the present case the expressions $G * 1$ occurring in (5.9.3) are given by (6.4.7).

The formal solution has poles in the complex ω-plane in particular at $\omega = 0, \omega = \pm k/\lambda$, and at the Poincaré frequencies $\omega_n(k)$ as indicated in Fig. 6.2.1. Suppressing the vertical mode index m and separating the Kelvin poles and the Poincaré poles from (5.9.3) and (6.4.7) we obtain

$$u(\omega, k, y) = \frac{u_*^2}{h}\left\{+\frac{i\lambda^2}{k(\omega^2\lambda^2 - k^2)} + \frac{4}{L\lambda^2}\sum_{n=0}^{\infty}\frac{\sin \varkappa_{n'}y}{\varkappa_{n'}\omega_{n'}^2(0)}\frac{1}{R^2}\left[\frac{-i\lambda^2}{k(\omega^2\lambda^2 - k^2)} + \frac{i}{k(\omega^2 - \omega_{n'}^2(k))}\right] + \frac{4}{L\lambda^2}\sum_{n=0}^{\infty}\frac{\cos \varkappa_{n'}y}{\omega_{n'}^2(0)} f\left[\frac{i\lambda^2}{\omega(\omega^2\lambda^2 - k^2)} - \frac{i}{\omega(\omega^2 - \omega_{n'}^2(k))}\right]\right\} \tag{6.9.3a}$$

$$v(\omega, k, y) = \frac{u_*^2}{h}\frac{\lambda}{R}\frac{4}{L\lambda^2}\sum_{n=0}^{\infty}\frac{\sin \varkappa_{n'}y}{\varkappa_{n'}}\frac{1}{k\omega(\omega^2 - \omega_{n'}^2(k))} \tag{6.9.3b}$$

$$p(\omega, k, y) = \frac{u_*^2}{h}\left\{\frac{i}{\omega(\omega^2\lambda^2 - k^2)} + \frac{\lambda}{R}\frac{4}{L\lambda^2}\sum_{n=0}^{\infty}\frac{\cos \varkappa_{n'}y}{\omega_{n'}^2(0)}\left[\frac{i}{k(\omega^2\lambda^2 - k^2)} - \frac{i}{k^2\lambda^2(\omega^2 - \omega_{n'}^2(k))}\right] + \frac{4}{L\lambda^2}\sum_{n=0}^{\infty}\frac{\sin \varkappa_{n'}y}{\varkappa_{n'}\omega_{n'}^2(0)}\cdot\frac{i}{R^2}\left[\frac{-1}{\omega(\omega^2\lambda^2 - k^2)} + \frac{1}{\omega\lambda^2(\omega^2 - \omega_{n'}^2(k))}\right]\right\} \tag{6.9.3c}$$

with $n' = 2n + 1$. Note that

$$\omega_{n'}^2(0) = f^2 + \frac{\varkappa_{n'}^2}{\lambda^2} = f^2(1 + R^2\varkappa_{n'}^2)$$

and

$$\omega_{n'}^2(k) = \omega_{n'}^2(0) + \frac{k^2}{\lambda^2}.$$

From (6.4.7) we easily verify that

$$\frac{4}{L\lambda^2}\sum_{n=0}^{\infty}\frac{\sin \varkappa_{n'} y}{\varkappa_{n'}\omega_{n'}^2(0)} = R^2 - R^2\,\frac{\cosh \dfrac{L}{2R}\left(1-\dfrac{2y}{L}\right)}{\cosh \dfrac{L}{2R}}. \tag{6.9.4a}$$

Differentiating this we get

$$\frac{4}{L\lambda^2}\sum_{n=0}^{\infty}\frac{\cos \varkappa_{n'} y}{\omega_{n'}^2(0)} = R\,\frac{\sinh \dfrac{L}{2R}\left(1-\dfrac{2y}{L}\right)}{\cosh \dfrac{L}{2R}}. \tag{6.9.4b}$$

With the aid of these expressions we may rewrite (6.9.3a) and (6.9.3c) to obtain

$$\begin{aligned} u = \frac{u_*^2}{h}\Bigg\{ &\frac{1}{2\cosh\dfrac{L}{2R}}\left[\frac{e^{\frac{L}{2R}-\frac{y}{R}}}{\omega^2}\left(\frac{i}{k}-\frac{i}{k-\omega\lambda}\right)+\frac{e^{-\frac{L}{2R}+\frac{y}{R}}}{\omega^2}\left(\frac{i}{k}-\frac{i}{k+\omega\lambda}\right)\right] \\ &+\frac{4}{L\lambda^2}\sum_{n=0}^{\infty}\left[\frac{\sin\varkappa_{n'}y}{\varkappa_{n'}\omega_{n'}^2(0)}\cdot\frac{i}{R^2k(\omega^2-\omega_{n'}^2(k))}-\frac{if\cos\varkappa_{n'}y}{\omega_{n'}^2(0)\,\omega(\omega^2-\omega_{n'}^2(k))}\right]\Bigg\} \end{aligned} \tag{6.9.5}$$

$$\begin{aligned} p = \frac{u_*^2}{h}\Bigg\{ &\frac{1}{2\cosh\dfrac{L}{2R}}\left[\frac{e^{\frac{L}{2R}-\frac{y}{R}}}{\omega^2\lambda}\left(\frac{i}{k}-\frac{i}{k-\omega\lambda}\right)-\frac{e^{-\frac{L}{2R}+\frac{y}{R}}}{\omega^2\lambda}\left(\frac{i}{k}-\frac{i}{k+\omega\lambda}\right)\right] \\ &+\frac{4}{L\lambda^2}\sum_{n=0}^{\infty}\left[-\frac{if\cos\varkappa_{n'}y}{k\omega_{n'}^2(0)\,(\omega^2-\omega_{n'}^2(k))}+\frac{if^2\sin\varkappa_{n'}y}{\varkappa_{n'}\omega_{n'}^2(0)\,\omega(\omega^2-\omega_{n'}^2(k))}\right]\Bigg\} \end{aligned} \tag{6.9.6}$$

with $n' = 2n + 1$.

Next we estimate tne Fourier inversion with respect to k. Taking (B18), (B20), and (B21) into account and using

$$\int_{-\infty}^{+\infty}\frac{(-i)\,e^{ikx}}{(k\pm\lambda(\omega+i\varepsilon))}\frac{dk}{2\pi} = \pm\theta(\pm x)\,e^{\pm i\omega\lambda x} \tag{6.9.7}$$

we arrive at

$$u(x,y,\omega) = \frac{u_*^2}{h}\Bigg\{\frac{1}{2\cosh\dfrac{L}{2R}}\left[\frac{e^{\frac{L}{2R}-\frac{y}{R}}}{\omega^2}(-\theta(x)+\theta(x)\,e^{i\omega\lambda x})-\frac{e^{-\frac{L}{2R}+\frac{y}{R}}}{\omega^2}(\theta(x)+\theta(-x)\,e^{-i\omega\lambda x})\right]$$

$$+\frac{4}{L}\sum_{n=0}^{\infty}\left[\frac{\sin \varkappa_{n'} y}{\varkappa_{n'}\omega_{n'}^2(0)\,R^2}\left(\frac{\theta(x)}{\lambda^2(\omega_{n'}^2(0)-\omega^2)}+\frac{\operatorname{sgn}(x)}{2\lambda^2}\cdot\frac{e^{i\sqrt{\omega^2-\omega_{n'}^2(0)}\,\lambda|x|}}{\omega^2-\omega_{n'}^2(0)}\right)\right.$$

$$\left.\left.-\frac{f\cos \varkappa_{n'} y\, e^{i\sqrt{\omega^2-\omega_{n'}^2(0)}\,\lambda|x|}}{2\lambda\omega_{n'}^2(0)\,\omega\sqrt{\omega^2-\omega_{n'}^2(0)}}\right]\right\} \tag{6.9.8 a}$$

$$v(x, y, \omega) = \frac{u_*^2}{h}\,\frac{i f\cdot 4}{L\omega}\sum_{n=0}^{\infty}\frac{\sin \varkappa_{n'} y}{\varkappa_{n'}(\omega^2-\omega_{n'}^2(0))}\left(\theta(x)-\frac{\operatorname{sgn}(x)}{2}\,e^{i\sqrt{\omega^2-\omega_{n'}^2(0)}\,\lambda|x|}\right) \tag{6.9.8 b}$$

$$p(x, y, \omega) = \frac{u_*^2}{h}\left\{\frac{1}{2\cosh\frac{L}{2R}}\left[\frac{e^{\frac{L}{2R}-\frac{y}{R}}}{\omega^2\lambda}\,\theta(x)\,(e^{i\omega\lambda x}-1)+\frac{e^{-\frac{L}{2R}+\frac{y}{R}}}{\omega^2\lambda}\,(\theta(x)+\theta(-x)\,e^{-i\omega\lambda x})\right]\right.$$

$$+\frac{4}{L}\sum_{n=0}^{\infty}\left[\frac{f\cos \varkappa_{n'} y}{\omega_{n'}^2(0)\,\lambda^2(\omega^2-\omega_{n'}^2(0))}\left(\theta(x)-\frac{\operatorname{sgn}(x)}{2}\,e^{i\sqrt{\omega^2-\omega_{n'}^2(0)}\,\lambda|x|}\right)\right.$$

$$\left.\left.-\frac{f^2\sin \varkappa_{n'} y}{\varkappa_{n'}\omega_{n'}^2(0)\,\omega}\cdot\frac{e^{i\sqrt{\omega^2-\omega_{n'}^2(0)}\,\lambda|x|}}{2\lambda\sqrt{\omega^2-\omega_{n'}^2(0)}}\right]\right\} \tag{6.9.8c}$$

with $n' = 2n + 1$.

These expressions may be used to discuss the Eulerian frequency spectra of the dynamical fields. Note the occurrence of peaks at the frequencies $\omega_{n'}(0)$, i.e., at those frequencies which correspond to zero group velocity of Poincaré waves.

It remains the inverse Fourier transformation of the set (6.9.8) concerning ω. The terms which involve the exponential with the argument $\sqrt{\omega^2-\omega_{n'}^2(0)}$ can be expressed in terms of the convolution integrals $C^{(1)}$ and $C^{(2)}$ which have been discussed in Sections 5.5. and 5.6. The inertial frequency f occurring in (5.6.1) and (5.6.2) must, however, be replaced by $\omega_{n'}(0)$. Then we obtain

$$u(x, y, t) = \frac{u_*^2}{h}\left\{\frac{1}{2\cosh\frac{L}{2R}}\left[e^{\frac{L}{2R}-\frac{y}{R}}\,\theta(x)\,(t\,\theta(t)-(t-\lambda x)\,\theta(t-\lambda x))\right.\right.$$

$$\left.+\,e^{-\frac{L}{2R}+\frac{y}{R}}\,(\theta(x)\,t\,\theta(t)+(t+\lambda x)\,\theta(-x)\,\theta(t+\lambda x))\right]$$

$$+\frac{4}{L}\sum_{n=0}^{\infty}\left[\frac{\sin \varkappa_{n'} y}{\varkappa_{n'}\omega_{n'}^2(0)}\,f^2\left(\theta(x)\,\theta(t)\,\frac{\sin(\omega_{n'}(0)\,t)}{\omega_{n'}(0)}-\frac{\operatorname{sgn}(x)}{2\omega_{n'}^2(0)}\,C_{tt}^{(2)}(|x|, t)\right)\right.$$

$$\left.\left.+\frac{f\cos \varkappa_{n'} y}{2\lambda\omega_{n'}^2(0)}\,C_t^{(1)}(|x|, t)\right]\right\} \tag{6.9.9 a}$$

$$v(x, y, t) = \frac{u_*^2}{h}\left\{-\frac{\theta(x)\,\theta(t)}{f}\left[1-\frac{\cosh\frac{L}{2R}\left(1-\frac{2y}{L}\right)}{\cosh\frac{L}{2R}}\right]\right.$$

$$\left.+\frac{4f}{L}\sum_{n=0}^{\infty}\frac{\sin \varkappa_{n'}y}{\varkappa_{n'}\omega_{n'}^2(0)}\left[\theta(x)\,\theta(t)\cos\omega_{n'}t+\frac{\operatorname{sgn}(x)}{2}C_t^{(2)}(|x|, t)\right]\right\} \quad (6.9.9\text{b})$$

$$p(x, y, t) = \frac{u_*^2}{h}\left\{\frac{1}{2\lambda\cosh\frac{L}{2R}}\left[\theta(x)\,(t\,\theta(t)-(t-\lambda x)\,\theta(t-\lambda x))\,e^{\frac{L}{2R}-\frac{y}{R}}\right.\right.$$

$$\left.-(t\,\theta(t)\,\theta(x)+\theta(-x)\,(t+\lambda x)\,\theta(t+\lambda x))\,e^{-\frac{L}{2R}+\frac{y}{R}}\right]$$

$$+\frac{4f}{L}\sum_{n=0}^{\infty}\left[\frac{\cos\varkappa_{n'}y}{\lambda^2\omega_{n'}^3(0)}\left(-\theta(x)\,\theta(t)\sin\omega_{n'}t+\frac{\operatorname{sgn}(x)}{2\omega_{n'}(0)}C_{tt}^{(2)}(|x|, t)\right)\right.$$

$$\left.\left.-\frac{f\sin\varkappa_{n'}y}{2\lambda\varkappa_{n'}\omega_{n'}^2(0)}C_t^{(1)}(|x|, t)\right]\right\} \quad (6.9.9\text{c})$$

with $n' = 2n + 1$. Let us indroduce the abbreviation

$$\begin{pmatrix} u \\ v \\ p \end{pmatrix} = \begin{pmatrix} u_{\mathrm{I}} \\ v_{\mathrm{I}} \\ p_{\mathrm{I}} \end{pmatrix} + \begin{pmatrix} u_{\mathrm{II}} \\ v_{\mathrm{II}} \\ p_{\mathrm{II}} \end{pmatrix} \quad (6.9.10)$$

where the terms with the index II refer to the Poincaré series occurring rightmostly in the set (6.9.9) while the terms with the index I denote the remaining parts of the solution.

First we discuss the Poincaré series. Using the asymptotic representations of $C^{(1)}$ and $C^{(2)}$ according to (5.5.12), (5.6.16), and (5.6.17) we get for $tf \gg 1$

$$u_{\mathrm{II}}(x, y, t) = \frac{u_*^2}{h}\frac{4}{L}\sum_{n=0}^{\infty}\left[\frac{f}{2}\frac{\sin\varkappa_{n'}y}{\varkappa_{n'}\omega_{n'}^2}\left(\frac{f\sin\omega_{n'}t}{\omega_{n'}}+\frac{x}{R}\left(\frac{2}{\pi\omega_{n'}t}\right)^{1/2}\sin\left(\omega_{n'}t+\frac{\pi}{4}\right)\right)\right.$$

$$\left.+\frac{\cos\varkappa_{n'}y}{2\lambda\omega_{n'}^3}\left(e^{-|x|\lambda\omega_{n'}}-\left(\frac{2}{\pi\omega_{n'}t}\right)^{1/2}\cos\left(\omega_{n'}t+\frac{\pi}{4}\right)\right)\right] \quad (6.9.11\text{a})$$

$$v_{\mathrm{II}}(x, y, t) = \frac{u_*^2}{h}\frac{4}{L}\sum_{n=0}^{\infty}\frac{\sin\varkappa_{n'}y}{2\varkappa_{n'}}\left[\frac{f\cos\omega_{n'}t}{\omega_{n'}^2}+\frac{x}{R}\left(\frac{2}{\pi\omega_{n'}t}\right)^{1/2}\frac{\cos\left(\omega_{n'}t+\frac{\pi}{4}\right)}{\omega_{n'}}\right.$$

$$\left.+\operatorname{sgn}(x)\frac{f}{\omega_{n'}^2}e^{-\omega_{n'}\lambda|x|}\right], \quad (6.9.11\text{b})$$

$$p_{\text{II}}(x, y, t) = -\frac{u_*^2}{h}\frac{4}{L}\sum_{n=0}^{\infty}\left[\frac{\cos \varkappa_{n'}y}{2\lambda^2\omega_{n'}^2}\left(\frac{f}{\omega_{n'}}\sin \omega_{n'}t + \frac{x}{R}\left(\frac{2}{\pi\omega_{n'}t}\right)^{1/2}\sin\left(\omega_{n'}t + \frac{\pi}{4}\right)\right)\right.$$
$$\left. + \frac{f^2 \sin \varkappa_{n'}y}{2\lambda\varkappa_{n'}\omega_{n'}^3}\left(e^{-\omega_{n'}\lambda|x|} - \left(\frac{2}{\pi\omega_{n'}t}\right)^{1/2}\cos\left(\omega_{n'}t + \frac{\pi}{4}\right)\right)\right] \qquad (6.9.11\,c)$$

where $n' = n + 1$.

With (6.9.9) to (6.9.11) we may describe the response in the inertial frequency range, i.e., for processes with frequencies exceeding the inertial frequency f. Immediatly after the onset of the forcing Poincaré cross-oscillations occur within the whole forcing area, i.e., $x > 0$. Moreover, from the line $x = 0$ Poincaré waves propagate in both negative and positive x-direction, governed by the convolutions $C^{(1)}$ and $C^{(2)}$. In the wake of these Poincaré waves there remain Poincaré oscillations which are no longer confined to the area $x > 0$, but having only half the amplitude of the initial cross-oscillations.

The alongshore propagating Poincaré waves decay algebraically, however, the cross-oscillations remain. This appears to be an important difference to the case of the semi-infinite ocean, where for large times the inertial motion has diminished completely.

Furthermore, in the wake of the Poincaré waves a current has developed close to the wind edge which is geostrophically balanced by the pressure, compare the third terms of (6.9.11 a, b, c). This portion of the solution rectifies the response in the vicinity of the wind discontinuity line $x = 0$.

The remaining parts of the response, in particular, u_{I} and p_{I} describes the generation of two Kelvin waves at the wind edge which are trapped at the coasts, i.e., at $y = 0$ and $y = L$, and which both propagate with the coast to the right.

In particular, for $R \ll L$ we may rewrite the corresponding terms of (6.9.9a, c) to obtain

$$u_{\text{I}}(x, y, t) = \frac{u_*^2}{h}\left[\theta(x)\,(t\,\theta(t) - (t - \lambda x)\,\theta(t - \lambda x))\,e^{-\frac{y}{R}}\right.$$
$$\left. + (t\,\theta(t)\,\theta(x) + \theta(-x)\,(t + \lambda x)\,\theta(t + \lambda x))\,e^{-\frac{L-y}{R}}\right] \qquad (6.9.12\,a)$$

$$p_{\text{I}}(x, y, t) = \frac{u_*^2}{\lambda h}\left[\theta(x)\,(t\,\theta(t) - (t - \lambda x)\,\theta(t - \lambda x))\,e^{-\frac{y}{R}}\right.$$
$$\left. - (t\,\theta(t)\,\theta(x) + \theta(-x)\,(t + \lambda x)\,\theta(t + \lambda x))\,e^{-\frac{L-y}{R}}\right]. \qquad (6.9.12\,b)$$

At the coast $y = 0$ the Kelvin wave propagates into the forcing area where it arrests the coastal jet, while the Kelvin waves at the coast $y = L$ radiate outside the forcing area. Thus the response is quite different at the two coasts, see Fig. 6.9.2.

At $y = 0$ the coastal jet appears to be arrested after the passage of the Kelvin fronts. Behind the front a pressure gradient develops which, close to the coast, balances the wind force. Moreover, there develops a coastal undercurrent in a manner outlined in Section 5.10.

At the opposite coast, however, the coastal jet develops as an accelerating downwind current which is not affected by the Kelvin waves.

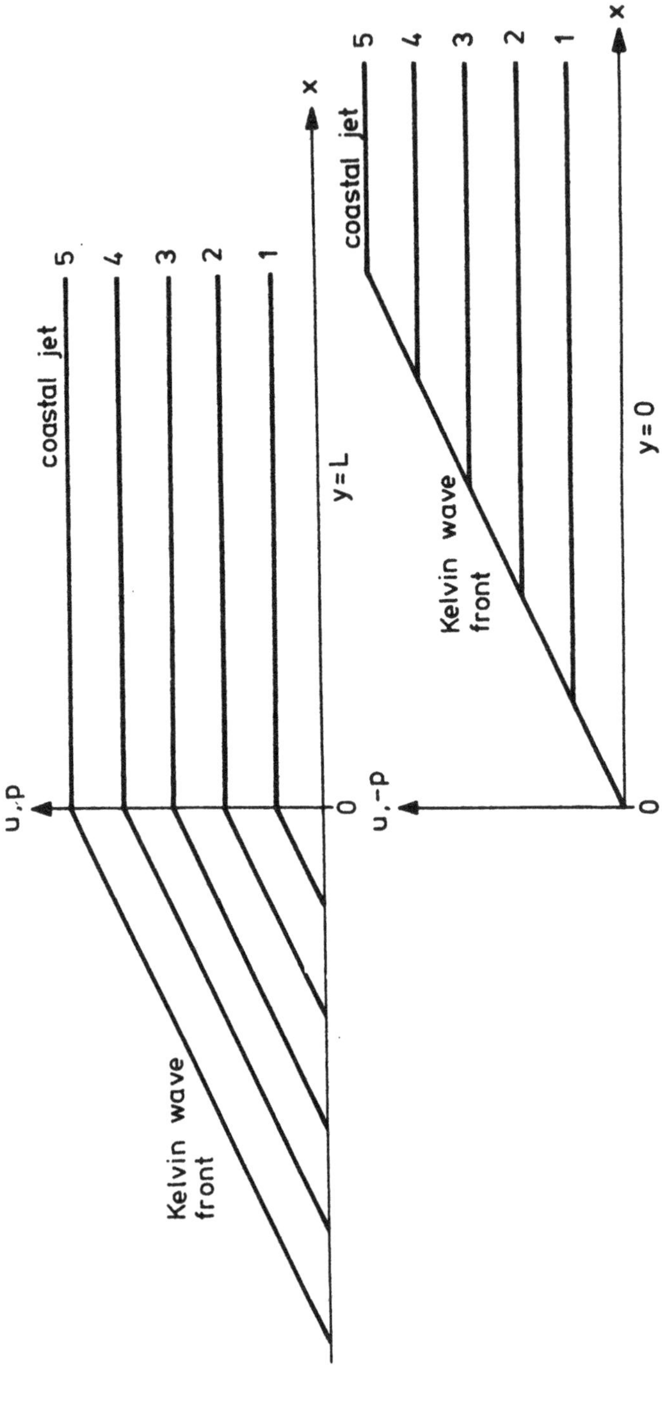

Fig. 6.9.2
Sketch of the response according to (6.9.12a, b). At the shore, $y = 0$, the Kelvin wave arrests the coastal jet, while at the opposite coast the coastal jet develops undisturbed and a Kelvin wave propagates outside the forcing region
Note that the behaviour at the two coasts corresponds to that shown in Figs. 5.9.4 and 5.9.5.

The contributions to v_I, see (6.9.9b) and (6.9.10), describe the usual Ekman current plus a coastal rectification.

Finally we derive the subinertial part of the modal vertical velocity defined by $w = -p_t$. From (6.9.12b) it follows

$$w_I(x, y, t) = -\frac{u_*^2}{h\lambda}\left\{\theta(x)\left[\theta(t) - \theta(t - \lambda x)\right] e^{-\frac{y}{R}} - \left[\theta(t)\,\theta(x) + \theta(-x)\,\theta(t + \lambda x)\right] e^{-\frac{L-y}{R}}\right\}. \tag{6.9.12c}$$

Thus, at the shore $y = 0$ downwelling associated with the coastal jet stops with the advent of the Kelvin wave front. At the opposite coast, i.e., at $y = L$, the Kelvin wave propagates upwelling outside the forcing region. Moreover, within the wind area the upwelling remains permanent at that coast.

In summary, the solution set (6.9.12) reveals a remarkable asymmetry of the response at the opposite shores of the channel.

6.10. FORCING THROUGH AN EXTERNAL BAROCLINIC PRESSURE GRADIENT

In this section we introduce an additional external force in terms of a baroclinic pressure gradient which acts on the channel model and we wish to study the associated response patterns. Such a case may be useful to model the dynamics of a channel-like connection between a large oceanic basin and an adjacent marginal sea where a baroclinic longitudinal pressure gradient may be sustained by the balance of freshwater outflow and salt-water inflow. As an example of such a system we mention the transition area between the North Sea and the Baltic Sea.

Let us consider an stationary external baroclinic pressure gradient $P_x(z)$ which varies linearly with depth, i.e., we put

$$X = -P_x(z) = -cz \tag{6.10.1}$$

where c is a constant. Note that z varies between 0 and $-H$. After Fourier transformation with respect to t and x, and expansion in vertical eigenfunctions we obtain

$$X_n = 2\pi\,\delta(\omega)\cdot 2\pi\,\delta(k)\,ca_n\,. \tag{6.10.2}$$

We use, just for simplicity, the eigenfunctions $F_n(z)$ which correspond to constant BVF, see Subsection 1.9.2. Then we get

$$a_0 = \frac{H^2}{2} \tag{6.10.3a}$$

$$a_n = (2)^{1/2}\frac{H^2}{(n\pi)^2}\left[(-1)^n - 1\right], \qquad n \geqq 1\,. \tag{6.10.3b}$$

Inserting (6.10.3) into the general formal solution (5.2.6) it follows

$$u_n = 2\pi\,\delta(\omega)\cdot 2\pi\,\delta(k)\,ca_n\frac{i}{\bar{\omega}}(1 + G * 1) \tag{6.10.4a}$$

$$v_n = 2\pi\,\delta(\omega)\cdot 2\pi\,\delta(k)\,ca_n\lambda_n^2 fG * 1 \tag{6.10.4b}$$

$$p_n = -2\pi\,\delta(\omega)\,2\pi\,\delta(k)\,ca_n\,\frac{if}{\bar{\omega}}\,G * 1\,. \tag{6.10.4c}$$

Here the dissipation parameter r (Rayleigh friction, Newtonian cooling) was introduced. That means all ω's were replaced by $\bar{\omega} = \omega + ir$ with $r \ll f$, except ω's which are the explicite variable of the forcing function, i.e., in the present case the ω in the delta function.

According to (6.4.7) we find

$$\delta(\omega)\,\delta(k)\,G * 1 = \delta(\omega)\,\delta(k)\,(-R_n^2)\left(1 - \frac{\cosh \frac{L}{2R_n}\left(1 - \frac{2y}{L}\right)}{\sinh \frac{L}{2R_n}}\right) \tag{6.10.5}$$

where

$$\alpha^2(ir, 0) = -\frac{1}{R^2}\left(1 + \frac{r^2}{f^2}\right) \approx -\frac{1}{R^2}$$

was used. The Fourier inversion of the set (6.10.4) is trivial because of the involved delta functions and so we get

$$u_n(y) = \frac{ca_n}{r}\,\frac{\cosh \frac{L}{2R_n}\left(1 - \frac{2y}{L}\right)}{\cosh\left(\frac{L}{2R_n}\right)} \tag{6.10.6a}$$

$$v_n(y) = -\frac{ca_n}{f}\left(1 - \frac{\cosh \frac{L}{2R_n}\left(1 - \frac{2y}{L}\right)}{\cosh\left(\frac{L}{2R_n}\right)}\right) \tag{6.10.6b}$$

$$p_n(y) = \frac{ca_n}{r\lambda_n}\,\frac{\sinh\left(\frac{L}{2R_n}\right)\left(1 - \frac{2y}{L}\right)}{\cosh\left(\frac{L}{2R_n}\right)}\,. \tag{6.10.6c}$$

The full solution follows after summation over all eigenfunctions $F_n(z)$, i.e.,

$$(u(y, z), v(y, z), p(y, z)) = \sum_{n=0}^{\infty} (u_n(y), v_n(y), p_n(y))\,F_n(z) \tag{6.10.7a}$$

and, moreover,

$$w(y, z) = -\frac{r}{N^2}\sum_{n=0}^{\infty} p_n(y)\,\frac{d}{dz}\,F_n(z)\,. \tag{6.10.7b}$$

In Fig. 6.10.2 the alongshore current u is plotted, where the numerical value $c = 7 \cdot 10^{-7}\,s^{-2}$ was adopted. The other parameters involved have the same values as in Section 6.5., see page 196. From Fig. 6.10.1 it can be seen that the alongshore current responses to

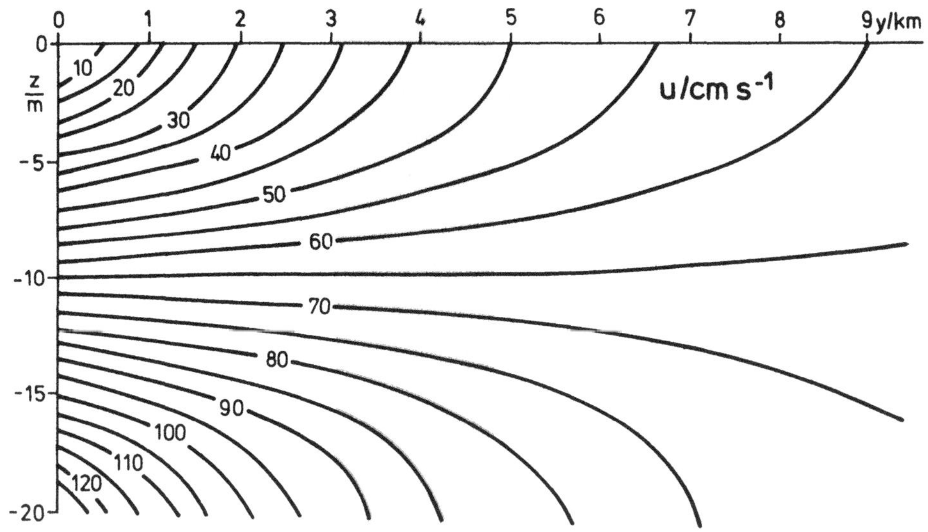

Fig. 6.10.1
Coastal jet in response to the baroclinic pressure gradient according to (6.10.7a)
Shown is the structure of the alongshore current which assumes its maximum values close to the bottom.

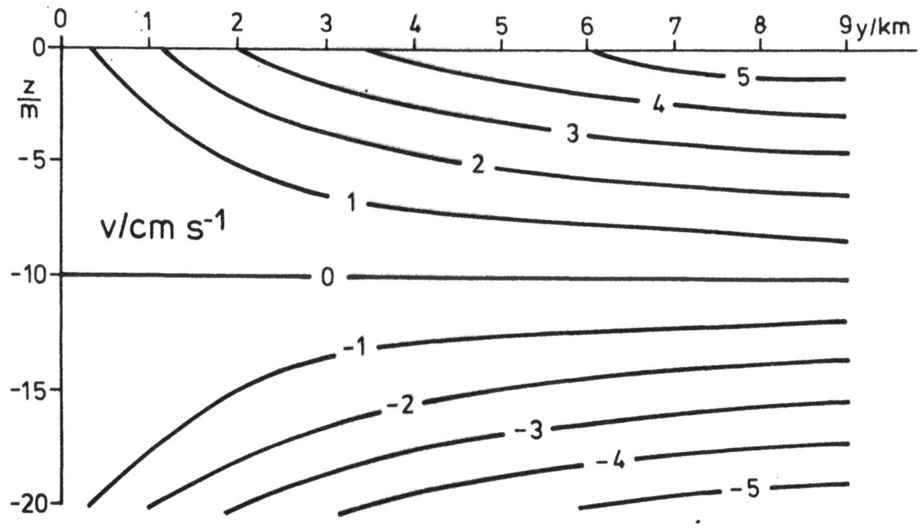

Fig. 6.10.2
The structure of the cross current associated with the coastal jet

the pressure gradient (6.10.1) with a coastal jet which, contrary to the wind driven coastal jets, has its maximum values near bottom. Fig. 6.10.2 shows the cross-shore current component $v(y, z)$, and in Fig. 6.10.3 the vertical velocity was plotted. Thus, the pressure gradient generates continuously upwelling at one coast and downwelling at the opposite coast.

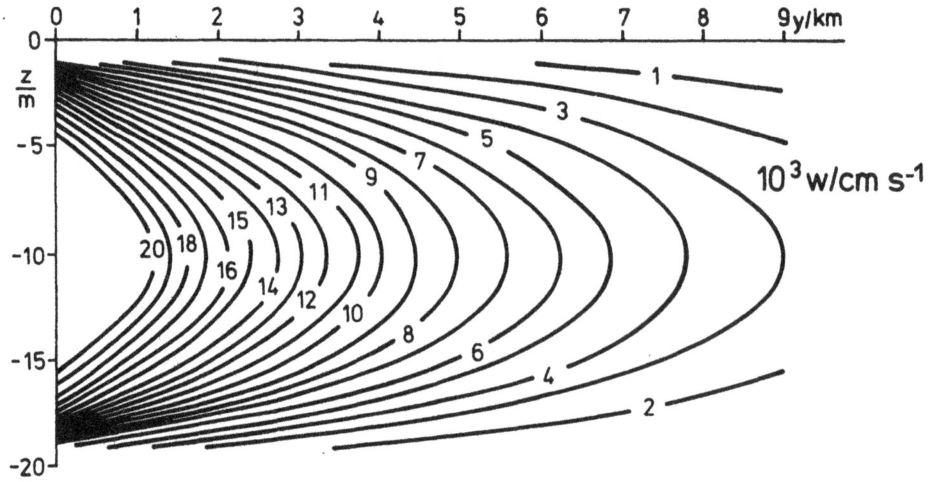

Fig. 6.10.3
The contours of the vertical velocity associated with the coastal jet

If we add an alongshore wind to the external pressure gradient we find the complete solution by adding the solution sets (6.5.1) and (6.10.6). If the wind blows in the direction of the negative pressure gradient the up- and downwelling contributions of both solutions have opposite sign and so only weak vertical velocities may result. Conversely, if the wind blows in opposite direction the up- and downwelling at the opposite coasts will increase since then the vertical velocities induced by the pressure gradient have the same sign as the wind-induced ones. Thus only one direction of the alongshore wind appears to be favourable to produce upwelling at one and downwelling at the opposite coast.

Moreover, if the wind blows in the direction of the negative pressure gradient the two coastal jet contributions form a resulting jet which affects the whole water column. If the wind blows in opposite direction, we have a downwind jet in the upperlayer and a pressure gradient induced countercurrent below, compare the Figs. 6.5.1 and 6.10.1.

CHAPTER 7

The midlatitudinal β-plane

7.1. GOVERNING EQUATIONS

In this chapter we take variation of the inertial frequency f with the meridional coordinate y into account, in particular, we use the β-plane approximation mentioned in Section 1.4.

Thus, we are dealing with the linear hydrostatic Boussinesq equations in the familar form

$$u_t - fv + p_x = X \tag{7.1.1a}$$

$$v_t + fu + p_y = Y \tag{7.1.1b}$$

$$w + \frac{p_{tz}}{N^2} = 0 \tag{7.1.1c}$$

$$u_x + v_y + w_z = 0 \tag{7.1.1d}$$

where, however, f varies with y according to $f = f_0 + \beta y$. The β-parameter has the typical value $\beta = 2 \cdot 10^{-11}$ (ms)$^{-1}$ ($\varphi \sim \pm 30°$).

First we put the basic equations in a form appropriate to the formulation of a solution theory. Let us derive an equation for the pressure p alone: Introducing the operator $\mathbf{L} = \frac{\partial^2}{\partial t^2} + f^2$ we obtain from (7.1.1a, b)

$$\mathbf{L}u = -fp_y - p_{xt} + fY + X_t \tag{7.1.2a}$$

$$\mathbf{L}v = fp_x - p_{yt} - fX + Y_t \tag{7.1.2b}$$

and from (7.1.1c, d) we get

$$u_x + v_y - \mathbf{Z}p_t = 0\,, \tag{7.1.2c}$$

where again the notation $\mathbf{Z} = \frac{\partial}{\partial z}\frac{1}{N^2}\frac{\partial}{\partial z}$ was used. Taking the relation

$$\frac{\partial}{\partial y}\mathbf{L}v = 2f\beta v + \mathbf{L}v_y \tag{7.1.3}$$

into account, and applying the operator $\mathbf{L}$ to (7.1.2c), and then using (7.1.2a, b) we arrive at

$$\Delta p_t + \mathbf{LZ}p_t - \beta p_x = f(Y_x - X_y) - \beta X + X_{xt} + Y_{yt} - 2\beta f v \tag{7.1.4}$$

with Δ being the horizontal Laplacian, $\Delta = \partial^2/\partial x^2 + \partial^2/\partial y^2$. In order to eliminate the v-term on the right-hand side of (7.1.4) we apply $\mathbf{L}$ again to this equation and use (7.1.2b). Then we obtain

$$\begin{aligned}\mathbf{L}(\Delta + \mathbf{LZ})\,p_t - \beta p_{xtt} + \beta f^2 p_x - 2\beta f p_{yt} &= \mathbf{L}[f(Y_x - X_y) - \beta X + X_{xt} + Y_{yt}]\\ &\quad + 2\beta f^2 X - 2\beta f Y_t\,.\end{aligned} \tag{7.1.5}$$

We expect that the β-effect affects only those processes which belong to the extreme subinertial domain, i.e., $\omega \ll f$. This implies that $\mathbf{L}$ can be replaced by f^2, and $p_{xtt} \ll f^2 p_x$, and, provided the x- and y-scales are comparable, $p_{yt} \ll f p_x$. Moreover, we may neglect βY_t. Then it follows

$$(\Delta + \mathbf{LZ})\,p_t + \beta p_x = f(Y_x - Y_y) + \beta X + X_{xt} + Y_{yt}\,, \tag{7.1.6}$$

where in the second term of the left-hand side $\mathbf{L}$ has been retained since then for $\beta \to 0$ (7.1.6) would yield the complete equation of the pressure on the f-plane, compare (3.2.13).

Since the variability of f is explicitly accounted for by the presence of β-terms in (7.1.6), it is a traditional approximation to consider f as constant when discussing the solutions of this equation. We note, however, that the explicit consideration of the y-dependence of f is of importance when discussing the effects of turning latitudes, see e.g. MUNK and PHILLIPS (1968), and GEISLER and DICKINSON (1972).

The horizontal currents u and v can be obtained to the same approximation from

$$u = -\frac{p_y}{f} - \frac{p_{xt}}{f^2} + \frac{Y}{f} + \frac{X_t}{f^2} \qquad v = \frac{p_x}{f} - \frac{p_{yt}}{f^2} - \frac{X}{f} + \frac{Y_t}{f^2}\,. \tag{7.1.7a), (7.1.7b}$$

In the absence of lateral boundaries and for very low frequencies we may approximate this by

$$u = \frac{1}{f}(-p_y + Y)\,, \qquad v = \frac{1}{f}(p_x - X)\,. \tag{7.1.8a, b}$$

To this approximation the currents consists of geostrophically adjusted currents plus surface layer trapped Ekman flows. The vertical velocity follows from (7.1.1c).

Dropping the time derivatives in (7.1.6) and inserting (7.1.8b) we arrive at

$$\beta v = (Y_x - X_y)\,. \tag{7.1.9}$$

We call this relation the Sverdrup balance because (7.1.9) corresponds to the classic depth averaged Sverdrup theory, see e.g. PEDLOSKY (1979).

7.2. GREEN'S FUNCTION FOR AN UNBOUNDED β-PLANE OCEAN AND ROSSBY WAVES

The first model we wish to consider is an unbounded β-plane ocean. In this case we may Fourier transform the equation (7.1.6) with repect to x, y, and t, and we obtain

$$(\mathbf{Z} - \tilde{\lambda}^2)\, p(\omega, k, \varkappa) = \frac{\mathrm{i}}{\omega(f^2 - \omega^2)} [f(\mathrm{i}kY - \mathrm{i}\varkappa X) + \beta X + \omega(kX + \varkappa Y)] \tag{7.2.1}$$

where

$$\tilde{\lambda}^2 = \left(k^2 + \varkappa^2 + \frac{k\beta}{\omega}\right) \frac{1}{f^2 - \omega^2}. \tag{7.2.2}$$

The boundary conditions are according to (3.2.14) and (3.2.15)

$$p_z = \frac{N^2}{g} p \quad \text{for} \quad z = 0 \tag{7.2.3 a}$$

and

$$p_z = 0 \quad \text{for} \quad z = -H. \tag{7.2.3 b}$$

Next we seek the corresponding Green's function $K(z, z')$ which is subject to

$$(\mathbf{Z} - \tilde{\lambda}^2)\, K(z, z') = \delta(z - z') \tag{7.2.4 a}$$

$$K_z(0, z') = -\frac{N^2}{g} K(0, z') \tag{7.2.4 b}$$

$$K_z(-H, z') = 0. \tag{7.2.4 c}$$

Let us, for simplicity, consider a constant BVF. Then the two linearly independent solutions of (7.2.4a) for $z \neq z'$ are

$$\varphi(z) = \cosh \tilde{\lambda} N z - \frac{N'}{\tilde{\lambda} g} \sinh \tilde{\lambda} N z \tag{7.2.5 a}$$

and

$$\psi(z) = \cosh \tilde{\lambda} N(z + H) \tag{7.2.5 b}$$

where φ and ψ fulfil the boundary condition (7.2.3a) and (7.2.3b), respectively. The Wronskian of φ and ψ is

$$W_r(\tilde{\lambda}) = -\tilde{\lambda} N \sinh \tilde{\lambda} N H - \frac{N^2}{g} \cosh \tilde{\lambda} N H, \tag{7.2.6}$$

and so we may construct the Green's function

$$K(z, z') = \theta(z - z')\, K^<(z, z') + \theta(z' - z)\, K^>(z, z') \tag{7.2.7 a}$$

with

$$K^<(z, z') = \frac{N^2 \varphi(z)\, \psi(z')}{W_r} = K^>(z', z). \tag{7.2.7 b}$$

As long as $\tilde{\lambda}^2$ is positive, i.e., $\omega < f$ or $\boldsymbol{k}^2 + \beta k/\omega > 0$, the Green's function and, therefore, the response of the ocean to external forcing is vertically localized. This follows from the occurrence of the hyperbolic functions in (7.2.5a, b) and (7.2.6). However, for negative $\tilde{\lambda}^2$, i.e., $\omega > f$ or $\mathbf{k}^2 + \beta k/\omega < 0$, the response oscillates sinusoidally in the vertical, and so the whole water column is affected.

Let $\tilde{\lambda}^2 < 0$ and put $\tilde{\lambda} = i\lambda$, where

$$\lambda^2 = \frac{1}{\omega^2 - f^2}\left(\mathbf{k}^2 + \frac{\beta k}{\omega}\right) \tag{7.2.8a}$$

then we have

$$\varphi(z) = \cos \lambda N z - \frac{N}{\lambda^2 g} \sin \lambda N z \tag{7.2.8b}$$

$$\psi(z) = \sin \lambda N(z + H) \tag{7.2.8c}$$

and

$$W_r(i\lambda) = \lambda N \sin \lambda NH - \frac{N^2}{g} \cos \lambda NH .$$

In this case the Green's function is formally equivalent to (3.3.5), (3.3.6). The Wronskian has zeros at

$$\tan \lambda_n NH = \frac{N}{g\lambda_n}, \qquad n = 0, 1, 2, \ldots$$

where the λ_n's are the vertical eigenvalues

$$\lambda_0 = (gH)^{-1/2}, \qquad \lambda_n = \frac{n\pi}{NH} \quad \text{for} \quad n \geqq 1 . \tag{7.2.9}$$

Thus, the zeros of the Wronskian determine the following dispersion relation

$$\lambda^2 = \lambda_n^2 = \frac{1}{\omega^2 - f^2}\left(\mathbf{k}^2 + \frac{\beta k}{\omega}\right) \tag{7.2.10}$$

or, equivalently,

$$\omega^3 - \left(f^2 + \frac{\mathbf{k}^2}{\lambda_n^2}\right)\omega - \frac{\beta k}{\lambda_n^2} = 0 . \tag{7.2.11}$$

It can be shown that the discriminant is always negative, i.e.,

$$\left(\frac{k\beta}{2\lambda_n}\right)^2 - \left[\frac{1}{3}\left(f^2 + \frac{\mathbf{k}^2}{\lambda_n^2}\right)\right]^3 < 0$$

implying that the equation (7.2.11) has three real roots.

Concerning the zonal wavenumber k (7.2.11) yields a more simple quadratic equation

$$k^2 + \frac{\beta k}{\omega} - \omega^2\lambda_n^2 + \varkappa^2 + \frac{1}{R_n^2} = 0$$

with $R_n = (\lambda_n f)^{-1}$ being the Rossby radius. This equation has the roots

$$k^{(1,2)} = -\frac{\beta}{2\omega} \pm \left[\frac{\beta^2}{4\omega^2} + \omega^2\lambda^2 - \left(\varkappa^2 + \frac{1}{R_n^2}\right)\right]^{1/2} \tag{7.2.12 a}$$

The radical in (7.2.12) may be written in the following manner

$$k^{(1,2)} = -\frac{\beta}{2\omega} \pm \frac{\lambda}{\omega}\sqrt{(\omega^2 - \omega_i^2)(\omega^2 - \omega_r^2)} \tag{7.2.12 b}$$

where

$$\omega_{i,r}^2 = \frac{f^2}{2} + \frac{\varkappa^2}{2\lambda_n^2} \pm \sqrt{\left(\frac{\varkappa^2}{2\lambda^2} + \frac{f^2}{2}\right)^2 - \frac{\beta^2}{4\lambda^2}}\,. \tag{7.2.12 c}$$

Since in the midlatitudes $f^2 \gg \beta/\lambda$ we find the approximate roots

$$\omega_i^2 \approx f^2 + \frac{\varkappa^2}{\lambda_n^2} = f^2(1 + R_n^2\varkappa^2) \gtrsim f^2 \tag{7.2.13 a}$$

$$\omega_r^2 \approx \frac{\beta^2}{4\left(\varkappa^2 + \frac{1}{R_n^2}\right)} \lesssim \left(\frac{\beta R_n}{2}\right)^2. \tag{7.2.13 b}$$

Thus the zonal wavenumber is real for $|\omega| > f$ and $|\omega| < \beta R_n/2$. For frequencies $\beta R_n/2 < |\omega| < f$, however, the $k_{1,2}$ are complex with the imaginary parts

$$\pm\frac{\lambda}{\omega}\sqrt{(\omega_i^2 - \omega^2)(\omega^2 - \omega_r^2)}\,.$$

This indicates the possibility of zonally localized response pattern. Since the inertial frequency f is much larger than the maximum frequency of Rossby waves, i.e., $f \gg \beta R_n/2$, there exists a large frequency gap between the inertial and subinertial wave processes. Thus in the high frequency domain we may neglect the β-term in (7.2.12), while in the low frequency range we may ignore the ω^2-term, and so we obtain the following three approximate roots

$$\omega_{1,2} = \pm\sqrt{f^2 + \frac{k^2}{\lambda_n^2}} \tag{7.2.14 a}$$

$$\omega_3 = -\frac{\beta k}{k^2 + \frac{1}{R_n^2}}\,. \tag{7.2.14 b}$$

Eq. (7.2.14a) defines the dispersion relation of inertial waves, or Poincaré waves, which have been discussed in Chapter 3, while (7.2.14b) yields a new dispersion relation due to planetary (Rossby) waves.

Next we discuss some properties of the Rossby waves inferred from the dispersion relation (7.2.14b). In Fig. 7.2.1 the dispersion relation (7.2.14b) is shown by means of an altitude chart. In Fig. 7.2.2 the dispersion relation was plotted for $\varkappa = 0$. The barotropic Rossby radius is typically 2000 km and the first baroclinic Rossby radius has a

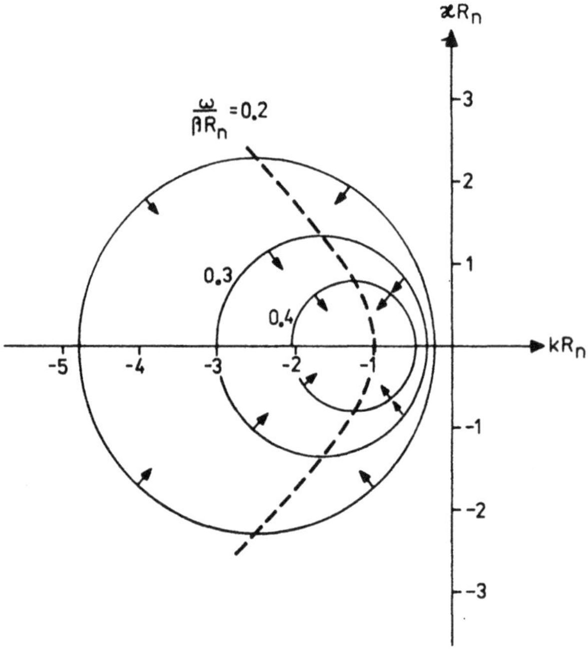

Fig. 7.2.1
Altitude chart of the dispersion relation. The circles indicates locations of fixed frequencies. The numerical values are given in terms of βR_n
The arrows refer to the direction of the group velocity (not to the magnitude). The broken line marks where the zonal component of the group velocity changes its sign.

typical value of 30 km. The maximum frequency, which is given by (7.2.13b), implies a minimum period of a week for the barotropic mode, and some months for the first baroclinic mode.

For the phase velocity we get

$$\boldsymbol{c} = \frac{\omega \boldsymbol{k}}{\boldsymbol{k}^2} = -\frac{\beta(k^2, k\varkappa)}{\boldsymbol{k}^2\left(\boldsymbol{k}^2 + \frac{1}{R_n^2}\right)^2}. \tag{7.2.15}$$

Thus $\boldsymbol{c}$ has always a negative zonal component. That is, the zonal phase propagation of Rossby waves is always westward.

The group velocity is defined by the wavenumber gradient of ω, i.e.,

$$\left(\frac{\partial}{\partial k}, \frac{\partial}{\partial \varkappa}\right)\omega = \boldsymbol{c}_{\mathrm{gr}} = \frac{\beta\left(k^2 - \varkappa^2 - \frac{1}{R_n^2}, 2k\varkappa\right)}{\left(\boldsymbol{k}^2 + \frac{1}{R_n^2}\right)^2}. \tag{7.2.16}$$

Let us put $\varkappa = 0$ and consider firstly two extreme cases, $k^2 \gg 1/R^2$ and $k^2 \ll 1/R_n^2$ which correspond to the short wave and to the long wave approximation, respectively. In the long wave approximation we obtain non-dispersive Rossby waves with

$$\omega = -R_n^2\beta k \tag{7.2.17a}$$

$$c = -R_n^2\beta \tag{7.2.17b}$$

$$c_{\mathrm{gr}} = \mathrm{c} \tag{7.2.17c}$$

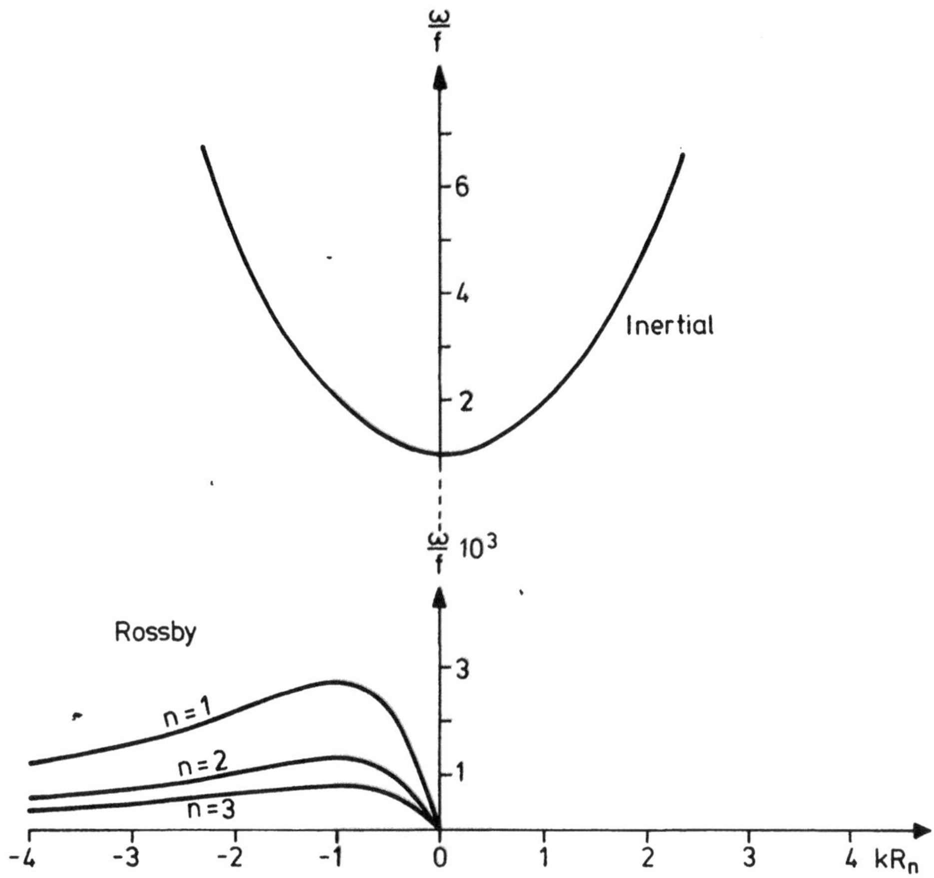

Fig. 7.2.2
The dispersion diagram for $\varkappa = 0$
Note that the ordinate is broken indicating a large frequency gap between inertial waves and Rossby waves. The label n refers to the first three vertical modes.

i.e., the phase speed equals the group velocity and both speeds are westward. Note that $c = -R^2\beta$ is the maximum phase speed. In the short wave approximation we have

$$\omega = -\frac{\beta}{k} \tag{7.2.18a}$$

$$c = -\frac{\beta}{k^2} \tag{7.2.18b}$$

$$c_{gr} = \frac{\beta}{k^2} = -c\,. \tag{7.2.18c}$$

Thus in the short wave domain, i.e., at large wavenumbers, the group velocity is eastward and, hence, opposite to the phase speed. This implies that information will be propagated eastward by the short Rossby waves and westward by the long Rossby waves.

Moreover, we consider the case with k being of the order of the inverse Rossby radius. The group velocity vanishes at $k = -R_n^{-1}$ and $\omega = +\beta R_n/2$, i.e., according to (7.2.13b), at the maximum frequency of the Rossby waves.

Furthermore, the group velocity possesses two maxima. The one is located at $\omega = 0$ and $k = 0$ with $c_{gr} = -\beta R_n^2$, and the other one at $\omega = \sqrt{3}\,\beta R_n/4$ and $k = -\sqrt{3}\,R_n^{-1}$ with $c_{gr} = \beta R_n^2/8$. Generally we may assume that the wave fronts of Rossby waves propagate with maximum group velocities from an initial disturbance. Behind the wave fronts waves with vanishing group velocity will remain, implying a typical wavelength of $2\pi R_n$ and a period of $4\pi/(\beta R_n)$. It appears that the Rossby wave fronts propagate eight times faster to the west than to the east. Moreover, due to the large barotropic Rossby radius the barotropic Rossby waves disperse much faster than the baroclinic ones.

With the methods outlined in Subsection 2.4.3 we may express the Green's function in terms of the vertical eigenfunctions, i.e.,

$$K(z, z') = \frac{2}{H} \sum_{n=0}^{\infty} \frac{\cos \lambda_n N z \cos \lambda_n N z'}{[1 + j_0(2\lambda_n NH)]\,(\lambda^2 - \lambda_n^2)}$$

$$= \frac{2}{H} \sum_{n=0}^{\infty} \frac{\cos(\lambda_n N z) \cos(\lambda_n N z')\, \omega(f^2 - \omega^2)}{[1 + j_0(2\lambda_n NH)]\, \lambda_n^2 (\omega - \omega_1)(\omega - \omega_2)(\omega - \omega_3)} \tag{7.2.19}$$

where SP $(2\lambda_n NH) \approx 1$ for $n = 0$ and zero for $n \geq 1$.

With the aid of $K(z, z')$ we can express the response of the unbounded midlatitude β-plane to external forcing in terms of convolutions of the body forces X and Y and of the Green's function. Introducing the notations
from (7.2.1) it follows

$$X * K = \int_{-H}^{0} dz'\, X(z')\, K(z, z'), \qquad Y * K = \int_{-H}^{0} dz'\, Y(z')\, K(z, z') \tag{7.2.20}$$

$$p = \frac{i}{\omega(f^2 - \omega^2)} [(ikf + \omega\varkappa)\, Y * K + (\beta - i\varkappa f + \omega k)\, X * K] \tag{7.2.21}$$

and from (7.1.8a, b) we have

$$u = \frac{1}{f}(Y - i\varkappa p), \qquad v = \frac{1}{f}(-X + ikp).$$

With (7.2.19) we may rewrite (7.2.21) as

$$p = \sum_{n=0}^{\infty} p_n(\omega, \boldsymbol{k})\, F_n(z) \tag{7.2.22}$$

where

$$p_n(\omega, \boldsymbol{k}) = \frac{i(ikf + \omega\varkappa)\, Y_n + (\beta - if\varkappa + \omega k)\, X_n}{\lambda_n^2 (\omega - \omega_1)(\omega - \omega_2)(\omega - \omega_3)}. \tag{7.2.23}$$

Here the ω_j $(j = 1, 2, 3)$ are the three real roots of (7.2.11). Approximate expressions of the ω_j's are given by (7.2.14). The X_n and Y_n are the projection coefficients of the forcing on the vertical eigenfunction $F_n(z)$. For $\beta = 0$ the expression (7.2.23) equals

(3.4.4a), thus (7.2.23) is the β-plane generalization of (3.4.4a). If the β-effect is of importance we may filter out the inertial waves assuming $\omega \ll f$ and we obtain

$$p_n(\omega, \boldsymbol{k}) = \frac{-\mathrm{i}}{\left(k^2 + \frac{1}{R_n^2}\right)(\omega - \omega_3(\boldsymbol{k}))} \left[(\mathrm{i}kf + \omega\varkappa)\, Y_n + (\beta - \mathrm{i}\varkappa f + \omega k)\, X_n\right]. \tag{7.2.24}$$

We have already mentioned that the Green's function is trapped at $z = z'$ provided ω is in the range $\beta R_n/2 < \omega < f$. The e-folding distance is according to (7.2.2) and (7.2.5)

$$L_v = (N\tilde{\lambda})^{-1} = \frac{(f^2 - \omega^2)^{1/2}}{N} \left(k^2 + \frac{\beta k}{\omega}\right)^{-1/2}. \tag{7.2.25}$$

This implies that it depends on the scales of the forcing whether the oceanic response is a barotropic or baroclinic one. Let us put $z' = -H_{mix}$ and $k = \varkappa$. The corresponding trapping scale is shown in Fig. 7.2.3 versus the time scale for different lenght scales. It appears that for $L \lesssim 250$ km the response is trapped near the surface for time scales ranging from the inertial period to 120 days, where the baroclinic Rossby waves come into play.

However, for the same range of time scales, but for length scales $L \gtrsim 500$ km the trapping scale is of the order of the depth of the ocean, and so the response to weather patterns with those scales is practically barotropic. Obviously, the only subinertial wave processes which may be excited with these scales are the barotropic Rossby waves.

This fact was utilized by WILLEBRAND et al. (1980) to investigate the response of the ocean to the fluctuating component of the wind stress as computed from twice daily weather maps.

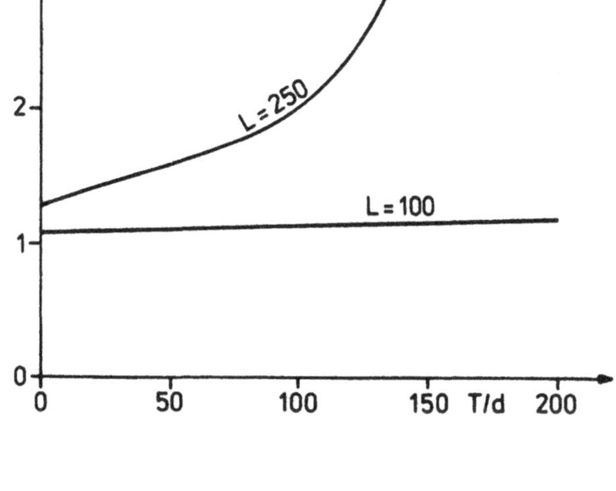

Fig. 7.2.3
The vertical scale L_v versus the time scale
The curves are labeled by different values of horizontal scales L in km.

7.3. FORCING OF AN UNBOUNDED β-PLANE OCEAN THROUGH ROTATIONAL WINDS

The response of an unbounded ocean on the β-plane to different wind patterns for time scales exceeding a pendulum day has been studied by several authors.

VERONIS and STOMMEL (1956) investigated the response to a meridional wind stress varying sinusoidally in zonal direction. They found that when a switched-on wind blows longer than an inertial period most of the energy goes into the barotropic and baroclinic Rossby waves, rather than into inertial waves. For a time-periodic wind with periods of the order of few weeks they could show that only the barotropic Rossby wave was excited. For increasing periods, however, the baroclinic modes come into play and, hence, the response becomes increasingly baroclinic.

This is in accordance with the general features of the forcing theory outlined in the previous section.

VERONIS (1958) and LONGUET-HIGGINS (1965) considered the response to initial disturbances of the windstress curl concentrated in a point, i.e., a problem close to the Green's function equation, and evaluated the wave patterns. Moreover, LONGUET-HIGGINS (1965) and LIGHTHILL (1967) estimated how moving storms radiate Rossby waves along their tracks.

An investigation of the quasi-geostrophic response of the ocean to stochastic forcing, in particular wind stress and atmospheric pressure, was presented by FRANKIGNOUL and MÜLLER (1979).

In the following two subsections we proceed to calculate the response to examples of the wind forcing on the basis of the Green's function and the formal solution as derived in the foregoing section.

7.3.1. Meridional wind

First we consider a meridional wind, switched on at $t = 0$, blowing from the north, and being confined to the half plane $x > 0$, see Fig. 7.3.1.

$$Y = -\frac{v_*^2}{H_{\text{mix}}}\,\theta(z + H_{\text{mix}})\,\theta(t)\,\theta(x) \tag{7.3.1}$$

$$X = 0\,. \tag{7.3.2}$$

Fourier transformation and expansion in vertical modes yield

$$Y_n(\omega, \boldsymbol{k}) = -\frac{v_*^2}{h_n}\,\frac{2\pi \mathrm{i}\,\delta(\varkappa)}{\omega + \mathrm{i}\varepsilon}\,\frac{(-\mathrm{i})}{k - \mathrm{i}\varepsilon}\,. \tag{7.3.3}$$

Since we wish to investigate the role of the β-effect we may filter out the inertial waves. Inserting (7.3.3) into (7.2.24) we get

$$p_n(\omega, \boldsymbol{k}) = -\frac{v_*^2}{h_n}\,\frac{2\pi\delta(\varkappa)\,f}{\omega(\omega - \omega_3)\left(k^2 + \dfrac{1}{R^2}\right)}\,. \tag{7.3.4}$$

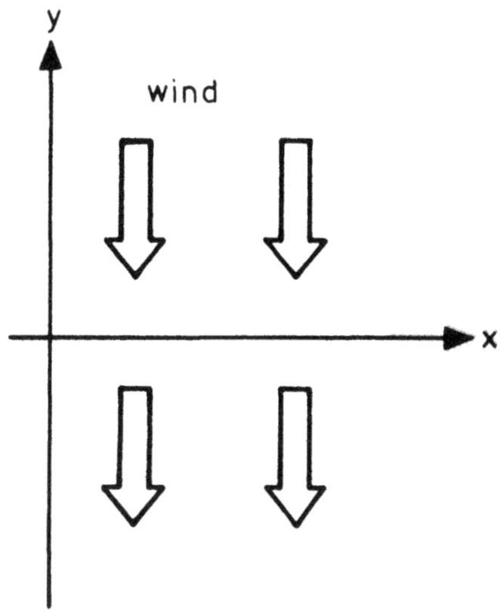

Fig. 7.3.1
Meridional wind blowing over a half plane

Using (7.2.14b) we may rewrite this to obtain

$$p_n(\omega, \boldsymbol{k}) = -\frac{v_*^2}{h_n} \frac{2\pi\, \delta(\varkappa) f}{\omega^2 (k - k_1)(k - k_2)} \tag{7.3.5}$$

where

$$k_{1,2} = -\frac{\beta}{2\omega} \pm \mathrm{i} \sqrt{\frac{1}{R^2} - \frac{\beta^2}{4\omega^2}} \,. \tag{7.3.6}$$

Next we perform the inverse Fourier transform with regard to k and $\varkappa$. The $\varkappa$-integration is trivial because of the involved delta function. To perform the k-integration we note that p_n has one pole in the upper and one in the lower half of the complex k-plane. Thus from the residues theorem we find

$$p_n(\omega, x) = -\frac{v_*^2}{h_n} \frac{\mathrm{i} f \,\mathrm{e}^{-\frac{\mathrm{i}\beta x}{2\omega}}}{\omega^2 \cdot 2\mathrm{i} \left(\frac{1}{R^2} - \frac{\beta^2}{4\omega^2} \right)^{1/2}} \left[\theta(x)\, \mathrm{e}^{-\sqrt{\frac{1}{R^2} - \frac{\beta^2}{4\omega^2}}\, x} + \theta(-x)\, \mathrm{e}^{\sqrt{\frac{1}{R^2} - \frac{\beta^2}{4\omega^2}}\, x} \right]. \tag{7.3.7}$$

We note that $p_n(\omega, x)$ yields the Eulerian frequency spectrum of the n-th modal component of the response. This spectrum has a peak at $\omega = \beta R/2$, that is, at that frequency which corresponds to zero group velocity.

Now we calculate the Fourier inversion of (7.3.7) concerning ω. Causality requires to close the integration contour in the lower half of the complex ω-plane. There exist two branch points, namely at $\omega = \pm\omega_b$, $\omega_b = \beta R/2$. We may choose the branch cuts from the points $\pm\omega_b$ to infinity parallel to the imaginary axis and deform the contour in the manner depicted in Fig. 7.3.2.

Then we have contributions from the residue of the pole at $\omega = 0$ and from the two line integrals along Γ_1 and Γ_3. We drop the index n and have

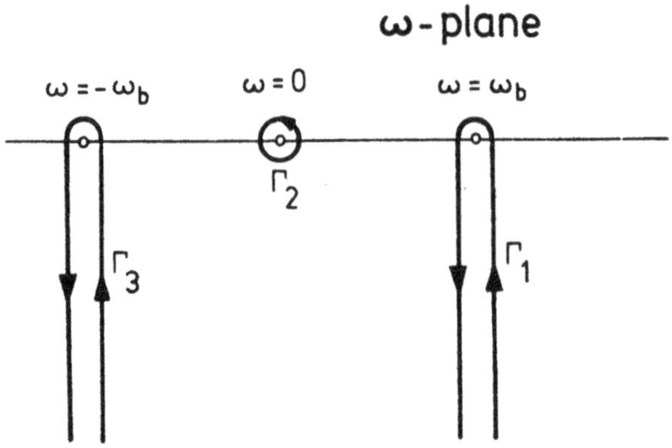

Fig. 7.3.2
The contours Γ_1, Γ_2, and Γ_3 in the ω-plane used in (7.3.8)

$$p(x,t) = -\frac{v_*^2}{h} f \int_{\Gamma_2} \frac{d\omega}{2\pi} \frac{e^{-i\omega t - \frac{i\beta x}{2\omega}}}{\omega i\beta \left(1-\left(\frac{\omega}{\omega_b}\right)^2\right)^{1/2}}$$

$$\times \left[\theta(x)\, e^{-i\frac{\beta x}{2\omega}\sqrt{1-\left(\frac{\omega}{\omega_b}\right)^2}} + \theta(-x)\, e^{+\frac{i\beta x}{2\omega}\sqrt{1-\left(\frac{\omega}{\omega_b}\right)^2}}\right]$$

$$+ \int_{\Gamma_1} \frac{d\omega}{2\pi} e^{-i\omega t}[\ldots] + \int_{\Gamma_3} \frac{d\omega}{2\pi} e^{-i\omega t}[\ldots] \tag{7.3.8}$$

where the brackets in the second and third term stand for the right side of (7.3.7). The first term yields two contributions, namely

$$\int_{\Gamma_2} \frac{d\omega}{2\pi i} \frac{e^{-i\omega t}}{\omega} \left[\theta(x)\, e^{-i\frac{\beta x}{\omega}} + \theta(-x)\, e^{-i\frac{\omega x}{\beta R^2}}\right]$$

$$= \theta(x) \int_{\Gamma_2} \frac{d\omega}{2\pi i} \frac{e^{-i\left(\omega t + \frac{\beta x}{\omega}\right)}}{\omega} - \theta(-x)\, \theta\left(t + \frac{x}{\beta R^2}\right)$$

where the term with the simple pole can easily be obtained. The other term, however, has an essential singularity at $\omega = 0$. That integral can be evaluated by expanding the exponential $\exp(-i\beta x/\omega)$ in a Taylor series and then calculating the residues term by term

$$\int_{\Gamma_2} \frac{d\omega}{2\pi i} \frac{e^{-i\omega t - i\frac{\beta x}{\omega}}}{\omega} = \int_{\Gamma_2} \frac{d\omega}{2\pi i} \sum_{n=0}^{\infty} (-i\beta x)^n \frac{e^{-i\omega t}}{\omega^{n+1}} = -\theta(t) \sum_{n=0}^{\infty} \frac{i^{2n}(\beta x t)^n}{(n!)^2} = -\theta(t)\, J_0(2\sqrt{\beta t x}). \tag{7.3.9}$$

Thus,

$$p(x, t) = +\frac{v_*^2 f}{h\beta}[\theta(-x)\,\theta(t + x/\beta R^2) + \theta(x)\,\theta(t)\,J_0(2\sqrt{\beta t x})] \qquad (7.3.10)$$
$$+ \text{ line integrals.}$$

The first term describes a Rossby wave propagating westward with the long wave group velocity βR^2. The second term describes the propagation of short Rossby waves having eastward group velocity, compare (7.2.18). That is, contrary to the f-plane case with a subinertial response being confined to the wind edge, long and short Rossby waves radiate westward and eastward, respectively, from the wind edge.

We proceed to analyze the line integrals starting with

$$\int_{\Gamma_1} \frac{d\omega}{2\pi} \frac{e^{-it}}{\omega^2\sqrt{\frac{1}{R^2} - \frac{\beta^2}{4\omega^2}}} \exp\left[-\sqrt{\frac{1}{R^2} - \frac{\beta^2}{4\omega^2}}\,|x| - \frac{i\beta x}{2\omega}\right]$$

$$= \frac{2\omega_b}{\beta}\int_{\Gamma_1} \frac{d\omega}{2\pi} \frac{\exp\left[-i\omega t - \frac{i\beta x}{2\omega} - \frac{\beta|x|}{2\omega\omega_b}\sqrt{(\omega-\omega_b)(\omega+\omega_b)}\right]}{\omega\sqrt{(\omega-\omega_b)(\omega+\omega_b)}}. \qquad (7.3.11)$$

Let $\omega = is + \omega_b$. This transformation maps the contour Γ_1 into Γ_1' on the s-plane, see Fig. 7.3.3, and (7.3.11) becomes

$$\frac{i\cdot 2\omega_b}{\beta}\,e^{-i\omega_b t}\int_{\Gamma_1'} \frac{ds}{2\pi} \frac{\exp\left[st - \frac{i\beta x}{2(is+\omega_b)} - \frac{\beta|x|\sqrt{is(is+2\omega_b)}}{2\omega(is+\omega_b)}\right]}{(is+\omega_b)\sqrt{is(is+2\omega_b)}}.$$

Moreover, transforming $\sqrt{s} = re^{i\varphi/2}$, with $\varphi = \pi$ above and $\varphi = -\pi$ below the cut, we obtain

$$\int_{\Gamma_1} = \frac{2\omega_b}{\pi\beta}\,e^{-i\omega_b t - i\frac{\pi}{4}}\int_{-\infty}^{+\infty} dr\,\frac{\exp\left[-r^2 t - \frac{i\beta x}{2(\omega_b - ir^2)} - \frac{i^{3/2}\beta\,|x|\,r\sqrt{2\omega_b - ir^2}}{2\omega_b(\omega_b - ir^2)}\right]}{(\omega_b - ir)^2\sqrt{2\omega_b - ir^2}}$$

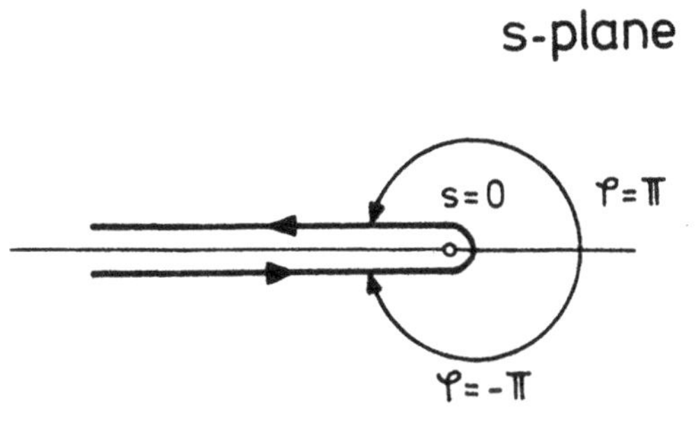

Fig. 7.3.3
The contour in the s-plane

where the symbol on the left-hand side represents the left-hand side of (7.3.11). For large times the leading contribution results from small r, i.e.,

$$\int_{\Gamma_1} = \frac{2\omega_b}{\pi\beta} \frac{e^{-i\omega_b t - \frac{i\pi}{4}}}{\sqrt{2\omega_b^3}} \int_{-\infty}^{+\infty} dr \exp\left[-r^2 t - \frac{i\beta x}{2\omega_b} - \frac{i\beta|x|\sqrt{i}\, r}{\sqrt{2\omega_b^3}} + O(r^2)\right]$$

$$= \frac{1}{\beta}\left(\frac{2}{\pi\omega_b t}\right)^{1/2} \exp\left[-i\left(\omega_b t + \frac{\beta x}{2\omega_b} + \frac{\beta^2 x^2}{8\omega_b^3 t} + \frac{\pi}{4}\right)\right] \tag{7.3.12}$$

Analoguously, the integration along Γ_3 yields for large times

$$\int_{\Gamma_3} = \frac{1}{\beta}\left(\frac{2}{\pi\omega_b t}\right)^{1/2} e^{i\left(\omega_b t + \frac{\beta x}{2\omega_b} + \frac{\beta^2 x^2}{8\omega_b^3 t} + \frac{\pi}{4}\right)} \tag{7.3.13}$$

Thus, with $\omega_b = \beta R/2$, the line integrals contribute for large times with

$$\int_{\Gamma_1} + \int_{\Gamma_3} = \frac{4}{\beta\sqrt{\pi\beta R t}} \cos\left(\frac{\beta R t}{2} + \frac{x}{R}\left(1 + \frac{x}{\beta R^2 t}\right) + \frac{\pi}{4}\right). \tag{7.3.14}$$

Supposing, moreover, $x/t \ll \beta R^2$ we get

$$\int_{\Gamma_1} + \int_{\Gamma_3} = \frac{4}{\beta\sqrt{\beta R t}} \cos\left(\frac{\beta R t}{2} + \frac{x}{R} + \frac{\pi}{4}\right). \tag{7.3.15}$$

This describes a wave process with frequency and wavenumber of vanishing group velocity, in accordance with the Eulerian spectral function (7.3.7) which has, as already mentioned, a prominent peak at the frequency of zero group velocity.

We note that the expression (7.3.15) may also be obtained by means of the method of stationary phases, compare e.g. CANE and SARACHIK (1976).

With (7.3.10) and (7.3.15) the complete result for large times reads

$$p_n(x,t) = +\frac{v_*^2}{h_n}\frac{f}{\beta}\left[\theta(-x)\,\theta\left(t + \frac{x}{\beta R_n^2}\right) + \theta(x)\, J_0(2\sqrt{\beta t x}) + \frac{2}{\sqrt{\beta R_n t \pi}} \cos\left(\frac{\beta R_n t}{2} + \frac{x}{R_n} + \frac{\pi}{4}\right)\right]. \tag{7.3.16a}$$

Moreover, to the approximation of (7.1.8a, b) we get

$$u_n(x,t) = -\frac{v_*^2}{h_n f}\,\theta(x) \tag{7.3.16b}$$

$$v_n(x,t) = \frac{v_*^2}{h_n\beta}\left[+\frac{\theta(-x)}{\beta R_n^2}\,\delta\left(t + \frac{x}{\beta R_n^2}\right) - \theta(x)\left(\frac{\beta t}{x}\right)^{1/2} J_1(2\sqrt{\beta x t}) - \frac{2}{R_n\sqrt{\pi\beta R_n t}} \sin\left(\frac{1}{2}\beta R_n t + \frac{x}{R_n} + \frac{\pi}{4}\right)\right]. \tag{7.3.16c}$$

In this set we have written the vertical mode index explicitly. The modal vertical velocity, $w_n = -p_n$, becomes

$$w_n(x, t) = \frac{v_*^2}{h_n} \frac{f}{\beta} \left[\theta(-x)\, \delta\left(t + \frac{x}{\beta R_n^2}\right) - \theta(x) \left(\frac{\beta x}{t}\right)^{1/2} J_1(2\sqrt{\beta x t}) \right.$$
$$\left. - \left(\frac{\beta R_n}{\pi t}\right)^{1/2} \sin\left(\frac{1}{2}\beta R_n t + \frac{x}{R_n} + \frac{\pi}{4}\right)\right] \qquad (7.3.16\text{d})$$

In the set (7.3.16) the terms with the Bessel functions depend only through h_n on the vertical mode index and, therefore, the corresponding sums over all vertical modes yield contributions confined to the upper layer.

Taking the aspects of the f-plane dynamics as outlined in Section 3.9. into account we may describe the complete response scenario of an unbounded ocean to a non-divergent rotational windstress over a half plane. After the radiation of inertial waves from the wind edge there remains an accelerating jet trapped within one Rossby radius at the wind edge, see Section 3.9.

From (7.3.6) it may be inferred that after times of the order $4\pi/(\beta R_n)$, which is the minimum period of Rossby waves, the β-dispersion comes into play and Rossby waves begin to propagate westward as well as eastward from the wind edge with their corresponding maximum group velocities. Thus the wind edge jet disperses into Rossby waves.

The asymptotic dynamical patterns are given by the set (7.3.16). On the western side the jet separates from the edge and propagates westward with the long wave group velocity. Since the barotropic mode travels very much faster than the baroclinic ones the separated branch of the jet is basically a baroclinic current, at least close to the wind edge. The eastern branch of the wind edge jet transforms into eastward propagating Rossby waves generating a series of alternating currents which are confined to the surface mixed layer. At any fixed position x these currents increase in time as $\sqrt{t}$ while their zonal width decreases with $1/t$. Between the separating wave fronts algebraically decaying Rossby waves with periods and wavelengths corresponding to vanishing group velocities remain.

7.3.2. Zonal wind

In this subsection we wish to investigate as a second example the response of the ocean to a zonal wind blowing from the east over the half plane $y < 0$ as sketched in Fig. 7.3.4.

$$X = -\frac{u_*^2}{H_{\text{mix}}}\, \theta(z + H_{\text{mix}})\, \theta(t)\, \theta(-y)\,, \qquad Y = 0\,. \qquad (7.3.17)$$

Fourier transform of (7.3.17) and expansion in terms of vertical eigenfunctions yield

$$X_n(\omega, \boldsymbol{k}) = \frac{u_*^2}{h_n} \frac{2\pi\, \delta(k)}{(\omega + \mathrm{i}\varepsilon)\,(\varkappa + \mathrm{i}\varepsilon)}\,,$$

and inserting this into (7.2.24) we have

$$p_n(\omega, \boldsymbol{k}) = -\frac{u_*^2}{h_n} \frac{(\mathrm{i}\beta + \varkappa f) \cdot 2\pi\delta(k)}{\omega^2 \varkappa \left(\varkappa^2 + \dfrac{1}{R_n^2}\right)}\,. \qquad (7.3.18$$

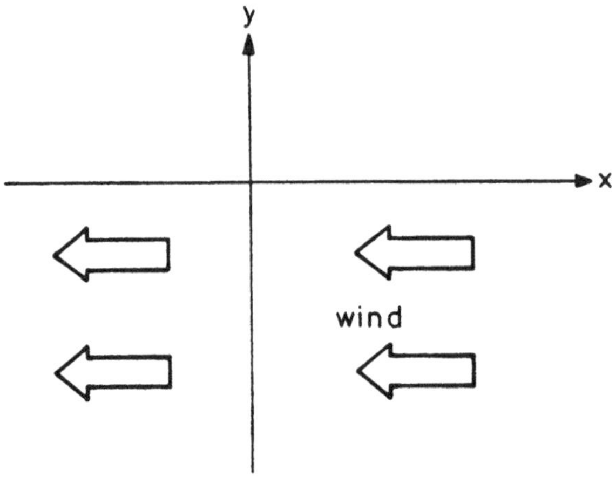

Fig. 7.3.4
Zonal wind blowing over a half plane

Here it was used that because of the involved delta function $\omega_3(k, \varkappa)$ vanishes, i.e., $\omega_3(0, \varkappa) = 0$. This implies that the forcing (7.3.17) can not excite Rossby waves. In order to perform the Fourier inversion of (7.3.18) into the $y - t$ domain we rewrite (7.3.18) in the following manner

$$p_n(\omega, \mathbf{k}) = -\frac{u_*^2}{h_n}\frac{2\pi\delta(k)}{\omega^2}\left[\frac{f}{\left(\varkappa^2 + \frac{1}{R_n^2}\right)} + \frac{\mathrm{i}\beta}{\varkappa\left(\varkappa^2 + \frac{1}{R_n^2}\right)}\right].$$

Then with the aid of the integrals (B 20) and (B 21) we get

$$p_n(y, t) = \frac{u_*^2}{h_n}\, t\,\theta(t)\left\{\frac{1}{2}R_n f\,\mathrm{e}^{-\frac{|y|}{R_n}} + R_n^2\beta\left[\theta(-y) + \frac{1}{2}\operatorname{sgn}(y)\,\mathrm{e}^{-\frac{|y|}{R_n}}\right]\right\}. \qquad (7.3.19\text{a})$$

We recognize that the first term represents the pressure field of a wind edge jet as already discussed in Section 3.9, compare in particular (3.9.5a). The important new result consists in the occurrence of the second term which is proportional to β.

Contrary to the response of a f-plane ocean this term yields a non-zero contribution within the forcing region, even far from the wind edge. For the corresponding horizontal currents we obtain to the geostrophic approximation (7.1.8a, b)

$$u_n(y, t) = +\frac{u_*^2}{h_n} f t\,\theta(t)\left[\frac{1}{2}\operatorname{sgn}(y)\,\mathrm{e}^{-\frac{|y|}{R_n}} + R_n\frac{\beta}{2f}\,\mathrm{e}^{-\frac{|y|}{R_n}}\right] \qquad (7.3.19\text{b})$$

$$v_n(y, t) = \frac{u_*^2}{h_n f}\,\theta(t)\,\theta(-y)\,. \qquad (7.3.19\text{c})$$

The modal vertical velocity, according to $w_n = -p_{nt}$, follows to be

$$w_n(y, t) = -\frac{u_*^2}{h_n}\,\theta(t)\left\{\frac{1}{2}R_n f\,\mathrm{e}^{-\frac{|y|}{R_n}} + R_n^2\beta\left[\theta(-y) + \frac{1}{2}\operatorname{sgn}(y)\,\mathrm{e}^{-\frac{|y|}{R_n}}\right]\right\}. \qquad (7.3.19\text{d})$$

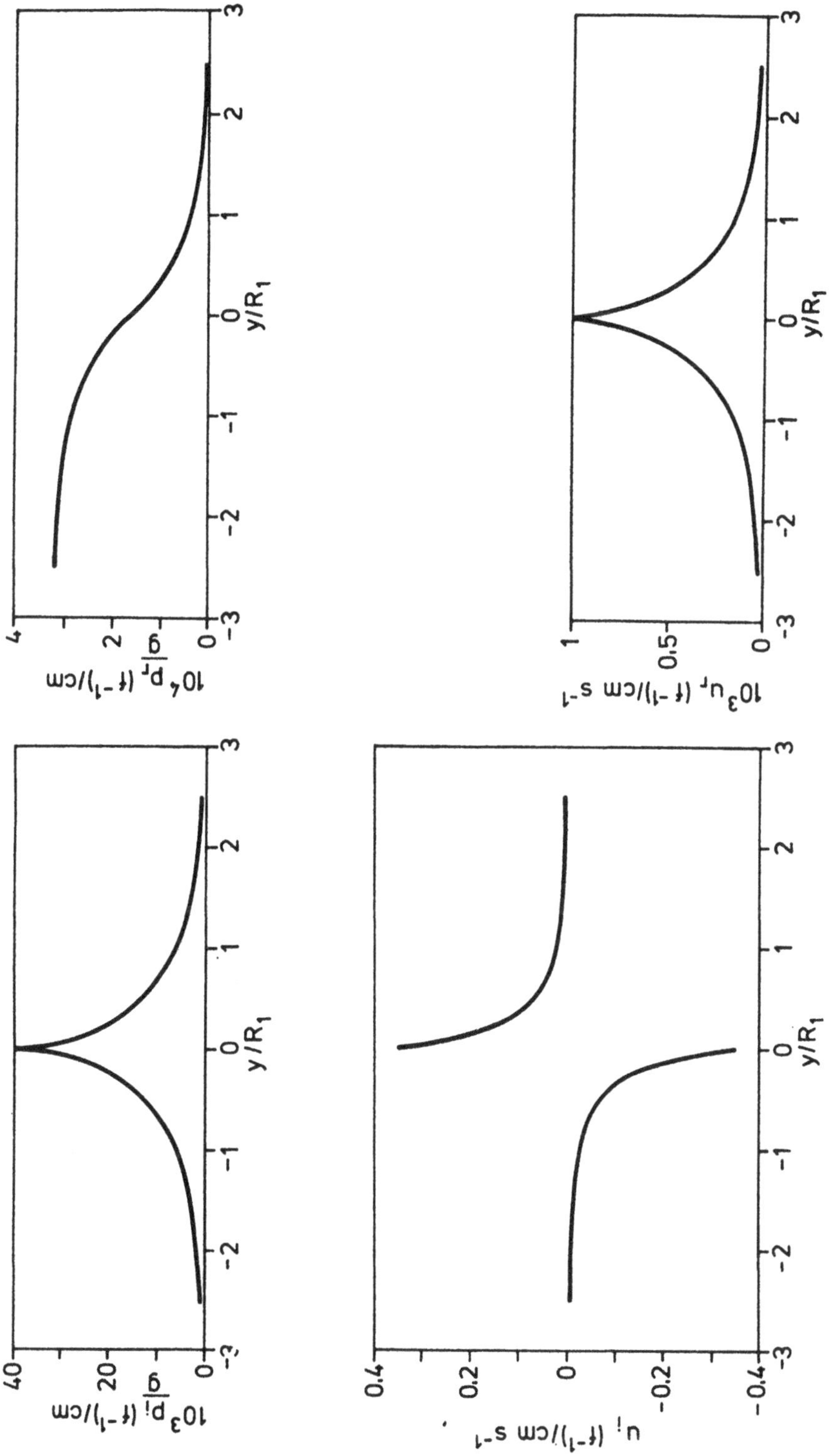

Fig. 7.3.5
The meridional structure of the different contributions to the response according to (7.3.19) at the time $t = f^{-1}$, the f-plane portion of the pressure p_i (*top left*) and of the geostrophic alongfront current u_i (*bottom left*), and the β-plane portion p_r (*top right*) and the associated current (*bottom right*)

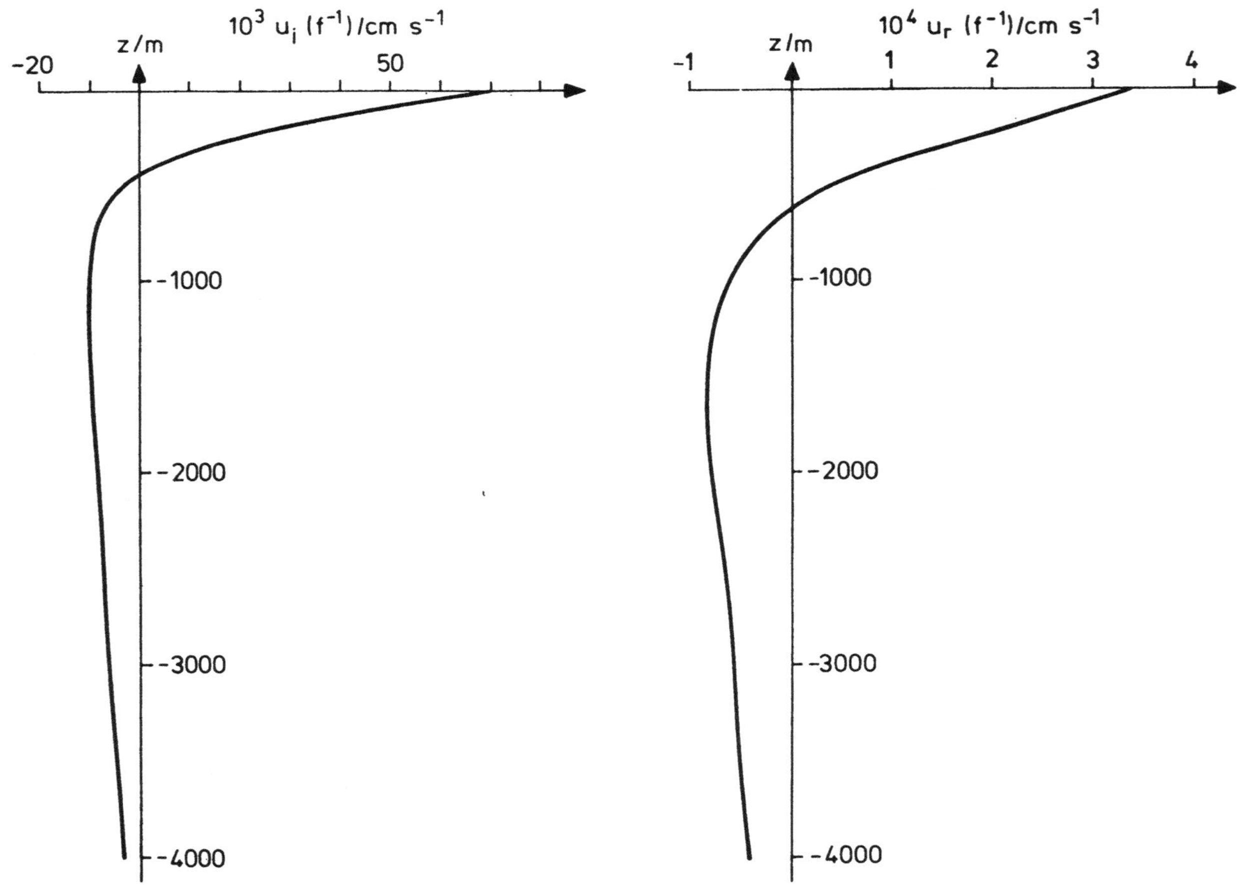

Fig. 7.3.6
The vertical structure of the baroclinic part u_i and u_r one Rossby radius R_1 away from the front, i.e., $y = R_1$
The sum includes the first 30 baroclinic modes.

The zonal current and the pressure consist of two contributions:

The first is governed by the f-plane dynamics and describes an accelerating downwind jet near the wind edge bordered by a countercurrent as discussed in Section 3.9. Such a regime implies the formation of a frontal zone. The current shear, u_y, is very large at the discontinuity line of the wind and, hence, the current system is likely to become unstable after a few inertial periods and may give rise to the formation of meanders and eddies.

The second is affected by the β-effect and describes an accelerating eastward jet which is not associated with such a strong shear and, hence, this along-front jet can be expected to be stable. The pressure increases within the forcing area, i.e., south of the front, indicating downwelling. Generally, the β-plane portion of the response contributes after a time of the order $t \gtrsim 2/(\beta R_n)$, i.e., it requires persistent wind fields.

The wind field (7.3.17) may serve as a crude model of the easterly trade winds in the North Pacific and in the North Atlantic. There exists observational evidence in both oceans, see e.g. RODEN (1975) and SIEDLER et al. (1985), that close to the subtropical fronts along-front jets, disturbed by some eddy motion, occur. Moreover, occurrences of downwelling in the forcing region are well documented. Thus, among other driving forces, the model outlined above may give a qualitative explanation of the observed response patterns.

In order to compute u and p we use the BVF-model (1.9.52) and the associated vertical eigenfunctions in the WKB treatment as discussed in Section 1.9.3. Moreover we adopt the value $u_* = 1$ cm/s.

Taking the first 30 baroclinic vertical eigenfunctions, $F_n(z)$, into account we obtain close to the surface the results shown in Fig. 7.3.5, where the contributions of the f-plane and of the β-plane dynamics are depicted separately. The pressure consists of a ridge, p_i, and of a step, p_r, within a transition zone of the width $2R_1 \approx 75$ km. It appears that the f-plane portion p_i develops within some 10 inertial periods, to the observed order of magnitude. The pressure step p_r, however, needs some 100 inertial periods to increase to comparable magnitudes. Thus the β-plane part of the response will be important for winds persistent for more than, say, 100 days. The associated currents u_i and u_r also shown in Fig. 7.3.5.

In Fig. 7.3.6 the vertical structures of u_i and u_r in a distance of R_1 from the front have been plotted.

The expressions (7.3.16) and (7.3.19) exhibit a remarkable difference in the response of the β-plane ocean to rotational forces over half planes, which depend on the orientation of the wind edges. In case of an east-west edge the associated jets disperse into Rossby waves. By contrast, for a zonally stretching wind edge Rossby waves are not excited, however, an additional along-front jet and a well-developed downwelling within the forcing region occur. The latter can be attributed to the divergence of the Ekman current (7.3.19c) which results from the meridional variation of f.

7.4. FORCING OF A β-PLANE OCEAN WITH AN EASTERN BOUNDARY

7.4.1. The Green's function

In the remainder of this chapter we analyze the influence of meridional boundaries on the response of a midlatitudinal β-plane ocean. First we consider the case of an eastern boundary, the coordinate system is depicted in Fig. 7.4.1. We start with the calculation of the Green's function.

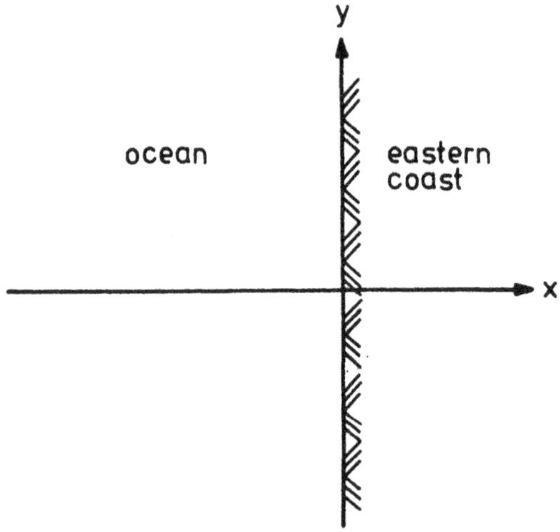

Fig. 7.4.1
The coordinate frame with an eastern boundary

Neglecting the higher time derivatives of (7.1.6), and after normal mode expansion in vertical eigenfunctions we get

$$\left(\Delta - \frac{1}{R_n^2}\right) p_{nt} + \beta p_{nx} = f(Y_{nx} - X_{ny}) + \beta X_n + X_{nxt} + Y_{nyt} \tag{7.4.1}$$

with

$$p(x, y, z, t) = \sum_n p_n(x, y, t)\, F_n(z)\,, \qquad (X, Y) = \sum_n (X_n, Y_n)\, F_n(z)\,.$$

The Fourier transform of (7.4.1) with respect to y and t yields

$$p_{nxx} - \left(\varkappa^2 + \frac{1}{R_n^2}\right) p_n + \frac{\mathrm{i}\beta}{\omega} p_{nx} = \frac{\mathrm{i}f}{\omega}(Y_{nx} - \mathrm{i}\varkappa X_n) + \frac{\mathrm{i}\beta}{\omega} X_n + X_{nx} + \mathrm{i}\varkappa Y_n\,. \tag{7.4.2a}$$

The presence of the coast at $x = 0$ requires the boundary condition

$$p_{nx} - \frac{\varkappa f p_n}{\omega} = \frac{\mathrm{i}fY_n}{\omega} + X_n \quad \text{for} \quad x = 0\,. \tag{7.4.2b}$$

Moreover, we postulate that p remains finite far from the coast, i.e.,

$$p_n < \infty \quad \text{for} \quad x \to -\infty\,. \tag{7.4.2c}$$

In the following we shall again drop the index n. The corresponding Green's function is subject to

$$K_{xx}(x, x') + \frac{i\beta}{\omega} K_x(x, x') - \left(\varkappa^2 + \frac{1}{R_n^2}\right) K(x, x') = \delta(x - x') \tag{7.4.3 a}$$

$$K_x(0, x') = \frac{\varkappa f}{\omega} K(0, x') \tag{7.4.3 b}$$

$$K(-\infty, x') = 0 \,. \tag{7.4.3 c}$$

Let for $x \neq x'$ the two linear independent solutions be $\Phi(x)$ and $\Psi(x)$ which fulfil

$$\Phi_x(0) = \frac{\varkappa f}{\omega} \Phi(0) \,, \qquad \Psi(-\infty) = 0 \,. \tag{7.4.4}$$

The occurrence of the x-derivative in (7.4.3 a) suggests the ansatz

$$(\Phi(x), \Psi(x)) = e^{-i\frac{\beta x}{2\omega}} (\varphi(x), \psi(x)) \,. \tag{7.4.5}$$

Then, for φ and ψ the equations

$$\varphi_{xx} - \alpha^2\varphi = 0 \,, \qquad \psi_{xx} - \alpha^2\psi = 0 \tag{7.4.6}$$

follow, with

$$\alpha^2 = \varkappa^2 + \frac{1}{R_n^2} - \frac{\beta^2}{4\omega^2} \,. \tag{7.4.7}$$

The boundary conditions (7.4.4) transform into

$$\varphi_x(0) = \left(\frac{i\beta}{2\omega} + \frac{\varkappa f}{\omega}\right) \varphi(0) \,, \qquad \psi(-\infty) = 0 \,. \tag{7.4.8}$$

The corresponding solutions are readily obtained

$$\varphi(x) = \frac{1}{\alpha}\left(\frac{i\beta}{2\omega} + \frac{\varkappa f}{\omega}\right) \sinh \alpha x + \cosh \alpha x \tag{7.4.9 a}$$

$$\psi(x) = e^{\alpha x} \,. \tag{7.4.9 b}$$

Next we estimate the Wronskian of Φ and Ψ

$$W_r(x) = \Phi_x\Psi - \Phi\Psi_x = e^{-i\frac{\beta x}{\omega}} \left(\frac{i\beta}{2\omega} - \alpha + \frac{\varkappa f}{\omega}\right) . \tag{7.4.10}$$

As a consequence of the occurrence of the first x-derivative in (7.4.3 a) the Wronskian depends explicitly on x.

Now we may construct the Green's function

$$K(x, x') = \theta(x - x')\, K^<(x, x') + \theta(x' - x)\, K^>(x, x') \tag{7.4.11 a}$$

with

$$K^{<}(x, x') = \frac{e^{-\frac{i\beta}{2\omega}(x-x')}\,\varphi(x)\,\psi(x')}{\frac{i\beta}{2\omega} - \alpha + \frac{\varkappa f}{\omega}} \tag{7.4.11 b}$$

$$K^{>}(x, x') = \frac{e^{-\frac{i\beta}{2\omega}(x-x')}\,\varphi(x')\,\psi(x)}{\frac{i\beta}{2\omega} - \alpha + \frac{\varkappa f}{\omega}}\,. \tag{7.4.11 c}$$

From (7.4.9) and (7.4.11) it can be seen that for real α the Green's function is localized at $x = x'$. However, for imaginary α, which implies according to (7.4.7) $\omega < \beta R/2$, $K(x, x')$ is oscillatory in x.

We may, e.g., $K^{<}(x, x')$ put into the form

$$K^{<}(x, x') = \frac{e^{-\frac{i\beta}{2\omega}(x-x')}\,\omega\left(i\,\frac{\beta}{2} + \varkappa f + \alpha\omega\right)\varphi(x)\,\psi(x')}{f^2(\varkappa - \varkappa_1)\,(\varkappa - \varkappa_2)}$$

where $\varkappa_{1,2} = -i\beta/(2f) \pm \lambda(\omega^2 - \beta^2 R^2/4)^{1/2}$. If the β-effect is negligible this expression becomes the subinertial limit of the f-plane Green's function discussed in Section 5.2., see, in particular, (5.2.14). Then the poles at $\omega^2\lambda^2 = \varkappa^2$ represents the ordinary Kelvin poles. On the β-plane, however, those poles are shifted into the complex $\varkappa$-plane and consequently the β-plane Kelvin waves are damped. For frequencies below the maximum frequency of Rossby waves, i.e., $\omega < \beta R/2$, the poles are located on the imaginary axis of the $\varkappa$-plane and then Kelvin waves do not longer exist.

7.4.2. The formal solution

In order to find the formal solution of the forcing problem in the frame of the considered model let us introduce the operator $\mathbf{D}(x)$ acting on x by

$$\mathbf{D}(x) = \frac{\partial^2}{\partial x^2} + \frac{i\beta}{\omega}\frac{\partial}{\partial x} - \left(\varkappa^2 + \frac{1}{R_n^2}\right). \tag{7.4.12}$$

Then (7.4.3a) reads

$$\mathbf{D}(x)\,K(x, x') = \delta(x - x')\,.$$

Let $f(x)$ and $g(x)$ be two arbitrary functions which vanish at some boundaries, say $x = a$ and $x = b$. Then the adjoint operator $\tilde{\mathbf{D}}(x)$ is defined by the relation

$$\int_a^b f(x)\,\mathbf{D}(x)\,g(x)\,\mathrm{d}x = \int_a^b g(x)\,\tilde{\mathbf{D}}(x)\,f(x)\,\mathrm{d}x\,,$$

implying

$$\tilde{\mathbf{D}}(x) = \frac{\partial^2}{\partial x^2} - \frac{i\beta}{\omega}\frac{\partial}{\partial x} - \left(\varkappa^2 + \frac{1}{R_n^2}\right). \tag{7.4.13}$$

Obviously the adjoint operator $\tilde{\mathbf{D}}(x)$ equals the complex conjugate of $\mathbf{D}(x)$, denoted by $\mathbf{D}^*(x)$, and, hence, $\mathbf{D}(x) = \tilde{\mathbf{D}}^*(x)$. Thus $\mathbf{D}(x)$ is a Hermitian operator.

By means of (7.4.13) and (7.4.11) we may readily verify that

$$\tilde{\mathbf{D}}(x')\, K(x, x') = \delta(x - x') . \tag{7.4.14}$$

Let us rewrite (7.4.2a) as

$$\mathbf{D}(x')\, p(x') = \frac{\mathrm{i}f}{\omega}\,[Y_{x'}(x') - \mathrm{i}\varkappa X(x')] + \frac{\mathrm{i}\beta}{\omega}\, X(x') + \mathrm{i}\varkappa Y(x') + X_{x'}(x') . \tag{7.4.15}$$

Multiply (7.4.15) by $K(x, x')$ and (7.4.14) by $p(x)$, integrate from $-\infty$ to 0, and substract the result to obtain

$$\int_{-\infty}^{0} [K(x, x')\, \mathbf{D}(x')\, p(x') - p(x')\, \tilde{\mathbf{D}}(x')\, K(x, x')]\, \mathrm{d}x'$$

$$= -p(x) + \frac{\mathrm{i}f}{\omega}\,(K * Y_{x'} - \mathrm{i}\varkappa K * X) + \frac{\mathrm{i}\beta}{\omega}\, K * X + K * X_{x'} + \mathrm{i}\varkappa K * Y$$

where

$$K * Y = \int_{-\infty}^{0} \mathrm{d}x'\, K(x, x')\, X(x') , \qquad K * Y = \int_{-\infty}^{0} \mathrm{d}x'\, K(x, x')\, Y(x') . \tag{7.4.16}$$

After twice integrating by parts and taking the boundary conditions (7.4.3b, c) and (7.4.2b, c) into account we obtain

$$\int_{-\infty}^{0} p(x')\, \tilde{\mathbf{D}}(x')\, K(x, x')\, \mathrm{d}x' = \int_{-\infty}^{0} K(x, x')\, \mathbf{D}(x')\, p(x')\, \mathrm{d}x' - \left[X(0) + \frac{\mathrm{i}f}{\omega}\, Y(0)\right] K(x, 0) . \tag{7.4.17}$$

Inserting this into (7.4.16) we arrive at

$$p(x) = -K(x, 0)\left[X(0) + \frac{\mathrm{i}f}{\omega}\, Y(0)\right] + \frac{\mathrm{i}f}{\omega}\,(K * Y_{x'} - \mathrm{i}\varkappa K * X)$$
$$+ \frac{\mathrm{i}\beta}{\omega}\, K * X + K * X_{x'} + \mathrm{i}\varkappa K * Y . \tag{7.4.18}$$

This is the formal solution of the forcing problem for the midlatidudinal β-plane with an eastern meridional boundary at $x = 0$. The remaining dynamical fields may be expressed in terms of p and of the external forces, compare (7.1.1). Thus in the subinertial region, $\omega \ll f$, we obtain

$$w = \frac{\mathrm{i}\omega}{N^2}\, p_z \tag{7.4.19a}$$

$$u = -\frac{\mathrm{i}\varkappa}{f}\, p + \frac{\mathrm{i}\omega}{f^2}\, p_x + \frac{Y}{f} \tag{7.4.19b}$$

$$v = \frac{p_x}{f} - \frac{\omega\varkappa}{f^2}\, p - \frac{X}{f} . \tag{7.4.19c}$$

7.4.3. Forcing parallel to the coast

In this and the following section we discuss some explicit examples. We start with a meridional wind blowing over a coastal strip of width l.

The wind may vary sinusoidally with the frequency ω_0, whereby $\omega_0 \ll f$

$$Y(x, z, t) = -\frac{v_*^2}{H_{\text{mix}}} \theta(z + H_{\text{mix}}) \theta(x + l) \sin \omega_0 t, \qquad X = 0. \tag{7.4.20}$$

After normal mode expansion and Fourier transforms with respect to t and y we get, compare (B 8),

$$Y_n(\omega, \varkappa, x) = -\frac{v_*^2}{h_n} \theta(x + l) \, i\omega_0 \operatorname{sgn}(\omega) \, \pi \, \delta(\omega^2 - \omega_0^2) \, 2\pi \, \delta(\varkappa). \tag{7.4.21}$$

We suppress the index n below. Inserting (7.4.21) in (7.4.18) it follows

$$p(\omega, \varkappa, x) = \frac{v_*^2}{h} \cdot 2\pi^2 \delta(\varkappa) \, i\omega_0 \operatorname{sgn}(\omega) \, \delta(\omega^2 - \omega_0^2) \frac{if}{\omega} [K(x, 0) - K(x, -l)]. \tag{7.4.22}$$

The expression consists of two contributions, one results from the presence of the boundary and the other comes from the wind curl Y_x which, in our case, is proportional to the delta function $\delta(l + x)$. Let us denote the first part by p^0 and the second one by p^l. We consider first p^0.

From (7.4.11) and (7.4.9) we obtain

$$p^0 = -\frac{v_*^2}{h} \cdot 2\pi^2 \, i\delta(\varkappa) \, \omega_0 \operatorname{sgn}(\omega) \, \delta(\omega^2 - \omega_0^2) \frac{if}{\omega} \left(\frac{i\beta}{2\omega} + \alpha\right) R^2 \, e^{\left(-\frac{i\beta}{2\omega} + \alpha\right) x} \tag{7.4.23}$$

where

$$\left(\frac{i\beta}{2\omega} - \alpha\right)\left(\frac{i\beta}{2\omega} + \alpha\right) = -\left(\frac{\beta^2}{4\omega^2} + \alpha^2\right) = -R^{-2}$$

was used.

Now we may consider the properties of the response for two frequency domains namely $\omega_0 > \beta R/2$ and $\omega_0 < \beta R/2$. For $\omega_0 > \beta R/2$ the Fourier inversion of (7.4.23) yields

$$p^0(x, t) = \frac{v_*^2}{h} f R^2 \, e^{\alpha x} \left[\frac{\beta}{2\omega_0^2} \sin\left(\omega_0 t + \frac{\beta x}{2\omega_0}\right) + \frac{\alpha}{\omega_0} \cos\left(\omega_0 t + \frac{\beta x}{2\omega_0}\right)\right]. \tag{7.4.24a}$$

With (7.4.24a) we obtain from (7.4.19b, c)

$$u(x, t) = -\frac{v_*^2}{hf}\left[\theta(x + l) \sin \omega_0 t - e^{\alpha x} \sin\left(\omega_0 t + \frac{\beta x}{2\omega_0}\right)\right] \tag{7.4.24b}$$

and

$$v(x, t) = \frac{v_*^2}{h\omega_0} e^{\alpha x} \cos\left(\omega_0 t + \frac{\beta x}{2\omega_0}\right). \tag{7.4.24 c}$$

As long as the β-dispersion is negligible the expressions (7.4.24a, c) describe a coastal jet whose off-shore scale is the Rossby radius. The jet is 90° out of phase with the wind. The zonal component consists, according to (7.4.24b), of the Ekman flow and a coastal rectification which ensures u to be zero at $x = 0$. This component is in phase with the wind.

With decreasing forcing frequency, i.e., ω_0 close to $\beta R/2$, the β-effect becomes increasingly important. Since α depends on the frequency through $\alpha = [1/R^2 - \beta^2/(4\omega_0^2)]^{1/2}$ the off-shore scale of the jet depends on the frequency or, equivalently, on the time scale of the forcing. Hence, it may considerably exceed the radius of deformation. Moreover, within a coastal region of the width α^{-1} there occurs marginally a wave propagation with the westward phase speed $c = -2\omega_0^2/\beta$.

Next we calculate the response pattern for $\omega_0 < \beta R/2$. In order to enter this frequency domain we have to go around the branch points, $\omega_b = \pm\beta R/2$, and so we must be carefully about + and − signs. In other words we have to stay on one Riemann's sheet of the complex ω-plane. This amounts to the choice

$$\alpha = \begin{cases} \sqrt{\dfrac{1}{R^2} - \dfrac{\beta^2}{4\omega_0^2}} & \text{for} \quad \omega_0 > \dfrac{1}{2}\beta R \\ \mathrm{i}\sqrt{\dfrac{\beta^2}{4\omega_0^2} - \dfrac{1}{R^2}} & \text{for} \quad 0 < \omega_0 < \dfrac{1}{2}\beta R \\ -\mathrm{i}\sqrt{\dfrac{\beta^2}{4\omega_0^2} - \dfrac{1}{R^2}} & \text{for} \quad -\dfrac{1}{2}\beta R < \omega_0 < 0 \\ \sqrt{\dfrac{1}{R^2} - \dfrac{\beta^2}{4\omega_0^2}} & \text{for} \quad \omega_0 < -\dfrac{1}{2}\beta R\,. \end{cases} \tag{7.4.25}$$

We introduce the notation $\tilde{\alpha} = \sqrt{\dfrac{\beta^2}{4\omega_0^2} - \dfrac{1}{R^2}}$. Then from (7.4.23) it follows

$$p^0(x, t) = \frac{v_*^2}{h}\frac{R^2 f}{\omega_0}\left(\frac{\beta}{2\omega_0} + \tilde{\alpha}\right)\sin\left(\omega_0 t + \frac{\beta x}{2\omega_0} - \tilde{\alpha}x\right) \tag{7.4.26 a}$$

For u^0 and v^0 we obtain

$$u^0(x, t) = -\frac{v_*^2}{hf}\left[\theta(x + l)\sin\omega_0 t + \sin\left(\omega_0 t + \frac{\beta x}{2\omega_0} - \tilde{\alpha}x\right)\right] \tag{7.4.26 b}$$

and

$$v^0(x, t) = \frac{v_*^2}{h\omega_0}\cos\left(\omega_0 t + \frac{\beta x}{2\omega_0} - \tilde{\alpha}x\right). \tag{7.4.26 c}$$

Let, moreover, $\omega_0 \ll \beta R/2$. Then we have approximately

$$\tilde{\alpha} = \frac{\beta}{2\omega_0}\left(1 - \frac{2\omega_0^2}{\beta^2 R^2}\right) = \frac{\beta}{2\omega_0} - \frac{\omega_0}{\beta R^2},$$

and, hence,

$$p^0(x, t) = \frac{v_*^2}{h}\frac{f\beta R^2}{\omega_0^2}\sin\left(\omega_0 t + \frac{\omega_0 x}{\beta R^2}\right) \tag{7.4.27a}$$

$$u^0(x, t) = -\frac{v_*^2}{hf}\left[\theta(x + l)\sin\omega_0 t - \sin\left(\omega_0 t + \frac{\omega_0 x}{\beta R^2}\right)\right] \tag{7.4.27b}$$

$$v^0(x, t) = \frac{v_*^2}{h\omega_0}\cos\left(\omega_0 t + \frac{\omega_0 x}{\beta R^2}\right). \tag{7.4.27c}$$

The set (7.4.27) describes forced Rossby waves with frequency ω_0 and wavenumber $\omega_0/(\beta R^2)$. These waves propagate westward with the long wave phase speed $c = -\beta R^2$, see (7.2.17b). That means for frequencies well below the maximum frequency of Rossby waves, $\beta R/2$, the coastal jet has dispersed into Rossby waves, which propagate westward with maximum phase speed.

Since near the eastern boundaries of both the Pacific and the Atlantic ocean the meridional components of the trade winds exhibit an annual cycle we may compare the minimum period of Rossby waves, $4\pi/(\beta R_n)$, with the time scale of one year. Using the vertical eigenvalues listed in Table 1.9.2 or 1.9.3 it appears that the barotropic, the first and, eventually, the second baroclinic mode can radiate as annual Rossby waves into the ocean. Thus according to (7.4.24) and (7.4.26) the vertical structure of the coastal jet oscillating with annual period is altered by the radiation of these Rossby waves.

In the wake of the westward propagating Rossby waves a complex system of equatorward and poleward flowing currents has been established whose zonal scales are the wavelengths of the annual Rossby waves. Due to the meridional variations of f, and, consequently, of the Rossby radius, the wavelength of the first mode Rossby wave varies from 400 km between 30° to 40° N to 1000 km between 20° to 30° N.

In spite of the high degree of idealization of the considered model these zonal scales agree well with observed scales of annual baroclinic Rossby waves, see e.g. WHITE (1977) and KANG and MAGAARD (1980).

Next we discuss the second contribution of (7.4.22), which arises from the wind curl. From (7.4.11) we obtain, after simple rearrangements, the following convenient form of the Green's function

$$K(x, -l) = \frac{1}{2\alpha}\,e^{-\frac{i\beta(x+l)}{2\omega}}\left[\frac{\left(\alpha + \frac{i\beta}{2\omega}\right)}{\left(\alpha - \frac{i\beta}{2\omega}\right)}e^{\alpha(x-l)} + e^{-\alpha|x+l|}\right]. \tag{7.4.28}$$

Thus, it follows from (7.4.22)

$$p^l(x, t) = -\frac{v_*^2 f}{h\cdot 2\omega_0\alpha}\left\{\cos\left(\omega_0 t + \frac{\beta}{2\omega_0}(x + l)\right)\left[e^{-\alpha|x+l|} + \left(1 - \frac{1}{2}\left(\frac{\beta R}{\omega_0}\right)^2\right)e^{\alpha(x-l)}\right]\right.$$
$$\left. + \sin\left(\omega_0 t + \frac{\beta}{2\omega_0}(x + l)\right)\frac{\beta\alpha R^2}{\omega_0}e^{\alpha(x-l)}\right\}. \tag{7.4.29a}$$

With $u^l = -p^l_{xt}/f^2$ and $v^l = p^l_x/f$ we get, moreover,

$$u^l(x,t) = -\frac{v_*^2}{h\cdot 2f\alpha}\left\{\frac{\beta}{2\omega_0}\cos\left[\omega_0 t + \frac{\beta}{2\omega_0}(x+l)\right]\left[e^{\alpha(x-l)} - e^{-\alpha|x+l|}\right]\right.$$
$$\left. - \alpha\sin\left[\omega_0 t + \frac{\beta}{2\omega_0}(x+l)\right]\left[e^{\alpha(x-l)} - \mathrm{sgn}\,(x+l)\, e^{-\alpha|x+l|}\right]\right\} \qquad (7.4.29\,\mathrm{b})$$

$$v^l(x,t) = \frac{v_*^2}{h\cdot 2\alpha\omega_0}\left\{\frac{\beta}{2\omega_0}\sin\left[\omega_0 t + \frac{\beta}{2\omega_0}(x+l)\right]\left[e^{\alpha(x-l)} - e^{-\alpha|x+l|}\right]\right.$$
$$\left. + \alpha\cos\left[\omega_0 t + \frac{\beta}{2\omega_0}(x+l)\right]\left[e^{\alpha(x-l)} - \mathrm{sgn}\,(x+l)\, e^{-\alpha|x+l|}\right]\right\}. \qquad (7.4.29\,\mathrm{c})$$

If the Rossby dispersion is negligible the expressions (7.4.29a, c) describe a wind edge jet which is 90° out of phase with the wind, and which has been considered in Section 3.9 for a switch-on forcing of a f-plane ocean. The zonal current u^l describes the rectification of the Ekman flow which was accounted for already in u^0, see (7.4.24b), in the vicinity of the wind edge.

For decreasing forcing frequency ω_0 the trapping scale depends on that frequency because then the β-dispersion comes into play and so the wind edge jet broadens. For frequencies below the maximum frequency of Rossby waves we find instead of (7.4.29)

$$p^l(x,t) = \frac{v_*^2 f}{h\beta}\left\{\frac{\beta^2R^2}{\omega_0^2}\sin\left[\omega_0 t + \frac{\beta l}{\omega_0} + \frac{\omega_0}{\beta R^2}(x+l)\right]\right.$$
$$\left. - \sin\left[\omega_0 t + \frac{\beta}{2\omega_0}(x+l+|x+l|) - \frac{\omega_0}{\beta R^2}|x+l|\right]\right\} \qquad (7.4.30\,\mathrm{a})$$

$$u^l(x,t) = \frac{v_*^2}{hf}\left\{\sin\left[\omega_0 t + \frac{\beta l}{\omega_0} + \frac{\omega_0}{\beta R^2}(x-l)\right] - \left[\theta(x+l) + \frac{\omega_0^2}{\beta R^2}\theta(-x-l)\right]\right.$$
$$\left. \times \sin\left[\omega_0 t + \frac{\beta}{2\omega_0}(x+l+|x+l|) - \frac{\omega_0}{\beta^2R^2}|x+l|\right]\right\} \qquad (7.4.30\,\mathrm{b})$$

$$v^l(x,t) = \frac{v_*^2}{h\omega_0}\left\{\cos\left[\omega_0 t + \frac{\beta l}{\omega_0} + \frac{\omega_0}{R}(x-l)\right] - \left[\theta(x+l) + \frac{\omega_0^2}{\beta^2R^2}\theta(-x-l)\right]\right.$$
$$\left. \times \cos\left[\omega_0 t + \frac{\beta}{2\omega_0}(x+l)\,\theta(x+l) - \frac{\omega_0}{\beta R^2}|x+l|\right]\right\} \qquad (7.4.30\,\mathrm{c})$$

where $\omega_0 \ll \beta R/2$ has been assumed. This set describes two kinds of Rossby waves radiating from the wind edge, namely, short Rossby waves which propagate with eastward group velocity into the forcing region, and long Rossby waves which propagate westward with the maximum group velocity $c = -\beta R^2$. The complete solution of the forcing problem, describing the response to (7.4.20) is given by the sum of the coastal and the wind curl contribution, i.e., the fields with the upper index "0" and "l".

Finally we append a calculation of the response to a steady wind blowing from north

$$Y = -\frac{v_*^2}{H_{mix}}\,\theta(z + H_{mix})\,. \tag{7.4.31}$$

Expansion of Y in vertical eigenfunctions and performing the Fourier transform yields

$$Y = -\frac{v_*^2}{h}(2\pi)^2\,\delta(\varkappa)\,\delta(\omega)\,.$$

In this case dissipation must be taken into account, i.e., we replace ω by $\bar\omega = \omega + ir$ in (7.4.18) and we find

$$p(x) = \frac{v_*^2}{h}(2\pi)^2\,\delta(\varkappa)\,\delta(\omega)\,\frac{if}{\bar\omega}\,K(x, 0)$$

or

$$p(x) = \frac{v_*^2}{h}\,(2\pi)^2\,\delta(\varkappa)\,\delta(\omega)\,\frac{f}{r^2}\,\frac{\beta\,e^{\left(\alpha - \frac{\beta}{2r}\right)x}}{2\alpha\left(\frac{\beta}{2r} - \alpha\right)}\,.$$

Assuming $r \ll \beta R/2$, for $\omega = 0$ we obtain

$$\alpha = \left(\frac{1}{R^2} + \frac{\beta^2}{4r^2}\right)^{1/2} \approx \frac{\beta}{2r} + \frac{r}{\beta R^2}$$

Hence, we get

$$p(x) = -\frac{v_*^2}{h}\,(2\pi)^2\,\delta(\varkappa)\,\delta(\omega)\,\frac{\beta R^2}{r^2}\,f\,e^{\frac{rx}{\beta R^2}}\,,$$

and the Fourier transform yields

$$p(x, t) = -\frac{v_*^2}{h}\,\frac{\beta^2 R^2}{r^2}\,e^{\frac{rx}{\beta R^2}}\,, \tag{7.4.32 a}$$

and so we get

$$v = \frac{p_x}{f} = -\frac{v_*^2}{hr}\,e^{\frac{rx}{\beta R^2}}\,. \tag{7.4.32 b}$$

This describes a steady eastern boundary current whose off-shore scale is given by $\beta R^2/r$, that is the distance the front of a long Rossby wave travels westward with the longwave group velocity in the time r^{-1}, i.e., the time scale at which dissipation becomes important. We may add (7.4.31) and (7.4.20) to model a meridional wind consisting of a steady part superimposed by an oscillation with seasonal or annual periods. Because of the linearity of the theory we may add the corresponding solutions to obtain a complex system of equatorward and poleward currents. Numerical simulations of such a regime were presented by several authors e.g. PHILANDER and YOON (1982), KRAUSS and WÜBBER (1982) and HOLLAND (1983).

7.4.4. Forcing by a zonal wind

The large-scale circulation patterns of the oceans consist of basin-wide anticyclonic gyres driven by the large scale wind field which in turn consists of the trade wind belts in the tropics and the west wind belts in the midlatitudes, see e.g. GILL (1982). Some aspects of the formation of the stationary patterns and of the associated time scales can be analyzed by considering the response of the ocean to a rotational wind. To this end we study in this subsection the response to a rotational zonal wind switched on at $t = 0$.

In particular we wish to analyze the response to a zonal wind, e.g.,

$$X(x, y, z, t) = \frac{u_*^2}{H_{\text{mix}}} \theta(z + H_{\text{mix}})\, \theta(t) \cos \varkappa_0 y\,, \qquad Y = 0 \tag{7.4.33}$$

which has a non-vanishing curl.

Expansion of X in vertical eigenfunctions and Fourier transforms with respect to t and y yields

$$X_n(x, \varkappa, \omega) = \frac{u_*^2}{h_n} \frac{\mathrm{i}}{\omega + \mathrm{i}\varepsilon} \pi[\delta(\varkappa + \varkappa_0) + \delta(\varkappa - \varkappa_0)]\,, \qquad Y = 0\,. \tag{7.4.34}$$

We suppress the vertical mode index below. Inserting (7.4.34) in the formal solution (7.4.18) yields

$$p(x) = \frac{u_*^2}{h} \frac{\mathrm{i}\pi}{\omega + \mathrm{i}\varepsilon} \left[-K(x, 0) + \left(\frac{f\varkappa}{\bar{\omega}} + \frac{\mathrm{i}\beta}{\bar{\omega}} \right) K * 1 \right] [\delta(\varkappa + \varkappa_0) + \delta(\varkappa - \varkappa_0)]\,. \tag{7.4.35}$$

The inverse Fourier transform of (7.4.35) with respect to the meridional wavenumber $\varkappa$ can readily be obtained

$$\begin{aligned} p(x, y, \omega) = \frac{u_*^2}{2h} \frac{\mathrm{i}}{\omega + \mathrm{i}\varepsilon} \Big\{ & -K(x, 0, \varkappa_0)\, \mathrm{e}^{\mathrm{i}\varkappa_0 y} - K(x, 0, -\varkappa_0)\, \mathrm{e}^{-\mathrm{i}\varkappa_0 y} \\ & + \frac{f\varkappa_0}{\bar{\omega}} [K(x, x', \varkappa_0) * 1\, \mathrm{e}^{\mathrm{i}\varkappa_0 y} - K(x, x', -\varkappa_0) * 1\, \mathrm{e}^{-\mathrm{i}\varkappa_0 y}] \\ & + \frac{\mathrm{i}\beta}{\bar{\omega}} [K(x, x', \varkappa_0) * 1\, \mathrm{e}^{\mathrm{i}\varkappa_0 y} + K(x, x', -\varkappa_0) * 1\, \mathrm{e}^{-\mathrm{i}\varkappa_0 y}] \Big\}\,. \end{aligned} \tag{7.4.36}$$

Next we have to calculate the convolution $K * 1$. With (7.4.11) it follows

$$\int_{-\infty}^{0} \mathrm{d}x'\, K(x, x', \varkappa_0) = \frac{\mathrm{e}^{-\frac{\mathrm{i}\beta}{2\bar{\omega}}}}{\frac{\mathrm{i}\beta}{2\bar{\omega}} - \alpha + \frac{f\varkappa_0}{\bar{\omega}}} \left[\varphi(x) \int_{-\infty}^{x} \mathrm{d}x'\, \mathrm{e}^{\left(\frac{\mathrm{i}\beta}{2\bar{\omega}} + \alpha\right)x'} + \mathrm{e}^{\alpha x} \int_{x}^{0} \mathrm{d}x'\, \mathrm{e}^{\frac{\mathrm{i}\beta}{2\bar{\omega}}x'} \varphi(x') \right] \tag{7.4.37}$$

where $\varphi(x)$ is given by (7.4.9a). Thus, we obtain

$$K * 1 = \frac{1}{\varkappa_0^2 + \frac{1}{R_n^2}} \left[\frac{\varkappa_0 f \, e^{\left(-\frac{i\beta}{2\bar{\omega}} + \alpha\right)x}}{i\frac{\beta}{2} + \varkappa_0 f - \alpha\bar{\omega}} - 1 \right] \tag{7.4.38}$$

This expression and consequently (7.4.36) has two branch points in the ω-plane at frequencies where $\alpha = [\varkappa_0^2 + R_n^{-2} - \beta^2/(4\omega^2)]^{1/2} = 0$. These frequencies correspond to Rossby waves with vanishing zonal group velocity. Moreover, there are singularities at $i\beta/2 + \varkappa_0 f = \alpha\bar{\omega}$ describing a damped coastal Kelvin wave and at $\omega = 0$ and at $\bar{\omega} = 0$ corresponding to non-oscillatory motion. In the following we confine ourselfes to the discussion of the asymptotic properties of (7.4.36) for large times or, equivalently, small frequencies.

For small ω we have approximately

$$\alpha = \frac{i\beta}{2\bar{\omega}}\left[1 - \frac{4\bar{\omega}^2}{\beta^2}(\varkappa_0^2 + R_n^{-2})\right]^{1/2} \approx \frac{i\beta}{2\bar{\omega}} - i\frac{\bar{\omega}}{\beta}(\varkappa_0^2 + R_n^{-2}). \tag{7.4.39}$$

Substituting (7.4.9b) in (7.4.11c) we obtain with (7.4.39)

$$K(x, 0, \varkappa_0) \approx \frac{\bar{\omega}\, e^{-i\frac{\bar{\omega}}{\beta}(\varkappa_0^2 + R_n^{-2})x}}{\varkappa_0 f + i\frac{\bar{\omega}^2}{\beta}(\varkappa_0^2 + R_n^{-2})} \approx \frac{\bar{\omega}}{\varkappa_0 f}\, e^{-i\frac{\bar{\omega}}{\beta}(\varkappa_0^2 + R_n^{-2})x}. \tag{7.4.40}$$

Inserting this into (7.4.36) it follows that the first two terms in (7.4.36) have no singularity in the ω-plane and, therefore, they do not contribute to the asymptotic regime. Similarly, we find from (7.4.38) for small ω

$$K * 1 = \frac{1}{\varkappa_0^2 + R_n^{-2}}\left[e^{-i\frac{\bar{\omega}}{\beta}(\varkappa_0^2 + R_n^{-2})x} - 1\right]. \tag{7.4.41}$$

Inserting this into (7.4.36) we obtain

$$p(x, y, \omega) = -\frac{u_*^2}{h}\frac{e^{-i\frac{\bar{\omega}}{\beta}(\varkappa_0^2 + R_n^{-2})x} - 1}{(\varkappa_0^2 + R_n^{-2})(\omega + i\varepsilon)(\omega + ir)}(f\varkappa_0 \sin\varkappa_0 y + \beta\cos\varkappa_0 y). \tag{7.4.42}$$

Inverting the Fourier transforms yields

$$p(x, y, t) = -\frac{u_*^2}{h}\frac{f\varkappa_0 \sin\varkappa_0 y + \beta\cos\varkappa_0 y}{\varkappa_0^2 + R^{-2}}\left[\theta(t)\frac{1 - e^{-rt}}{r} - \theta\left(t - \frac{x}{c}\right)\frac{e^{-\frac{x}{c}r} - e^{-tr}}{r}\right] \tag{7.4.43 a}$$

with

$$c = -\frac{\beta R^2}{1 + \varkappa_0^2 R^2}$$

being the maximum group velocity of long Rossby waves.

For a vanishing friction parameter r this reduces to

$$p(x, y, t) = -\frac{u_*^2}{h}\,\frac{f\varkappa_0 \sin\varkappa_0 y + \beta\cos\varkappa_0 y}{\varkappa_0^2 + R^{-2}}\left[\theta(t)\,t - \theta\left(t - \frac{x}{c}\right)\left(t - \frac{x}{c}\right)\right] \tag{7.4.43 b}$$

Substituting (7.4.43a, b) in (7.1.8a, b) we can calculate the horizontal current components. For the zonal current it follows

$$u(x, y, t) = \frac{u_*^2\varkappa_0}{hf}\,\frac{f\varkappa_0\cos\varkappa_0 y - \beta\sin\varkappa_0 y}{\varkappa_0^2 + R^{-2}}\left[\theta(t)\,\frac{1 - e^{-rt}}{r} - \theta\left(t - \frac{x}{c}\right)\frac{e^{-\frac{x}{c}r} - e^{-tr}}{r}\right] \tag{7.4.44 a}$$

in the viscid case, and

$$u(x, y, t) = \frac{u_*^2\varkappa_0}{hf}\,\frac{f\varkappa_0\cos\varkappa_0 y - \beta\sin\varkappa_0 y}{\varkappa_0^2 + R^{-2}}\left[\theta(t)\,t - \theta\left(t - \frac{x}{c}\right)\left(t - \frac{x}{c}\right)\right] \tag{7.4.44 b}$$

in the inviscid case, $r \to 0$.

For the meridional component we have analogously

$$\begin{aligned} v(x, y, t) = -\frac{u_*^2}{h}\Bigg\{&\theta(t)\,\frac{\cos\varkappa_0 y}{f} + \frac{\varkappa_0}{f^2}\,\frac{\varkappa_0 f\cos\varkappa_0 y - \beta\sin\varkappa_0 y}{\varkappa_0^2 + R^{-2}}\left[\theta(t) - \theta\left(t - \frac{x}{c}\right)e^{-tr}\right] \\ &- \theta\left(t - \frac{x}{c}\right)\left(\frac{\cos\varkappa_0 y}{f} + \frac{\varkappa_0}{\beta}\sin\varkappa_0 y\right)e^{-\frac{x}{c}r}\Bigg\} \end{aligned} \tag{7.4.45 a}$$

and

$$\begin{aligned} v(x, y, t) = -\frac{u_*^2}{h}\Bigg\{&\theta(t)\,\frac{\cos\varkappa_0 y}{f} + \frac{\varkappa_0}{f^2}\,\frac{\varkappa_0 f\cos\varkappa_0 y - \beta\sin\varkappa_0 y}{\varkappa_0^2 + R^{-2}}\left[\theta(t) - \theta\left(t - \frac{x}{c}\right)\right] \\ &- \theta\left(t - \frac{x}{c}\right)\left(\frac{\cos\varkappa_0 y}{f} + \frac{\varkappa_0}{\beta}\sin\varkappa_0 y\right)\Bigg\}. \end{aligned} \tag{7.4.45 b}$$

The modal vertical component can be obtained from $w = i\bar{\omega}p$, i.e. we get

$$w(x, y, t) = \frac{u_*^2}{h}\,\frac{f\varkappa_0\sin\varkappa_0 y + \beta\cos\varkappa_0 y}{\varkappa_0^2 + R^{-2}}\left[\theta(t) - \theta\left(t - \frac{x}{c}\right)e^{-\frac{x}{c}r}\right] \tag{7.4.46 a}$$

and

$$w(x, y, t) = \frac{u_*^2}{h}\,\frac{f\varkappa_0\sin\varkappa_0 y + \beta\cos\varkappa_0 y}{\varkappa_0^2 + R^{-2}}\left[\theta(t) - \theta\left(t - \frac{x}{c}\right)\right]. \tag{7.4.46 b}$$

The solutions (7.4.43) to (7.4.46) characterize two different physical regimes which we discuss with the aid of the inviscid solutions, just for simplicity. The initial phase is governed by a stationary upwelling driven by the wind stress curl as described by (7.4.46 b). This upwelling generates zonally stretching pressure ridges and troughs with accelerating amplitude as given by (7.4.43 b). The zonal current consists of alternating bands of accelerating zonal flows, see (7.4.44 b), while the meridional current is dominated by the Ekman flow.

After the arrival of the Rossby wave front the upwelling vanishes and the pressure as well as the zonal current stop growing, that is, the solutions become stationary. From (7.4.45b) it can be seen that the Sverdrup regime

$$\beta v = \frac{u_*^2}{h} \varkappa_0 \sin \varkappa_0 y = -X_y \tag{7.4.47}$$

has established. Thus, similar to the arresting of a coastal jet by the arrival of Kelvin waves, see Section 5.9., the accelerating response changes into a Sverdrup regime after the advent of the Rossby waves.

The associated time scale is given by the travel time of long Rossby waves. For example, for the Atlantic a long Rossby wave needs about 10 days for the barotropic mode and some years for the first baroclinic mode to cross the ocean. After this time the solutions (7.4.43) to (7.4.46) can be expected to describe a part of a gyre in the interior and the eastern part of the ocean.

For a more detailed discussion we refer to ANDERSON and GILL (1975).

7.5. FORCING OF A β-PLANE OCEAN WITH A WESTERN BOUNDARY

7.5.1. The Green's function and the formal solution

In this and the following subsections we wish to analyze the effects of a western boundary on the oceanic response. Similar to subsection 7.4.1. we take the x-axis as meridional boundary as shown in Fig. 7.5.1. This implies some minor modifications of the formalism outlined in Subsections 7.4.1 and 7.4.2, which will be summarized in this section. Instead of the requirements of the boundary conditions (7.4.2a), (7.4.3c), (7.4.4b), and (7.4.8b) in the present case the corresponding functions are postulated to remain finite for $x \to +\infty$, i.e.,

$$p(x), K(x, x'), \Psi(x), \Phi(x) < \infty \quad \text{for} \quad x \to \infty\,. \tag{7.5.1}$$

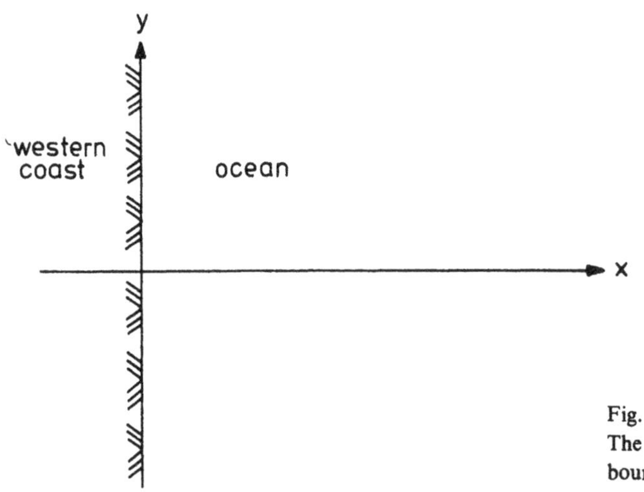

Fig. 7.5.1
The coordinate frame with a western boundary

Thus instead of (7.4.9b) we get

$$\psi(x) = \mathrm{e}^{-\alpha x} \tag{7.5.2}$$

where x is confined to the positive domain, $x > 0$. Note that the shape of $\varphi(x)$ according to (7.4.9a) is not affected. The Wronskian modifies to

$$W_r(x) = -\left(\frac{\mathrm{i}\beta}{2\omega} + \alpha + \frac{\varkappa f}{\omega}\right)\mathrm{e}^{-\frac{\mathrm{i}\beta}{\omega}x}. \tag{7.5.3}$$

With (7.4.9a), (7.5.2) and (7.5.3) we obtain

$$K(x, x') = \theta(x - x')\, K^{>}(x, x') + \theta(x' - x)\, K^{<}(x, x') \tag{7.5.4a}$$

where

$$K^{>}(x, x') = -\frac{\varphi(x')\,\psi(x)}{\dfrac{\mathrm{i}\beta}{2\omega} + \dfrac{\varkappa f}{\omega} + \alpha}\,\mathrm{e}^{-\frac{\mathrm{i}\beta}{2}(x-x')} \tag{7.5.4b}$$

and

$$K^{<}(x, x') = -\frac{\varphi(x)\,\psi(x')}{\dfrac{\mathrm{i}\beta}{2\omega} + \dfrac{\varkappa f}{\omega} + \alpha}\,\mathrm{e}^{-\frac{\mathrm{i}\beta}{2\omega}(x-x')}. \tag{7.5.4c}$$

We may rewrite this as

$$K(x, x') = -\frac{1}{2\alpha}\,\mathrm{e}^{-\frac{\mathrm{i}\beta}{2\omega}(x-x')}\left[\mathrm{e}^{-\alpha|x-x'|} - \frac{\dfrac{\mathrm{i}\beta}{2\omega} + \dfrac{\varkappa f}{\omega} - \alpha}{\dfrac{\mathrm{i}\beta}{2\omega} + \dfrac{\varkappa f}{\omega} + \alpha}\,\mathrm{e}^{-\alpha(x+x')}\right]. \tag{7.5.5}$$

With (7.5.4) from (7.4.2a, b) and (7.5.1) we may readily derive a source representation of p in the manner outlined in Subsection 7.4.2.

$$p(x) = K(x, 0)\left(X(0) + \frac{\mathrm{i}f}{\omega}\,Y(0)\right) + \frac{\mathrm{i}f}{\omega}(K * Y_{x'} - \mathrm{i}\varkappa K * X) + K * X_{x'} + \mathrm{i}\varkappa K * Y$$
$$+ \frac{\mathrm{i}\beta}{\omega} K * X. \tag{7.5.6}$$

7.5.2. Response to meridional forcing

Next we study the response of the ocean to a forcing by a meridional wind. We start with a switch-on forcing of the form

$$Y = \frac{v_*^2}{H_{\mathrm{mix}}}\,\theta(t)\,\theta(z + H_{\mathrm{mix}}) \tag{7.5.7}$$

whose Fourier transform is after projection onto vertical eigenfunctions

$$v = \frac{v_*^2}{h} \cdot 2\pi\,\delta(\varkappa)\,\frac{\mathrm{i}}{\omega + \mathrm{i}\varepsilon}\,. \tag{7.5.8}$$

From (7.5.6) we get with (7.5.8)

$$p(\omega, \varkappa, x) = \frac{v_*^2}{h}\,\frac{\mathrm{i}}{\omega + \mathrm{i}\varepsilon}\,2\pi\,\delta(\varkappa)\,\frac{\mathrm{i}f}{\omega}\,K(x, 0) \tag{7.5.9}$$

where according to (7.5.5)

$$K(x, 0) = -\frac{\mathrm{e}^{-\frac{\mathrm{i}\beta}{2\omega}x - \alpha x}}{\dfrac{\mathrm{i}\beta}{2\omega} + \alpha}\,.$$

The estimation of the inverse Fourier transform of (7.5.9) amounts to the calculation of the integral

$$p(x, t) = +\frac{v_*^2}{h}\,fR^2 \int_{-\infty}^{+\infty} \frac{\mathrm{d}\omega}{2\pi}\,\frac{1}{\omega^2}\left(-\frac{\mathrm{i}\beta}{2\omega} + \alpha\right)\mathrm{e}^{-\frac{\mathrm{i}\beta}{2\omega} - \alpha x - \mathrm{i}\omega t} \tag{7.5.10}$$

with $\alpha = [1/R^2 - \beta^2/(4\omega^2)]^{1/2}$. This integral can be evaluated in the manner outlined in Subsection 7.3.1, see in particular (7.3.11) to (7.3.12). The choice of the integration contour is depicted in Fig. 7.3.2. The asymptotic result follows as

$$p(x, t) = \frac{v_*^2}{h}\left\{-J_0(2\sqrt{\beta x t}) + \sqrt{\frac{2f}{\pi\omega_\mathrm{b}}}\,\frac{1}{(\omega_\mathrm{b} t)^{3/2}}\left[x\cos\left(\omega_\mathrm{b} t + \frac{x}{R} + \frac{\pi}{4}\right)\right.\right.$$
$$\left.\left.- R\sin\left(\omega_\mathrm{b} t + \frac{x}{R} + \frac{\pi}{4}\right)\right]\right\} \tag{7.5.11 a}$$

where $\omega_\mathrm{b} = \beta R/2$.

Moreover, from $v = p_x/f$ and $u = -p_{xt}/f^2 + Y/f = -v_t/f + Y/f$ we obtain

$$v(x, t) = \frac{v_*^2}{h}\left[\left(\frac{t}{x\beta}\right)^{1/2} J_1(2\sqrt{x\beta t}) - \left(\frac{2}{\pi}\right)^{1/2}\frac{1}{\omega_\mathrm{b}}\,\frac{1}{(\omega_\mathrm{b} t)^{3/2}}\,\frac{x}{R}\cos\left(\omega_\mathrm{b} t + \frac{x}{R} + \frac{\pi}{4}\right)\right]$$

and (7.5.11 b)

$$u(x, t) = \frac{v_*^2}{hf}\left[1 - J_0(2\sqrt{x\beta t}) - \left(\frac{2}{\pi}\right)^{1/2}\frac{1}{(\omega_\mathrm{b} t)^{3/2}}\,\frac{x}{R}\cos\left(\omega_\mathrm{b} t + \frac{x}{R} + \frac{\pi}{4}\right)\right]. \tag{7.5.11 c}$$

With the results of Section 5.4 and the set (7.5.11) we may describe the complete response scenario. In the initial phase, i.e., time scales exceeding an inertial period, the response consists of off-shore radiating inertial waves and the formation of a coastal jet, see (5.4.13) and (5.4.14). For larger time scales, particularly $t \gtrsim 4\pi/(R\beta)$, the β-dispersion comes into play and the coastal jet disperses into Rossby waves. Behind a front of short Rossby waves propagating eastward with maximum group velocity the asymptotic regime governed by the set (7.5.11 c) establishes. That is, the coastal jet has dispersed into a band of alternating jets, which are described by the Bessel functions. According to (7.5.11b) these jets increase in time as $\sqrt{t}$ and being confined to the surface layer. We may define the off-shore scale L_b of the jet close to the coast by the first zero of $J_0(2\sqrt{x\beta t})$. From $2\sqrt{L_b\beta t} = 2.4$ we get $L_b = 1.4/(\beta t)$. Thus, the jet becomes progressively thinner.

The remaining terms of (7.5.11) describe Rossby waves of zero group velocity which decay algebraically as $t^{-3/2}$. Thus after sufficiently long time the regime of the western boundary layer may be described by

$$p(x, z, t) = -\frac{v_*^2}{H_{mix}} J_0(2\sqrt{x\beta t})\, \theta(z + H_{mix}) \tag{7.5.12a}$$

$$v(x, z, t) = \frac{v_*^2}{H_{mix}} \left(\frac{t}{\beta x}\right)^{1/2} J_1(2\sqrt{x\beta t})\, \theta(z + H_{mix}) \tag{7.5.12b}$$

$$u(x, z, t) = \frac{v_*^2}{H_{mix}} \frac{1}{f}\, \theta(z + H_{mix}) \left[1 - J_0(2\sqrt{x\beta t})\right] \tag{7.5.12c}$$

where the sum over all vertical modes was performed according to (3.8.20).

Boundary layers of the form (7.4.4) have been discussed by several authors. Thus, e.g., ANDERSON and GILL (1975) studied the spin-up of the ocean taking the formation of boundary layers into account. ANDERSON and ROWLANDS (1976) analyzed the response of an equatorial β-plane near the west coast to local forcing by a longshore wind in order to model the Somali current.

Clearly, for very long time frictional effects have to be taken into account since otherwise the boundary layer (7.5.12) diminishes. In order to study this problem we consider a steady forcing which may be considered as a switch-on forcing a very long time after the switch-on process, i.e.,

$$Y = \frac{v_*^2}{H_{mix}} \theta(z + H_{mix}) . \tag{7.5.13}$$

After Fourier transformation and expansion in vertical modes we get

$$Y = \frac{v_*^2}{h} (2\pi)^2\, \delta(\omega)\, \delta(\varkappa) . \tag{7.5.14}$$

Replacing in our general solution (7.5.6) ω by $\bar{\omega} = \omega + ir$ and inserting (7.5.14) we get

$$p = -\frac{v_*^2}{h} (2\pi)^2\, \delta(\omega)\, \delta(\varkappa) \frac{if}{\bar{\omega}} \frac{e^{-\left(\alpha + \frac{\beta}{2\bar{\omega}}\right)x}}{\frac{i\beta}{2\bar{\omega}} + \alpha} .$$

Let us assume that $r \ll \beta R/2$. Then we obtain

$$p(x, z, t) = -\frac{v_*^2}{H_{mix}} \frac{f}{\beta} e^{-\frac{\beta}{r}x}\, \theta(z + H_{mix}) \tag{7.5.15a}$$

where the sum over all vertical modes was performed. Moreover, we have

$$v(x, z, t) = \frac{v_*^2}{H_{\text{mix}}} \frac{f}{r} e^{-\frac{\beta}{r}x} \theta(z + H_{\text{mix}}). \tag{7.5.15 b}$$

This describes a steady western boundary current whose off-shore scale is r/β, i.e., the distance of the front of a short Rossby wave travels eastward with its short wave group velocity in the time r^{-1}. Contrary to the eastern frictional boundary current (7.4.32), the western one becomes narrower as r decreases. We may assume that the forcing (7.5.13) is the limit case of (7.5.7) a very long time after the onset. Then (7.5.15) means, that the thinnening of the western boundary layer described by (7.5.12) will be stopped by frictional processes after a time of the order r^{-1}. Moreover, the alternating system of currents and counter currents in (7.5.7) has been damped by the dissipation.

CHAPTER 8

The equatorial β-plane

8.1. INTRODUCTIONAL REMARKS

The response of the equatorial ocean to wind forcing differs significantly in both the transient and equilibrium state from that of the midlatitudes.

Contrary to the midlatitudes, where a spectral gap between inertial and Rossby waves exists, the transient response of the equatorial ocean to changing wind includes all time scales and is confined to a small latitudinal belt centered at the equator, referred to as the equatorial wave guide. Thus signals may propagate rapidly along the equator without large losses in amplitude by meridional dispersion.

The stationary response of equatorial oceans to trade winds over the Pacific and the Atlantic, and to the Monsoons over the Indian Ocean consists in the central parts of the oceans of intense, predominately zonal currents located at the surface and within the thermocline, and of strong boundary currents near the coast.

The equatorial response affects the climate of the earth. The seasonal and interannual changes in the heat content of the upper equatorial ocean depend not only on the local fluxes of heat from the atmosphere to the oceans but also on the large scale response of the oceans to variable winds. A striking example of the intensity of the interaction between atmosphere and the tropical oceans is the so called "El Nino-Southern Oscillation (ENSO)" phenomenon, see e.g. PHILANDER (1983).

In order to explain such phenomena it is important to understand the response of the ocean to changing winds — see, for example, the survey article of McCREARY (1985). The present chapter is devoted to this problem. Again, we try to give a systematic account of the response of the equatorial ocean to different wind patterns by means of the Green's function methods. We confine ourselves to an unbounded ocean. With regard to phenomena such as reflection of equatorial waves at meridional boundaries and the formation of coastal boundary layers of the equatorial ocean we refer to, e.g., MOORE and PHILANDER (1977), and ANDERSON and ROWLANDS (1976).

8.2. THE GOVERNING EQUATIONS

The most important dynamical property of the equatorial ocean is the vanishing of the inertial frequency on the equator, which amounts to the equatorial β-plane approximation

$$f = \beta y . \tag{8.2.1}$$

This approximation will be valid as long as the solutions are strongly enough localized around the equator. Inserting (8.2.1) in the set (1.8.1) the resulting linear Boussinesq equations on an equatorial β-plane are

$$u_t + ru - \beta y v + p_x = X \tag{8.2.2}$$

$$v_t + rv + \beta y u + p_y = Y \tag{8.2.3}$$

$$p_z - b = 0 \tag{8.2.4}$$

$$u_x + v_y + w_z = 0 \tag{8.2.5}$$

$$b_t + rb + N^2(z)\, w = 0 . \tag{8.2.6}$$

The same relaxation parameter r has been used for both the Rayleigh friction and the Newtonian cooling mechanism. This assumption has not much influence on the transient solutions. It was shown by YAMAGATA and PHILANDER (1985) that it is not the Rayleigh damping but the Newtonian cooling that changes the final state of response drastically compared with the inviscid case.

The chosen model is close to LIGHTHILL (1969). We adopt an exponential BVF profile and assume a vanishing vertical velocity component at the flat bottom and at the rigid surface. The latter assumption filters out the barotropic mode. It appears, that the meridional scale being associated with the barotropic mode is too large and would violate the equatorial β-plane approximation.

First we expand u, v, p, X, and Y in terms of vertical eigenfunctions

$$(u, v, p, X, Y) = \sum_m (u_m, v_m, p_m, X_m, Y_m)\, F_m(z) . \tag{8.2.7}$$

Eliminating b and w by (8.2.4) and (8.2.6), inserting (8.2.7) in the remaining equations, and performing the Fourier transforms with respect to t and x we get the set

$$-\mathrm{i}\bar{\omega} u_m - \beta y v_m + \mathrm{i}k p_m = X_m \tag{8.2.8}$$

$$-\mathrm{i}\bar{\omega} v_m + \beta y u_m + p_{my} = Y_m \tag{8.2.9}$$

$$\mathrm{i}k u_m + v_{my} - \mathrm{i}\bar{\omega}\lambda_m^2 p_m = 0 \tag{8.2.10}$$

with $\bar{\omega} = \omega + \mathrm{i}r$. From this set we find in the usual manner the following equation for v_m alone. Dropping the vertical mode index m we get

$$v_{ss}(\omega, k, s) + \left(q + \frac{1}{2} - \frac{s^2}{4}\right) v(\omega, k, s) = \mathrm{i}\, \frac{(\bar{\omega}^2\lambda^2 - k^2)\, Y}{2\lambda\beta\bar{\omega}} + \sqrt{\frac{\lambda}{2\beta}}\, \frac{s}{2}\, X + \frac{k}{\bar{\omega}\sqrt{2\lambda\beta}}\, X_s \tag{8.2.11}$$

where the abbreviations $s = \sqrt{2\lambda\beta}\, y$, and

$$2q + 1 = \frac{(\bar{\omega}^2\lambda^2 - k^2)}{\lambda\beta} - \frac{k}{\lambda\bar{\omega}} \tag{8.2.12}$$

have been introduced.

Moreover, we have the boundary conditions

$$v(\omega, k, s) \to 0 \quad \text{for} \quad s \to \pm\infty . \tag{8.2.13}$$

The equations (8.2.11) and (8.2.12) form an inhomogeneous boundary value problem which can be solved in terms of the corresponding Green's function. The remaining dynamic variables u and p in (8.2.8) to (8.2.10) can be expressed in terms of the meridional component v and the external forces X and Y. This will be done in the following section. For integer q ($q = 0, 1, 2, ...$) we mention that the relation (8.2.12) is the well known dispersion relation of free equatorial trapped waves, henceforth denoted as ETW.

8.3. THE GREEN'S FUNCTION AND THE GENERAL FORMAL SOLUTION

The Green's function is subject to

$$G_{ss}(q, s; s') + \left(q + \frac{1}{2} - \frac{s^2}{4}\right) G(q, s; s') = \delta(s - s') \tag{8.3.1}$$

and

$$G(q, s, s') \to 0 \quad \text{for} \quad s \to \pm\infty. \tag{8.3.2}$$

The two linearly independent solutions of the homogeneous version of (8.3.1), i.e., $s \neq s'$, are parabolic cylinder functions $D_q(s)$ and $D_q(-s)$, see e.g. GRADSTEIN and RYSHIK (1971). The one, $D_q(s)$, fulfils the boundary condition at $s \to \infty$, and the other one, $D_q(-s)$, fulfils the boundary condition at $s \to -\infty$. The Wronskian of $D_q(s)$ and $D_q(-s)$ is given by

$$W_r = D(s)\frac{\mathrm{d}}{\mathrm{d}s} D_q(-s) - D_q(-s)\frac{\mathrm{d}}{\mathrm{d}s} D_q(s) = -\frac{\sqrt{2\pi}}{\Gamma(-q)} \tag{8.3.3}$$

where $\Gamma(-q)$ is Euler's gamma function. Now we can construct the Green's function and obtain

$$G(q, s; s') = \theta(s - s')\, G^{>}(q, s; s') + \theta(s' - s)\, G^{<}(q, s; s') \tag{8.3.4}$$

with

$$G^{>}(q, s; s') = -\frac{\Gamma(-q)}{\sqrt{2\pi}} D_q(s)\, D_q(-s') = G^{<}(q, s'; s) . \tag{8.3.5}$$

Let us discuss the properties of the Green's function. First we note that for integer q, i.e., $q = n$ with $n = 0, 1, 2, \ldots$, the parabolic cylinder functions equal the Hermite functions

$$D_n(s) = \frac{1}{\sqrt{2^n}}\, e^{-\frac{s^2}{4}} H_n\left(\frac{s}{\sqrt{2}}\right) \tag{8.3.6}$$

where $H_n(x)$ is the Hermite polynomial of n-th order. We note the first few $H_n(x)$:

$$H_0 = 1 , \qquad H_1(x) = 2x , \qquad H_2(x) = 4x^2 - 2 , \qquad H_3(x) = 8x^3 - 12x , \ \ldots$$

Moreover, the singularities of the Green's function (8.3.5) are determined by the poles of the gamma function $\Gamma(-q)$, which has poles at negative integers, i.e., at $q = n$. Expanding the gamma function around its poles yields

$$\Gamma(-q) \approx \frac{(-1)^n}{n!(n-q)}$$

where $(-1)^n/n!$ is the residue of the pole at $q = n$. A partial fraction expansion of (8.3.4) and (8.3.5) with regard to the poles of $\Gamma(-q)$ yields the bilinear representation of the Green's function in terms of Hermite functions

$$G(q, s; s') = \sum_{n=0}^{\infty} \frac{e^{-\frac{(s^2+s'^2)}{4}}}{2^n n! \sqrt{2\pi}(q-n)} H_n\left(\frac{s}{\sqrt{2}}\right) H_n\left(\frac{s'}{\sqrt{2}}\right). \tag{8.3.7}$$

Thus we have the two equivalent representations (8.3.4) and (8.3.5), and (8.3.7) of the Green's function. Note that the scaled meridional distance is $s = \sqrt{2\beta\lambda}y$. The scale $(\beta\lambda)^{-1/2}$ is called the equatorial Rossby radius which corresponds to the midlatitude Rossby radius $(\lambda f)^{-1}$. From (8.3.7) it can be seen that the Green's function is localized meridionally at the equator within a Rossby radius. This area is denoted as the equatorial wave guide. Now we readily obtain the formal solution of (8.2.11) in terms of a source representation. Introducing the following convolution integrals

$$G * Y = \int_{-\infty}^{+\infty} \mathrm{d}s'\, G(s; s')\, Y(s') \tag{8.3.8 a}$$

$$G * s'X = \int_{-\infty}^{+\infty} \mathrm{d}s'\, s'G(s; s')\, X(s') \tag{8.3.8 b}$$

$$G * X_{s'} = \int_{-\infty}^{+\infty} \mathrm{d}s'\, G(s; s') \frac{\partial}{\partial s'} X(s') \tag{8.3.8 c}$$

we get

$$v(\omega, k, s) = \mathrm{i}\frac{(\bar{\omega}^2\lambda^2 - k^2)}{2\lambda\beta\bar{\omega}} G * Y + \sqrt{\frac{\lambda}{2\beta}} G * \frac{s'}{2} X + \frac{k}{\bar{\omega}\sqrt{2\lambda\beta}} G * X_{s'}\,. \tag{8.3.9 a}$$

The remaining variables u and p can easily be expressed in terms of v, and of the external forcing components X and Y.

The results are

$$u(\omega, k, s) = \sqrt{\frac{\lambda}{2\beta}} \left(\frac{k}{\bar{\omega}\lambda}\frac{\partial}{\partial s} - \frac{s}{2}\right) G * Y$$
$$+ \frac{\mathrm{i}\bar{\omega}\lambda^2}{\bar{\omega}^2\lambda^2 - k^2}\left[X + \left(\frac{s}{2} - \frac{k}{\bar{\omega}\lambda}\frac{\partial}{\partial s}\right)\left(G * \frac{s'}{2} X + \frac{k}{\bar{\omega}\lambda} G * X_{s'}\right)\right] \tag{8.3.9 b}$$

$$p(\omega, k, s) = \frac{1}{\sqrt{2\lambda\beta}}\left(\frac{\partial}{\partial s} - \frac{k}{\bar{\omega}\lambda}\frac{s}{2}\right) G * Y$$
$$+ \frac{ik}{\bar{\omega}^2\lambda^2 - k^2}\left[X + \left(\frac{s}{2} - \frac{\bar{\omega}\lambda}{k}\frac{\partial}{\partial s}\right)\left(G * \frac{s'}{2}X + \frac{k}{\bar{\omega}\lambda} G * X_{s'}\right)\right] \tag{8.3.9 c}$$

The set (8.3.9) represents already a general formal solution of the forcing problem of an unbounded equatorial ocean. From the mathematical point of view the physics of the response is governed by the analytical properties of the these expressions in the complex ω-plane. The poles near the origin of the ω-plane correspond to steady motion, where friction balances the external forces. The poles of the Green's function, see (8.3.5), determine the dispersion relation (8.2.12) of free ETW, and the pole at $\omega = k/\lambda$, which occurs only in (8.3.9b) and (8.3.9c), corresponds to the equatorial Kelvin wave. The role of the pole $\omega = -k/\lambda$ will be discussed below. Now we proceed with the discussion of the dispersion relation in more detail. As noted above the Green's function (8.3.7) or (8.3.4), (8.3.5) has poles at $q = n$. Inserting this in (8.2.12) we obtain the dispersion relation for the free ETW, which may be written as a polynomial in $\bar{\omega}$,

$$\bar{\omega}^3 - \left(\frac{k^2}{\lambda^2} + \frac{\beta}{\lambda}(2n + 1)\right)\bar{\omega} - \frac{\beta k}{\lambda^2} = 0\,. \tag{8.3.10}$$

It can be shown that the discriminant is always negative, i.e.,

$$\left(\frac{\beta k}{2\lambda}\right)^2 - \left[\frac{1}{3}\left(\frac{k^2}{\lambda^2} + \frac{\beta}{\lambda}(2n + 1)\right)\right]^3 < 0\,,$$

implying that the equation (8.3.10) has three real roots. However, these roots can only approximately be estimated.

Writing (8.3.10) as a polynomial in zonal wavenumber k we get a simple quadratic equation

$$k^2 + \frac{\beta}{\bar{\omega}}k - \bar{\omega}^2\lambda^2 + \lambda\beta(2n + 1) = 0\,. \tag{8.3.11}$$

The two roots of (8.3.11) can be estimated exactly and are

$$k_n^{(1,2)} = -\frac{\beta}{2\bar{\omega}} \pm \sqrt{\frac{\beta^2}{4\bar{\omega}^2} + \lambda^2\bar{\omega}^2 - \lambda\beta(2n + 1)}\,, \tag{8.3.12 a}$$

or, equivalently,

$$k_n^{(1,2)} = -\frac{\beta}{2\bar{\omega}} \pm \frac{\lambda}{\bar{\omega}}\sqrt{(\bar{\omega}^2 - \omega_i^2)(\bar{\omega}^2 - \omega_r^2)} \tag{8.3.12 b}$$

where

$$\omega_{i,r} = \sqrt{\frac{\beta}{\lambda}}\left(\sqrt{\frac{n+1}{2}} \pm \sqrt{\frac{n}{2}}\right) \tag{8.3.12 c}$$

A graphical representation of the dispersion relation of the free ETW and the equatorial Kelvin wave are given in Fig. 8.3.1. Comparing the dispersion relation of the unbounded ocean on an equatorial β-plane with that one of the unbounded, midlatitudinal β-plane described in Section 7.2. we observe both significant differences and striking similarities.

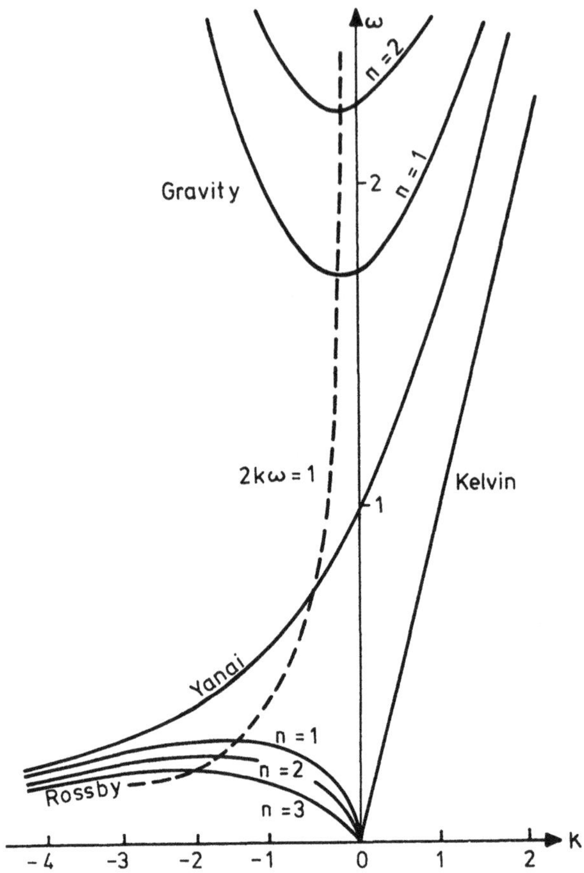

Fig. 8.3.1
Dispersion diagram of free equatorial trapped waves showing the different roots of the dispersion relation (8.3.12) for the first few meridional modes
The frequency ω and the zonal wavenumber k are scaled with $\sqrt{\lambda/\beta}$ and $1/\sqrt{\lambda\beta}$, respectively. The Kelvin branch is no root of (8.3.12) rather it corresponds to the Kelvin poles occurring in (8.3.9b, c).

The important difference is that the large frequency gap between inertial and Rossby waves found in the midlatitudinal ocean is filled out in the equatorial ocean by the equatorial Kelvin wave and by the $n = 0$ mode of the ETW, which is called the Yanai wave or mixed Rossby-gravity wave. The dispersion relation of the Yanai mode can be readily obtained from (8.3.12) by setting $n = 0$, which yields

$$k_0^{(1)} = \lambda\bar{\omega} - \frac{\beta}{\bar{\omega}}, \qquad k_0^{(2)} = -\lambda\bar{\omega}. \tag{8.3.13}$$

It will be shown below that $k_0^{(2)}(\omega)$ is not really a pole of the Green's function of the unbounded equatorial β-plane. We note, however, that $k_0^{(2)}$ is the dispersion relation of the "anti"-Kelvin wave, which occurs in an equatorial ocean bounded meridionally by one or two coasts, see e.g. CANE and SARACHIK (1977). The dispersion relation of the Yanai wave $k_0^{(1)}$ behaves for low frequencies like a Rossby wave with westward phase propagation, and for high frequencies like an inertial-gravity wave with eastward phase propagation. The group velocities are

$$c_{gr} = \lambda^{-1} \tag{8.3.14a}$$

for the Kelvin wave, and

$$c_{gr} = \left(\lambda + \frac{\beta}{\omega^2}\right)^{-1} \tag{8.3.14b}$$

for the Yanai wave. Both group velocities are eastward with amplitudes of the order of 2—3 ms^{-1}, if the numerical values of λ listed in Table 1.9.3 are used.

This implies that wave fronts propagate rapidly eastward over the equatorial ocean. For example, it takes the wave front approximately one month to cross the equatorial Pacific. Compared to the response time of the midlatitudinal ocean, which is of the order of decades, it becomes evident that the interaction between the ocean and the atmosphere is much more intense in the equatorial area than in midlatitudes.

Similar to the dispersion relation of the unbounded ocean on a midlatitudinal β-plane the dispersion relation of the ETW splits off into the high frequency inertial-gravity waves and the low frequency Rossby waves. Comparing the dispersion relations (7.2.11) and (7.2.12) with (8.3.10) and (8.3.12) it is evident that both have the same analytical form, in particular if we put

$$f^2 = \frac{\beta}{\lambda}(2n + 1)\,, \qquad R^{-2} = \lambda\beta(2n + 1)\,. \tag{8.3.15}$$

Using the numerical values of the vertical eigenvalues given in Table 1.9.3 from (8.3.12c) it follows that, the maximum period of the inertial-gravity waves of the ETW, i.e., $n = 1$, is for the first vertical mode of the order of five days. The minimum period of the first mode Rossby wave is according to (8.3.12c) of the order of 30 days.

Neglecting the term $\beta k \lambda^{-2}$ in (8.3.10) we obtain an approximate dispersion relation for the inertial-gravity wave branch of the ETW

$$\omega_{1,2} = \pm \sqrt{\frac{k^2}{\lambda^2} + \frac{\beta}{\lambda}(2n + 1)}\,. \tag{8.3.16}$$

The associated approximate group speed is

$$\frac{\partial \omega}{\partial k} = c_{gr} = \pm \frac{1}{\lambda} \frac{1}{\sqrt{1 + \frac{\beta\lambda}{k^2}(2n + 1)}}\,. \tag{8.3.17}$$

Thus the gravity waves can propagate as well eastward as westward. The maximum group velocity equals $\pm\lambda^{-1}$ which means that inertial wave fronts travel as fast as the Kelvin wave, however, by contrast, in both directions of the wave guide. The approximate group velocity vanishes at $k = 0$ according to (8.3.17). However, the exact dispersion relation (8.3.12) reveals that the group velocity of gravity waves vanishes at $k = -\beta/(2\omega_i)$. We will see below that this small difference may have important consequences for that part of the response, which consists of inertial gravity waves.

Neglecting ω^3 in (8.3.10) yields the third approximate root of this equation, which defines the dispersion relation of equatorial trapped Rossby waves

$$\omega_3 = -\frac{\beta k}{k^2 + \lambda\beta(2n + 1)}\,. \tag{8.3.18}$$

This implies that the zonal phase propagation of the Rossby waves is always westward. For the group velocity we get from (8.3.18)

$$c_{gr} = \beta \frac{k^2 - \lambda\beta(2n+1)}{k^2 + \lambda\beta(2n+1)^2} . \tag{8.3.19}$$

The group velocity of the Rossby waves vanishes at $k = -\beta/(2\omega_r)$ and possesses two maxima. The one is located at $\omega = 0$ and $k = 0$, $c_{gr} = -\lambda^{-1}/(2n+1)$, and belongs to the long wave domain of the Rossby waves, the other one is located at $\omega = \sqrt{3\beta/[16\lambda(2n+1)]}$ and $k = -\sqrt{3\lambda\beta(2n+1)}$, i.e., $c_{gr} = \lambda^{-1}/[8(2n+1)]$. Assuming that wave fronts propagate with maximum group velocities from an initial disturbance it appears that, similar to the midlatitudinal case, Rossby wave fronts propagate eight times faster to the west than to the east. However, the westward maximum group velocity of the gravest Rossby mode, $n = 1$, is one third of the group velocity of the Kelvin wave, i.e., about 1 m/s for the first vertical mode. That means the wave front needs approximately three months to cross the equatorial Pacific from east to west. This indicates a much shorter response time of the equatorial ocean compared to the midlatitudinal oceans.

Finally, we note that the general formal solution given above is equivalent to the formalism of Cane and Sarachik (1976). This can be seen using the bilinear expansion (8.3.7) of the Green's function. Then the calculation of the convolutions, i.e., $G * X$, reduces to the estimation of the coefficients of the Hermite expansion of the forcing functions, and the expressions given by Cane and Sarachik (1976) follow.

8.4. THE EQUATORIAL KELVIN WAVE

In this section we investigate the meaning of the poles $\omega = \pm k/\lambda$ and the associated motions. In the general solution (8.3.9) both poles occur in the expressions of the zonal current and the pressure field, in particular in those terms which are associated with the zonal component X of the wind forcing. Moreover, from (8.3.13) follows that $\omega = -k/\lambda$ is a pole inherent in the Green's function (8.3.4), (8.3.5), namely for $q = 0$. Hence, we have to study the meaning of both poles for all terms of the general solution (8.3.9).

Using the identities

$$\frac{k}{\omega^2 - \left(\frac{k}{\lambda}\right)^2} = \frac{\lambda}{2}\left[\frac{1}{\omega - \left(\frac{k}{\lambda}\right)} - \frac{1}{\omega + \left(\frac{k}{\lambda}\right)}\right] \tag{8.4.1a}$$

$$\frac{\omega}{\omega^2 - \left(\frac{k}{\lambda}\right)^2} = \frac{1}{2}\left[\frac{1}{\omega - \frac{k}{\lambda}} + \frac{1}{\omega + \frac{k}{\lambda}}\right] \tag{8.4.1b}$$

$$\frac{1}{\omega}\,\frac{1}{\omega^2 - \left(\frac{k}{\lambda}\right)^2} = -\frac{\lambda^2}{k^2}\left[\frac{1}{\omega} - \frac{1}{2}\left(\frac{1}{\omega - \frac{k}{\lambda}} + \frac{1}{\omega - \frac{k}{\lambda}}\right)\right] \tag{8.4.1c}$$

we may rewrite in (8.3.9b), (8.3.9c) the terms associated with the zonal forcing

$$u(\omega, k, s) = \sqrt{\frac{\lambda}{2\beta}}\left(\frac{k}{\lambda\bar{\omega}}\frac{\partial}{\partial s} - \frac{s}{2}\right)G * Y + \frac{i}{2}\frac{X *}{\left(\bar{\omega} - \frac{k}{\lambda}\right)}[\delta(s - s') + \Lambda^-(s)\,\Lambda^-(s')\,G(s, s')]$$
$$+ \frac{i}{2}\frac{X *}{\bar{\omega} + \frac{k}{\lambda}}[\delta(s - s') + \Lambda^+(s)\,\Lambda^+(s')\,G(s, s')] + \frac{i}{\bar{\omega}}\frac{\partial}{\partial s}G * X_{s'} \tag{8.4.2a}$$

$$p(\omega, k, s) = \frac{1}{\sqrt{2\beta\lambda}}\left(\frac{\partial}{\partial s} - \frac{k}{\bar{\omega}\lambda}\frac{s}{2}\right) G * Y + \frac{i}{2}\frac{X *}{\left(\bar{\omega} + \frac{k}{\lambda}\right)}[\delta(s - s') + \Lambda^-(s)\,\Lambda^-(s')\,G]$$
$$+ \frac{i}{2}\frac{X *}{\left(\bar{\omega} + \frac{k}{\lambda}\right)}[\delta(s - s') + \Lambda^+(s)\,\Lambda^+(s')\,G(s, s')] - \frac{i}{\bar{\omega}\lambda}\frac{s}{2}G * X_{s'}. \tag{8.4.2b}$$

Here the operators

$$\Lambda^{\pm}(s) = \left(\frac{d}{ds} \pm \frac{s}{2}\right)$$

have been introduced. Using well-known recurrence relations of the parabolic cylinder functions, see e.g. GRADSTEIN and RYSHIK (1971),

$$\Lambda^+(s)\,D_q(\pm s) = \pm q D_{q-1}(\pm s) \tag{8.4.3a}$$
$$\Lambda^-(s)\,D_q(\pm s) = \mp D_{q+1}(\pm s) \tag{8.4.3b}$$

we easily verify that

$$\Lambda^+(s)\,G_q(s, s') = -\frac{q}{2}\Gamma(-q)\,[\theta(s - s')\,D_{q-1}(s)\,D_q(-s') - \theta(s' - s)\,D_{q-1}(-s)\,D_q(s')] \tag{8.4.4}$$

and, taking into account the Wronskian (8.3.3),

$$\Lambda^+(s)\,\Lambda^+(s')\,G_q(s, s') = -\delta(s - s') + qG_{q-1}(s, s') \tag{8.4.5a}$$
$$\Lambda^-(s)\,\Lambda^-(s')\,G_q(s, s') = -\delta(s - s') + (q + 1)\,G_{q+1}(s, s')\,. \tag{8.4.5b}$$

Here the subscript q is meant to be an index and does not refer to a differentiation. Substituting (8.4.5) in (8.4.2) we find an equivalent formulation for the zonal current component u and the pressure perturbation p

$$u(\omega, k, s) = \sqrt{\frac{\lambda}{2\beta}}\left(\frac{k}{\bar{\omega}\lambda}\frac{\partial}{\partial s} - s\right) G_q * Y + \frac{i}{\bar{\omega}}\frac{\partial}{\partial s}G_q * X_{s'}$$
$$+ \frac{i}{2}\left[\frac{(q + 1)\,G_{q+1} * X}{\bar{\omega} - \frac{k}{\lambda}} + \frac{qG_{q-1} * X}{\bar{\omega} + \frac{k}{\lambda}}\right] \tag{8.4.6a}$$

$$p(\omega, k, s) = \frac{1}{\sqrt{2\lambda\beta}}\left(\frac{\partial}{\partial s} - \frac{k}{\bar{\omega}\lambda}\frac{s}{2}\right) G_q * Y - \frac{\mathrm{i}}{\lambda\bar{\omega}}\frac{s}{2} G_q * X_{s'}$$

$$+ \frac{\mathrm{i}}{2\lambda}\left[\frac{(q+1)\, G_{q+1} * X}{\bar{\omega} - \frac{k}{\lambda}} - \frac{q G_{q-1} * X}{\bar{\omega} + \frac{k}{\lambda}}\right]. \qquad (8.4.6\,\mathrm{b})$$

In these expressions the poles at $\bar{\omega} = 0$, and at $\bar{\omega} = \pm k/\lambda$ are separated.

It is convenient for the following to put the dispersion relation (8.2.12) in the form

$$2q = \frac{\lambda^2\bar{\omega}^2 - k^2}{\beta} - \frac{\lambda\bar{\omega} + k}{\bar{\omega}\lambda} = \frac{\bar{\omega}\lambda + k}{\lambda\beta}\left(\bar{\omega}\lambda - k - \frac{\beta}{\bar{\omega}}\right) \qquad (8.4.7)$$

For the Kelvin pole, $\bar{\omega} = k/\lambda$, (8.4.7) implies that $q = -1$. Taking the relation $-(q+1)\,\Gamma(-(q+1)) = \Gamma(-q)$ into account we find with (8.3.4), (8.3.5)

$$\lim_{q \to -1} (q+1)\, G_{q+1}(s, s') * X = \Gamma(1)\, D_0(s)\, D_0(s') * X(s') = \exp\left(-\frac{s^2 + s'^2}{4}\right) * X(s'). \qquad (8.4.8)$$

Thus the Kelvin pole corresponds in fact to a forced equatorial wave trapped in the equatorial wave guide. Because the Kelvin pole occurs only in those terms of u and p which are associated with the zonal forcing X we may conclude that Kelvin waves are only excited by zonal winds.

Using (8.4.8), it can readily be shown that the pressure and the zonal velocity of the equatorial Kelvin wave in (8.4.6) are in geostrophic balance similarly to the coastal Kelvin wave described in Section 5.3.

Finally we discuss the properties of the general solution (8.3.9) at the pole $\bar{\omega} = -k/\lambda$, which according to (8.4.7) implies $q = 0$. Putting $q = 0$ and $\bar{\omega}\lambda = -k$ in all terms of (8.3.9) except in the denominator we obtain

$$v = \mathrm{i}\,\frac{\bar{\omega}\lambda + k}{\beta}\, G_0 * X + \sqrt{\frac{\lambda}{2\beta}}\, \Lambda^+ G_0 * X \qquad (8.4.9\,\mathrm{a})$$

$$u = -\sqrt{\frac{\lambda}{2\beta}}\, \Lambda^+(s)\, G_0 * Y - \frac{\mathrm{i}\lambda X *}{2(\bar{\omega}\lambda + k)}\left[\delta(s - s') + \Lambda^+(s)\, \Lambda^+(s')\, G_0\right] \qquad (8.4.9\,\mathrm{b})$$

$$p = \frac{\Lambda^+(s)\, G_0 * Y}{\sqrt{2\beta\lambda}} - \frac{\mathrm{i}X *}{2(\bar{\omega}\lambda + k)}\left[\delta(s - s') + \Lambda^+(s)\, \Lambda^+(s')\, G_0\right]. \qquad (8.4.9\,\mathrm{c})$$

Substituting (8.4.5a) in (8.4.9b) and (8.4.9c) yields

$$u = -\sqrt{\frac{\lambda}{2\beta}}\, \Lambda^+(s)\, G_0 * Y + \lim_{q \to 0} \frac{\mathrm{i}\lambda}{2(\bar{\omega}\lambda + k)}\, q G_{q-1} * X \qquad (8.4.10\,\mathrm{a})$$

$$p = \frac{1}{\sqrt{2\beta\lambda}}\, \Lambda^+(s)\, G_0 * Y - \lim_{q \to 0} \frac{\mathrm{i}}{2(\bar{\omega}\lambda + k)}\, q G_{q-1} * X\,. \qquad (8.4.10\,\mathrm{b})$$

Obviously, the pole $\bar{\omega}\lambda = -k$ which occurs in G_0, will be reduced in the first term of (8.4.9a) by the numerator. Moreover, using (8.4.4) and the relation

$$\lim_{q \to 0} (-q)\,\Gamma(-q) = \Gamma(-q+1) = \Gamma(1) = 1$$

we get

$$\Lambda^+(s)\,G_0(s, s') = \theta(s - s')\,D_{-1}(s)\,D_0(-s') - \theta(s' - s)\,D_{-1}(-s)\,D_0(s')\,.$$

Thus the poles $\bar{\omega}\lambda = -k$ vanish in the second term of (8.9a) and in the first terms of (8.4.10a) and (8.4.10b). Finally, substituting (8.4.7) in the second terms of (8.4.10) reveals that the pole $\bar{\omega}\lambda = -k$ is eliminated completely in the general solution of the response of an unbounded equatorial ocean.

It appears, however, that in a meridionally bounded ocean this pole corresponds to the coastal Kelvin wave which propagates within one Rossby radius along the zonal coast from east to west, see PHILANDER (1978).

8.5. THE RESPONSE OF THE EQUATORIAL OCEAN TO A ZONALLY CONSTANT WIND

On the basis of the general formal solution derived in the foregoing sections we shall consider several explicit examples of wind patterns acting upon an equatorial ocean. We start with a switched-on forcing which is zonally constant and meridionally bell-shaped

$$X(x, y, z, t) = \frac{u_*^2}{H_{\text{mix}}}\,\theta(z + H_{\text{mix}})\,\theta(t)\,e^{-\frac{y^2}{2L_a^2}} \tag{8.5.1 a}$$

$$Y(x, y, z, t) = \frac{v_*^2}{H_{\text{mix}}}\,\theta(z + H_{\text{mix}})\,\theta(t)\,e^{-\frac{y^2}{2L_a^2}}\,. \tag{8.5.1 b}$$

Expansion in vertical eigenfunctions and Fourier transforms with respect to x and t yield

$$X_m(\omega, k, y) = \frac{u_*^2}{h_m}\,\frac{2\pi\,\mathrm{i}\delta(k)}{\omega + \mathrm{i}\varepsilon}\,e^{-\frac{y^2}{2L_a^2}} \tag{8.5.2 a}$$

$$Y_m(\omega, k, y) = \frac{v_*^2}{h_m}\,\frac{2\pi\,\mathrm{i}\delta(k)}{\omega + \mathrm{i}\varepsilon}\,e^{-\frac{y^2}{2L_a^2}}\,. \tag{8.5.2 b}$$

Again we drop the vertical mode index m and introduce the variable $s = \sqrt{2\beta\lambda}\,y$. Thus we replace $e^{-y^2/(2L_a^2)}$ by $e^{-\zeta s^2/4}$, with ζ being the squared ratio of the meridional scale of the atmospheric force to the oceanic Rossby radius, i.e., $\zeta = \lambda\beta L_a^2$.

Substituting (8.5.2) in (8.3.9) and using $k\,\delta(k) = 0$ we get

$$v(\omega, k, s) = 2\pi\delta(k)\left[-\frac{v_*^2\lambda\bar{\omega}}{2\beta h(\omega + \mathrm{i}\varepsilon)}\,G_q * e^{-\frac{\zeta s'^2}{4}} + \frac{u_*^2}{h}\sqrt{\frac{\lambda}{2\beta}}\,\frac{\mathrm{i}}{\omega + \mathrm{i}\varepsilon}\,G_q * \frac{s'}{2}\,e^{-\frac{\zeta s'^2}{4}}\right] \tag{8.5.3 a}$$

$$u(\omega, k, s) = 2\pi\delta(k)\left[\sqrt{\frac{\lambda}{2\beta}}\frac{\mathrm{i}}{\omega+\mathrm{i}\varepsilon}\frac{v_*^2}{h}\frac{s}{2}G_q * \mathrm{e}^{-\frac{\zeta s'^2}{4}} - \frac{u_*^2}{\bar{\omega}h}\frac{1}{\omega+\mathrm{i}\varepsilon}\left(\mathrm{e}^{-\frac{\zeta s^2}{4}} + \frac{s}{2}G_q * \frac{s'}{2}\mathrm{e}^{-\frac{\zeta s'^2}{4}}\right)\right] \tag{8.5.3 b}$$

$$p(\omega, k, s) = \frac{v_*^2}{h}\frac{1}{\sqrt{2\beta\lambda}}\frac{2\pi\,\mathrm{i}\delta(k)}{\omega+\mathrm{i}\varepsilon}\frac{\partial}{\partial s}G_q * \mathrm{e}^{-\frac{\zeta s'^2}{4}} + \frac{u_*^2}{\lambda h\bar{\omega}}\frac{2\pi\delta(k)}{\omega+\mathrm{i}\varepsilon}\frac{\partial}{\partial s}G_q * \frac{s'}{2}\mathrm{e}^{-\frac{\zeta s'^2}{4}}. \tag{8.5.3 c}$$

Moreover, the relation (8.2.12) reduces to

$$2q = \frac{\bar{\omega}^2\lambda}{\beta} - 1. \tag{8.5.4}$$

In order to calculate the two types of convolutions which occur in (8.5.3) we use the bilinear representation of the Green's function (8.3.7). Then we get

$$G_q * \mathrm{e}^{-\frac{\zeta s'^2}{4}} = \sum_{n=0}^{\infty}\frac{\mathrm{e}^{-\frac{s^2}{4}}H_n\left(\frac{s}{\sqrt{2}}\right)}{2^n n!\sqrt{\pi}(q-n)}\int_{-\infty}^{+\infty}\frac{\mathrm{d}s'}{\sqrt{2}}\mathrm{e}^{-(1+\zeta)\frac{s'^2}{4}}H_n\left(\frac{s'}{\sqrt{2}}\right) = \frac{2\beta}{\lambda}\sum_{n=0}^{\infty}\frac{I_n\psi_n(\eta)}{\bar{\omega}^2 - \frac{\beta}{\lambda}(2n+1)} \tag{8.5.5}$$

and

$$G_q * \frac{s'}{2}\mathrm{e}^{-\frac{\zeta s'^2}{4}} = \frac{\beta}{\lambda}\sum_{n=0}^{\infty}\frac{\sqrt{n+1}\,I_{n+1} + \sqrt{n}\,I_{n-1}}{\bar{\omega}^2 - \frac{\beta}{\lambda}(2n+1)}\psi_n(\eta) \tag{8.5.6}$$

with $\eta = s/\sqrt{2}$ and $\psi_n(\eta)$ being the Hermite function. Here I_n denotes the projection coefficients of the meridional shape of the forcing on the Hermite functions, i.e.,

$$I_n = \int_{-\infty}^{+\infty}\mathrm{d}\eta\,\psi_n(\eta)\,\mathrm{e}^{-\frac{\eta^2}{2}\zeta} \tag{8.5.7}$$

and in (8.5.6) the recurrence relation

$$\eta\psi_n(\eta) = \sqrt{\frac{n+1}{2}}\psi_{n+1}(\eta) + \sqrt{\frac{n}{2}}\psi_{n-1}(\eta)$$

has been used. The index n is due to the meridional modes and should not be confused with the dropped vertical mode index. The expressions (8.5.5) and (8.5.6) have poles at

$$\bar{\omega} = \sqrt{\frac{\beta}{\lambda}(2n+1)} - \mathrm{i}r = \omega_n - \mathrm{i}r \tag{8.5.8 a}$$

and

$$\omega = -\sqrt{\frac{\beta}{\lambda}(2n+1)} - ir = -\omega_n - ir \tag{8.5.8 b}$$

which correspond to oscillations in the equatorial wave guide.

The frequencies are determined by the intersections of the dispersion curves of the inertia-gravity waves and the ordinate $k = 0$ in the dispersion diagram, compare Fig. 8.3.1. The response of the equatorial wave guide in terms of standing gravity waves resembles the response of a channel, expressed by standing Poincaré modes as discussed in Chapter 6. Moreover, we will see that the non-oscillatory portion of the response, which is determined by the poles at $\omega = -i\varepsilon$ and $\omega = -ir$ in (8.5.3), consists of jets similar to the coastal jets which are, for example, described in Section 5.4.

Next we estimate the projection coefficients I_n introduced by (8.5.7). Due to the properties of the Hermite functions the odd projection coefficients vanish while the even coefficients are

$$I_{2n} = \pi^{1/4}\sqrt{\frac{2}{1+\zeta}}\left(\frac{1-\zeta}{1+\zeta}\right)^n \sqrt{(2n)!}\,\frac{1}{2^n n!} \tag{8.5.9}$$

From (8.5.9) follows readily the recurrence relation

$$I_{2(n+1)} = \left(\frac{1-\zeta}{1+\zeta}\right)\sqrt{\frac{2n+1}{2(n+1)}}\, I_{2n}\,. \tag{8.5.10}$$

To proceed further we consider the response to a zonal and to a meridional wind separately. Since we deal with a linear theory both solutions can be added in order to get the complete response to (8.5.1 a, b).

We start with a meridional wind, i.e., we put $X = 0$ and substitute (8.5.9) in (8.5.5) and the resulting expression in (8.5.3). Then we invert the Fourier transforms and get

$$v(\eta, t) = \theta(t)\,\frac{v_*^2}{h}\,e^{-rt}\sum_{n=0}^{\infty} I_{2n}\,\frac{\sin\omega_{2n}t}{\omega_{2n}}\,\psi_{2n}(\eta) \tag{8.5.11 a}$$

$$u(\eta, t) = \theta(t)\,\frac{v_*^2}{h}\sqrt{\frac{\lambda}{\beta}}\sum_{n=0}^{\infty} 2I_{2n}\sqrt{\frac{2n+1}{2}}$$

$$\times\left[\frac{4n+3}{(4n+1)(4n+5)} - e^{-rt}\left(\frac{\cos\omega_{2n}t}{4n+1} + \frac{\cos\omega_{2(n+1)}t}{4n+5}\right)\right]\psi_{2n+1}(\eta) \tag{8.5.11 b}$$

$$p(\eta, t) = \theta(t)\,\frac{v_*^2}{h}\,\frac{1}{(\lambda\beta)^{1/2}}\sum_{n=0}^{\infty} I_{2n}\sqrt{\frac{2n+1}{2}}$$

$$\times\left[\frac{4}{(4n+1)(4n+5)} + e^{-rt}\left(\frac{\cos\omega_{2n+2}t}{4n+5} - \frac{\cos\omega_{2n}t}{4n+1}\right)\right]\psi_{2n+1}(\eta) \tag{8.5.11 c}$$

Here the relations $r \ll \omega_n$ and $\zeta \ll 1$ were assumed, i.e., a small friction and a meridional scale of the forcing much larger than the equatorial Rossby radius $(\lambda\beta)^{-1/2}$. The ω_{2n} are defined by (8.5.8 a, b) in obvious manner.

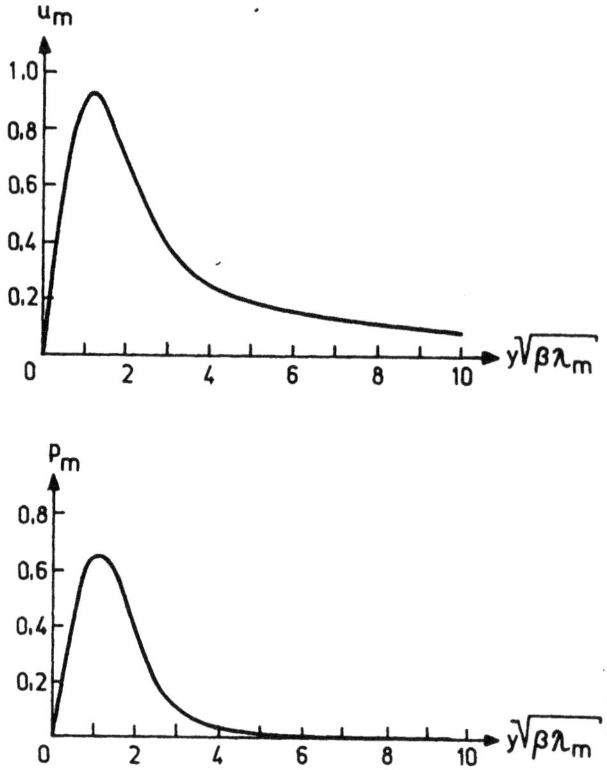

Fig. 8.5.1
The meridional structure of the m-th vertical mode of the steady part of the zonal velocity generated by an uniform wind blowing toward the north (a) and the corresponding pressure perturbation (b)
The velocity and the pressure are scaled with $(v_*^2/h)\sqrt{\lambda/\beta}$ and $(v_*^2/h)\,(1/\sqrt{\lambda\beta})$, respectively.

The first phase of this response scenario is governed by transients consisting of inertial oscillations with periods of, say, ten days and less. When these oscillations are damped out there remains a stationary response in the u and p field which is sketched in Fig. 8.5.1. The pressure disturbance is confined to the equator within few Rossby radii.

The associated pattern of the zonal current component, also shown in Fig. 8.5.1, consists of an eastward current north of the equator, $y > 0$ and a countercurrent south of the equator, $y < 0$. Outside the equatorial belt a pure Ekman current remains.

Finally we calculate the response to a pure zonal wind, i.e., we put $Y = 0$ in (8.3.9). Then we obtain in the same manner

$$v(\eta, t) = -2\theta(t)\,\frac{u_*^2}{h}\sqrt{\frac{\lambda}{\beta}}\sum_{n=0}^{\infty} I_{2n}\sqrt{\frac{2n+1}{2}}\;\frac{1-\mathrm{e}^{-rt}\cos\omega_{2n+1}t}{4n+3}\,\psi_{2n+1}(\eta) \tag{8.5.12 a}$$

$$u(\eta, t) = \frac{u_*^2}{h}\,\theta(t)\left\{\frac{1-\mathrm{e}^{-rt}}{r}\left[\mathrm{e}^{-\frac{\zeta\eta^2}{2}} - 2\eta\sum_{n=0}^{\infty}\sqrt{\frac{2n+1}{2}}\,\frac{I_{2n}}{4n+3}\,\psi_{2n+1}(\eta)\right]\right.$$

$$\left. + 2\eta\,\mathrm{e}^{-rt}\sum_{n=0}^{\infty}\sqrt{\frac{2n+1}{2}}\;\frac{I_{2n}}{4n+3}\;\frac{\sin\omega_{2n+1}t}{\omega_{2n+1}}\,\psi_{2n+1}(\eta)\right\} \tag{8.5.12 b}$$

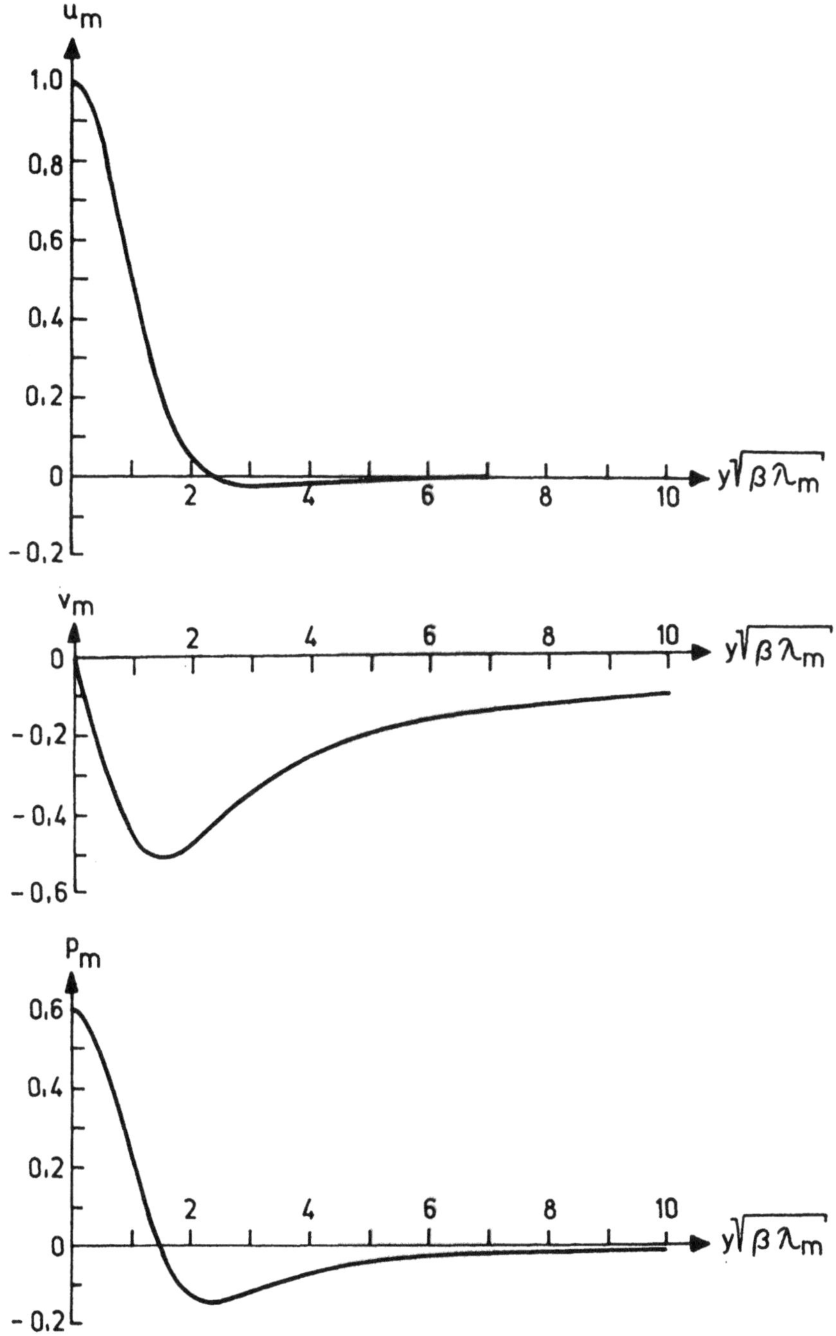

Fig. 8.5.2
The meridional structure of the m-th vertical mode of the equatorial Yoshida jet generated by an uniform eastward blowing wind (a), the corresponding meridional velocity (b), and the corresponding pressure perturbation (c)
u, v, and p are scaled with $u_*/(hr)$, $u_*^2\sqrt{\lambda/\beta}$, and $u_*^2/(\lambda rh)$, respectively.

$$p(\eta, t) = \frac{u_*^2}{\lambda h}\left\{\frac{2(1-e^{-rt})}{r}\sum_{n=0}^{\infty}\left[\frac{2n+1}{4n+3}I_{2n}\psi_{2n}(\eta) - \eta\sqrt{\frac{2n+1}{2}}\frac{I_{2n}}{4n+3}\psi_{2n+1}(\eta)\right]\right.$$

$$\left. -e^{-rt}\sum_{n=0}^{\infty}\left(\frac{2n+1}{4n+3}\frac{\sin\omega_{2n+1}t}{\omega_{2n+1}} - \frac{2n}{4n-1}\frac{\sin\omega_{2n-1}t}{\omega_{2n-1}}\right)I_{2n}\psi_{2n}(\eta)\right\}. \quad (8.5.12\,c)$$

Again the response consists of a transient part, governed by inertial oscillations, and a stationary part. However, the meridional symmetry properties of the meridional and zonal current have been exchanged. The non-oscillatory part of the response is sketched in Fig. 8.5.2. The zonal current consist of an equatorially trapped jet.

Such a equatorial jet, which was first described by YOSHIDA (1959), is equivalent to the coastal jet as discussed, for example, in Chapter 5.

We note a remarkable difference in the steady responses of the equatorial ocean to the midlatitudinal ocean. The former responds with intense zonal jets to large scale zonal winds, whereas the latter responds to large scale forcing only with Ekman transport as shown in Section 3.7.

Outside the equatorial wave guide the meridional current consists of a meridional Ekman current which is confined to the surface layer. The associated Ekman mass flux causes Ekman pumping in the equatorial wave guide and so the equatorial oceans which are exposed to the prevailing SE trade winds are favoured upwelling areas. This has important consequences for the climatic interaction with the atmosphere as well for the ecosystem of the equatorial oceans. The amplitudes of the vertical modes of the vertical current component can be calculated from (8.5.12c) by $w = -(p_t + rp)$ which yields for the stationary part

$$w(\eta, t) = -2\theta(t)\frac{u_*^2}{\lambda h}\sum_{n=0}^{\infty}\left[\frac{2n+1}{4n+3}I_{2n}\psi_{2n}(\eta) - \eta\sqrt{\frac{2n+1}{2}}\frac{I_{2n}}{4n+3}\psi_{2n+1}(\eta)\right] \quad (8.5.13)$$

The Ekman pumping into the equatorial wave guide accumulates mass there causing a pressure anomaly which in turn causes the geostrophically adjusted zonal jet.

8.6. THE RESPONSE TO A SPATIALLY BOUNDED FORCING

In the previous section it was shown that the equatorial ocean responds to large scale zonal winds with inertial-gravity oscillations and intense zonal jets. From the studies of the midlatitudinal oceans we know that the response to spatially bounded wind patterns may be severely modified by waves which radiate from the edges of the wind distribution. It is the aim of the following two subsections to study these effects in the response of the equatorial ocean. In particular we consider zonal winds which are zonally bounded. In Subsection 8.6.1. we calculate the complete solution in the Fourier domain, in Subsection 8.6.2. we discuss the asymptotic properties of the response in the physical space, and in Subsection 8.6.3. the Kelvin wave and the equatorial undercurrent shall be studied.

8.6.1. Solution in the Fourier domain

We wish to study the response of the equatorial ocean to a zonal wind, switched on at $t = 0$, which blows over the area between the lines $x = -a$ and $x = a$. That is, we assume

$$X(t, x, y, z) = -\frac{u_*^2}{H_{mix}} \theta(t)\, \theta(a - |x|)\, \theta(z + H_{mix}), \qquad Y = 0. \tag{8.6.1a}$$

The spatial distribution of this forcing is sketched in Fig. 8.6.1.

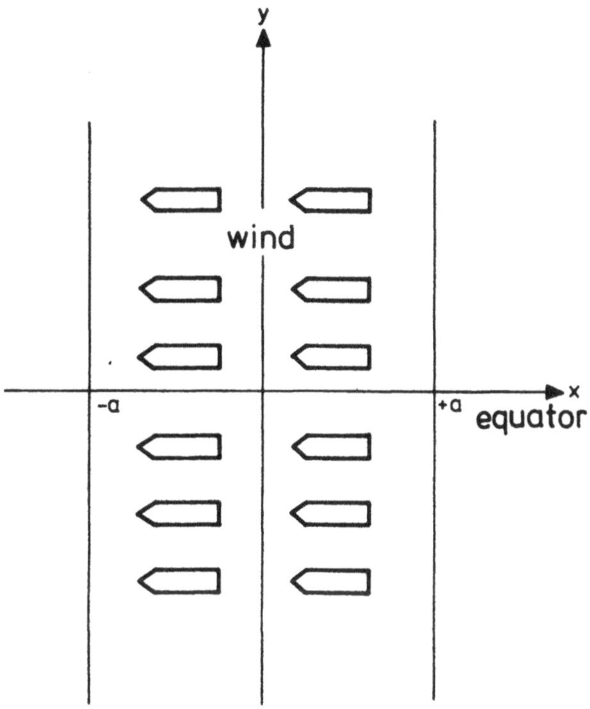

Fig. 8.6.1
Sketch of the wind patch according to (8.6.1a)

After vertical mode expansion and Fourier transforms with respect to t and x, see appendix (B. 23) and (B. 12), we get

$$X(\omega, k, y) = -\frac{2\sin(ka)}{k} \frac{i}{\omega + i\varepsilon} \frac{u_*^2}{h_m}. \tag{8.6.1b}$$

Again we drop the vertical mode number below.

In order to calculate the response it is convenient to start with the equations (8.3.9a), (8.4.6a), and (8.4.6b) which in this case assume the form

$$v(\omega, k, s) = \sqrt{\frac{\lambda}{2\beta}}\, G_q(s, s') * \frac{s'}{2} X(\omega, k, s') \tag{8.6.2a}$$

$$u(\omega, k, s) = \frac{i}{2}\left[\frac{(q+1)}{\bar{\omega} - \frac{k}{\lambda}} G_{q+1}(s, s') * X + \frac{q}{\bar{\omega} + \frac{k}{\lambda}} G_{q-1}(s, s') * X\right] \quad (8.6.2\text{ b})$$

$$p(\omega, k, s) = \frac{i}{2\lambda}\left[\frac{(q+1)}{\bar{\omega} - \frac{k}{\lambda}} G_{q+1}(s, s') * X - \frac{q}{\bar{\omega} + \frac{k}{\lambda}} G_{q-1}(s, s') * X\right]. \quad (8.6.2\text{ c})$$

We calculate the convolutions occurring in the set (8.6.2) with the aid of the bilinear representation of the Green's function (8.3.7).

The projection of the forcing (8.6.1) on the Hermite modes can easily be obtained from (8.5.9) and (8.5.10) for $L_a \to \infty$, i.e., $\zeta = 0$,

$$I_{2n} = \frac{\pi^{1/4}\sqrt{2(2n)!}}{2^n n!}. \quad (8.6.3)$$

Using the recurrence relation

$$\eta\psi_n(\eta) = \sqrt{\frac{n+1}{2}}\,\psi_{n+1}(\eta) + \sqrt{\frac{n}{2}}\,\psi_{n-1}(\eta)$$

from (8.6.2) we get the representation of the response in terms of Hermite functions

$$v(\omega, k, \eta) = -2i\,\frac{u_*^2}{h}\,\frac{2\sin ka}{(\omega + i\varepsilon)\,k}\sqrt{\frac{\lambda}{\beta}}\sum_{n=0}^{\infty}\sqrt{\frac{2n+1}{2}}\,\frac{I_{2n}}{2q+1-(4n+3)}\,\psi_{2n+1}(\eta) \quad (8.6.4\text{ a})$$

$$u(\omega, k, n) = \frac{u_*^2}{h}\,\frac{2\sin ka}{(\omega + i\varepsilon)\,k}\,\frac{\lambda}{2}\sum_{n=0}^{\infty}\left\{\frac{2(q+1)}{(\bar{\omega}\lambda - k)\,[2(q+1) - 4n]} + \frac{2q}{(\bar{\omega}\lambda + k)\,[2(q-1) - 4n]}\right\} I_{2n}\psi_{2n}(\eta) \quad (8.6.4\text{ b})$$

$$p(\omega, k, \eta) = \frac{u_*^2}{h}\,\frac{2\sin ka}{(\omega + i\varepsilon)\,k}\,\frac{1}{2}\sum_{n=0}^{\infty}\left\{\frac{2(q+1)}{(\bar{\omega}\lambda - k)\,[2(q+1) - 4n]} - \frac{2q}{(\bar{\omega}\lambda + k)\,[2(q-1) - 4n]}\right\} I_{2n}\psi_{2n}(\eta). \quad (8.6.4\text{c})$$

From the first sum in (8.6.4b) and (8.6.4c) we split off the $n = 0$ terms and renumber the indices in the remaining sums. Then with (8.2.12) we obtain

$$u(\bar{\omega}, k, \eta) = \frac{1}{2}\,\frac{u_*^2}{h}\,\frac{2\sin ka}{(\omega + i\varepsilon)\,k}$$

$$\times\left\{\frac{\lambda I_0}{(\bar{\omega}\lambda - k)}\psi_0(\eta) + \sum_{n=0}^{\infty}\frac{(\bar{\omega}^2\lambda + \omega k + \beta)\sqrt{2n+1}\,\psi_{2(n+1)}(\eta) + (\bar{\omega}^2\lambda - \bar{\omega}k - \beta)\sqrt{2(n+1)}\,\psi_{2n}(\eta)}{\beta\bar{\omega}[2q+1-(4n+3)]\sqrt{2(n+1)}}\,I_{2n}\right\}$$

(8.6.5 a)

$$p(\omega, k, \eta) = \frac{-1}{2\lambda} \frac{u_*^2}{h} \frac{2 \sin ka}{(\omega + i\varepsilon) k}$$

$$\times \left\{ \frac{\lambda I_0}{(\bar{\omega}\lambda - k)} \psi_0(\eta) + \sum_{n=0}^{\infty} \frac{(\bar{\omega}^2\lambda + \bar{\omega}k + \beta)\sqrt{2n+1}\,\psi_{2(n+1)}(\eta) - (\bar{\omega}^2\lambda - \bar{\omega}k - \beta)\sqrt{2(n+1)}\,\psi_{2n}(\eta)}{\beta\bar{\omega}[2q + 1 - (4n+3)]\sqrt{2(n+1)}} I_{2n} \right\}. \tag{8.6.5 b}$$

From (8.6.4a) it follows that the response of the meridional component consists of inertial waves and Rossby waves associated with odd Hermite modes, while according to (8.6.4b, c) in the response of the zonal current and of the pressure these waves are associated with even Hermite modes. Moreover, in u and p there occurs an eastward propagating Kelvin wave having the meridional structure of the lowest Hermite mode ψ_0.

Let us study the response pattern of the meridional current component in more detail. In order to proceed we first perform the Fourier transform of (8.6.4a) with respect to k. Using (8.2.12) we can express the denominator of (8.6.4a) as

$$2q + 1 - (4n + 3) = -\frac{1}{\lambda\beta}(k - k_1)(k - k_2) \tag{8.6.6}$$

where k_1 and k_2 are given by (8.3.12). Since the k_1 and k_2 are complex for frequencies within the interval $\omega_r \leqq \omega \leqq \omega_i$ we get from (8.6.6) the prescription how to go around the poles in the complex k-plane when inverting the Fourier transforms. It is convenient to introduce the integral

$$\hat{I}(\bar{\omega}, x) = \int_{-\infty}^{+\infty} \frac{dk}{2\pi} \frac{2\sin(ka)\,e^{ikx}}{k(k - k_1)(k - k_2)}. \tag{8.6.7}$$

Obviously the meridional current component in the ω, x space can be expressed by $\hat{I}(\bar{\omega}, x)$ as

$$v(\bar{\omega}, x, \eta) = +\frac{2\lambda u_*^2}{h}\sqrt{\lambda\beta}\sum_{n=0}^{\infty}\sqrt{\frac{2n+1}{2}}\,I_{2n}\frac{i}{\omega + i\varepsilon}\hat{I}(\bar{\omega}, x)\,\psi_{2n+1}(\eta). \tag{8.6.8}$$

Below we will show that u and p, and, hence, the complete response can also be expressed through $\hat{I}(\bar{\omega}, x)$. Therefore, we denote $\hat{I}(\bar{\omega}, x)$ as basic integral.

In order to evaluate $\hat{I}(\bar{\omega}, x)$ we use the fact that (8.6.7) can be written as a convolution integral

$$\hat{I}(\bar{\omega}, x) = \int_{-\infty}^{+\infty} dx'\,\theta(a - |x'|)\,\tilde{I}(x - x') \tag{8.6.9}$$

where

$$\tilde{I}(x) = \int_{-\infty}^{+\infty} \frac{dk}{2\pi} \frac{e^{ikx}}{(k - k_1)(k - k_2)}. \tag{8.6.10}$$

The poles k_1 and k_2 are given by (8.3.12b).

Let us choose k_1 and k_2 according to

$$k_{1,2} = -\frac{\beta}{2\bar{\omega}} \pm i\frac{\lambda}{\bar{\omega}}[(\omega_i^2 - \bar{\omega}^2)(\bar{\omega}^2 - \omega_r^2)]^{1/2}.$$

Then k_1 is located in the upper and k_2 in the lower half of the complex k-plane. Thus $\tilde{I}(x)$ can readily be calculated with the aid of the residues technique. Closing the contour in the upper half-plane for $x > 0$ and in the lower half-plane for $x < 0$, respectively, we get

$$\tilde{I}(x) = \frac{i}{k_1 - k_2}\left[\theta(x)\, e^{ik_1 x} + \theta(-x)\, e^{ik_2 x}\right].$$

Substituting this in (8.6.9) yields explicitly

$$\hat{I}(\bar{\omega}, x) = -\frac{1}{k_1 k_2(k_1 - k_2)}\{\theta(a - |x|)(k_2 - k_1) + k_2[\theta(x - a)\, e^{ik_1(x-a)} - \theta(x + a)\, e^{ik_1(x+a)}]$$

$$+ k_1[\theta(-(x - a))\, e^{ik_2(x-a)} - \theta(-(x + a))\, e^{ik_2(x+a)}]\}. \qquad (8.6.11)$$

We emphasize that this calculation is rigorous. Performing first the Fourier transform of (8.6.4a) with respect to ω would have the drawback that, due to the cumbersome algebraic problems related with the dispersion relation in the form (8.3.10), the result can only be obtained approximately, see CANE and SARACHIK (1976). The basic integral (8.6.11) describes how waves generated at the edges of the wind field propagate from these edges. Note that k_1 represents the branch of the dispersion relation with eastward group velocity, and k_2 that one with westward group velocity.

With the relations

$$k_1 - k_2 = 2\frac{\lambda}{\bar{\omega}}\sqrt{(\bar{\omega}^2 - \bar{\omega}_i^2)(\bar{\omega}^2 - \bar{\omega}_r^2)} \quad \text{and} \quad k_1 k_2 = -\lambda^2\left[\bar{\omega}^2 - \frac{\beta}{\lambda}(4n + 3)\right] \qquad (8.6.12)$$

which follows from (8.3.12) we can discuss the analytical properties of $\hat{I}(\bar{\omega}, x)$. It follows that $\hat{I}(\bar{\omega}, x)$ has branch points at $\bar{\omega} = \pm\omega_i$ and $\bar{\omega} = \pm\omega_r$, simple poles at $\bar{\omega} = \pm\beta(4n + 3)/\lambda$, and an essential singularity at $\bar{\omega} = 0$. Moreover, the basic integral (8.6.11) is bounded at both small and high frequencies. Particularly for $\bar{\omega} \to \infty$, we obtain $\hat{I}(\bar{\omega}, x) \sim (\lambda\bar{\omega})^{-2}$.

It can readily be shown that $\hat{I}(\bar{\omega}, x)$ has vanishing residues at the simple poles, $\pm\beta(4n + 3)/\lambda$, because k_1 vanishes at these frequencies. This implies that, the response of the equatorial ocean to a wind patch lacks pure inertial oscillations in contrast to the response to a zonally unbounded wind as described in the previous section.

Obviously, the expression (8.6.8) with (8.6.11) can be considered to be the Eulerian frequency spectrum of the meridional current component at any location x, y, note that $\eta = s/\sqrt{2} = \sqrt{\lambda\beta}\, y$. The spectrum is increasing at very low frequencies, and has peaks at frequencies of zero group velocities for both the Rossby and inertial wave modes. For high frequencies it decays as $(\lambda\bar{\omega})^{-3}$. Moreover, pure inertial oscillations do not occur. Next we estimate the response of the zonal current component by performing the Fourier transforms of (8.6.5a, b) with respect to the zonal wavenumber. The frequency spectra of u and p can be expressed in terms of the basic integral (8.6.11) and of the Kelvin wave contribution

$$u(\omega, x, \eta) = -\frac{u_*^2}{h}\frac{i\lambda}{2}\frac{i}{(\omega + i\varepsilon)}$$

$$\times\left\{\frac{I_0\psi_0(\eta)}{\bar{\omega}\lambda}[\theta(a - |x|) + \theta(x - a)\, e^{i\lambda\bar{\omega}(x-a)} - \theta(x + a)\, e^{i\lambda\bar{\omega}(x+a)}]\right.$$

$$-\sum_{n=0}^{\infty} \frac{I_{2n}}{\sqrt{2(n+1)}} \left[\left(\left(\bar{\omega}\lambda + \frac{\beta}{\bar{\omega}}\right) \hat{I}(x, \bar{\omega}) - \mathrm{i} \frac{\mathrm{d}}{\mathrm{d}x} \hat{I}(x, \bar{\omega})\right) \sqrt{2n+1}\, \psi_{2(n+1)}(\eta)\right.$$

$$\left. + \left(\left(\bar{\omega}\lambda - \frac{\beta}{\bar{\omega}}\right) \hat{I}(x, \bar{\omega}) + \mathrm{i} \frac{\mathrm{d}}{\mathrm{d}x} \hat{I}(x, \bar{\omega})\right) \sqrt{2(n+1)}\, \psi_{2n}(\eta)\right]\Bigg\} \tag{8.6.13}$$

and

$$p(\omega, x, \eta) = -\frac{\mathrm{i}}{2} \frac{u_*^2}{h} \frac{\mathrm{i}}{(\omega + \mathrm{i}\varepsilon)}$$

$$\times \Bigg\{ \frac{I_0 \psi_0(\eta)}{\bar{\omega}\lambda} \left[\theta(a - |x|) + \theta(x - a)\, \mathrm{e}^{\mathrm{i}\bar{\omega}\lambda(x-a)} - \theta(x + a)\, \mathrm{e}^{\mathrm{i}\bar{\omega}\lambda(x+a)}\right]$$

$$-\sum_{n=0}^{\infty} \frac{I_{2n}}{\sqrt{2(n+1)}} \left[\left(\left(\bar{\omega}\lambda + \frac{\beta}{\bar{\omega}}\right) \hat{I}(x, \bar{\omega}) - \mathrm{i} \frac{\mathrm{d}}{\mathrm{d}x} \hat{I}(x, \bar{\omega})\right) \sqrt{2n+1}\, \psi_{2(n+1)}(\eta)\right.$$

$$\left. - \left(\left(\bar{\omega}\lambda - \frac{\beta}{\bar{\omega}}\right) \hat{I}(x, \bar{\omega}) + \frac{\mathrm{d}}{\mathrm{d}x} I(x, \bar{\omega})\right) \sqrt{2(n+1)}\, \psi_{2n}(\eta)\right]\Bigg\}. \tag{8.6.14}$$

Differentiating the basic integral (8.6.11) yields

$$\frac{\mathrm{d}\hat{I}}{\mathrm{d}x} = \frac{-\mathrm{i}}{k_1 - k_2} \{\theta(x - a)\, \mathrm{e}^{\mathrm{i}k_1(x-a)} - \theta(x + a)\, \mathrm{e}^{\mathrm{i}k_1(x+a)} + \theta(-(x - a))\, \mathrm{e}^{\mathrm{i}k_2(x-a)}$$

$$- \theta(-(x + a))\, \mathrm{e}^{\mathrm{i}k_2(x+a)}\}. \tag{8.6.15}$$

From (8.6.13), (8.6.14) with (8.6.15) it follows that the spectra of u and p have peaks at the frequencies of vanishing group velocities of both the inertial gravity and Rossby waves. For high frequencies, $\omega \to \infty$, these spectra are "red", for $\omega \to 0$ they increase as $\bar{\omega}^{-2}$.

In contrast to the case of the large scale forcing considered in Section 8.5., where the peak frequencies are determined by the inertial oscillations, in the present case the spectral peaks are due to those frequencies which correspond to zero group velocity.

8.6.2. Asymptotic solution of the response in the physical space

In this subsection we shall study the response of an equatorial ocean to the forcing (8.6.1) in the physical space, i.e., in the x, y, t domain.

In order to calculate the response in the time domain we have to invert the Fourier transforms of (8.6.8), (8.6.13), and (8.6.14) with respect to ω.

From the method of stationary phases, see, for example, PEDLOSKY (1979), it can be inferred that the Fourier kernels developed at frequencies of maximum group velocities, describe the propagation of the wave fronts which move with these maximum group velocities.

According to (8.3.17) the maximum group velocities of inertial gravity waves are $\pm\lambda^{-1}$. For the Rossby waves we find from (8.3.19) the eastward and westward maximum group velocities to be $[8(4n + 3)\lambda]^{-1}$ and $-[(4n + 3)\lambda]^{-1}$, respectively. Thus the Rossby wave fronts move eight times faster to the west than to the east. Between the fronts, which propagate from the edges of the wind patch, waves with smaller group

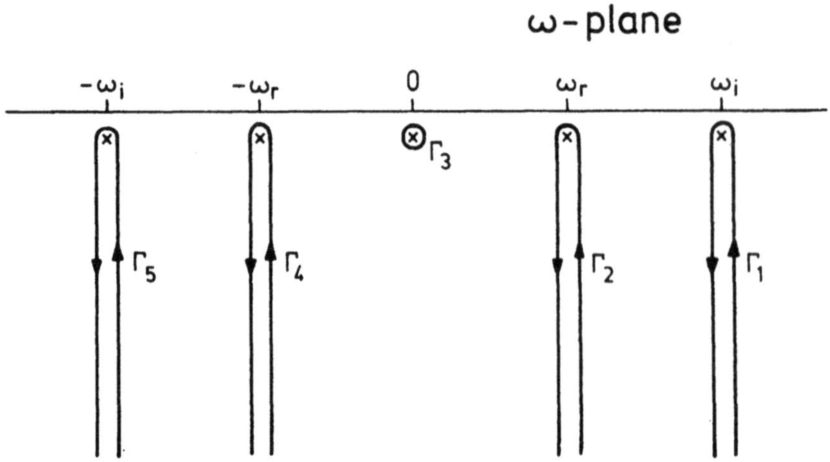

Fig. 8.6.2
The contours Γ_1 to Γ_5 used for the calculation of the asymptotic properties of the dynamical fields in the equatorial wave guide

velocities establish. In the following we wish to analyze these wave processes. This amounts to the investigation of Fourier integrals of (8.6.8), (8.6.13), and (8.6.14) for large times.

The analytical properties of the basic integral $\hat{I}(\varpi, x)$ (8.6.11) are already discussed in the foregoing section, i.e., there are branch points at $\pm\omega_i$ and $\pm\omega_r$, and an essential singularity at $\varpi = 0$.

Moreover, in expressions (8.6.8), (8.6.13), and (8.6.14) occur poles at $\omega = 0$. Thus we can use the integration contour shown in Fig. 8.6.2. The asymptotic expression for large times can be obtained in the same manner as outlined in Section 7.3.

We start with the asymptotic evaluation of the meridional current v. The contribution along the contours Γ_1 and Γ_5 yield for large t.

$$v^{(1,5)}(x, \eta, t) = u_*^2 \left(\frac{\beta}{\lambda}\right)^{1/2} \sum_{n=0}^{\infty} \left(\frac{2n+1}{2}\right)^{1/2} \frac{I_{2n}}{h}$$
$$\times \frac{\psi_{2n+1}(\eta)}{\omega_i \sqrt{\omega_i^2 - \omega_r^2}\sqrt{\pi\omega_i t}\left(4n + 3 - \frac{\lambda}{\beta}\omega_i^2\right)}$$
$$\times \left[\cos\left(\omega_i t + \frac{\beta}{2\omega_i}(x-a) + \frac{\pi}{4}\right) - \cos\left(\omega_i t + \frac{\beta}{2\omega_i}(x+a) + \frac{\pi}{4}\right)\right]. \tag{8.6.16}$$

It is interesting to note that this result cannot be obtained if first the Fourier inversion with respect to ω, using the approximate roots (8.3.16), would be performed, see CANE and SARACHIK (1976). Then the exact root ω_i would be approximated by the inertial frequency, which corresponds to $k = 0$. In that case the phase information in (8.6.16) is lost and the cosine-terms in (8.6.16) would cancel each other.

From (8.6.16) it follows that the asymptotic regime is described by inertial waves with vanishing group velocities. Note that ω_i is the frequency, and $\beta/2\omega_i$ the associated wavenumber of zero group velocity. The inertial waves decay algebraically as $(\omega_i t)^{-1/2}$.

In the same way we obtain from the integrals along Γ_2 and Γ_4

$$v^{(2,4)}(x,\eta,t) = -u_*^2\left(\frac{\beta}{\lambda}\right)^{1/2}\sum_{n=0}^{\infty}\left(\frac{2n+1}{2}\right)^{1/2}\frac{I_{2n}}{h}\,\frac{\psi_{2n+1}(\eta)}{\omega_r\sqrt{\omega_i^2-\omega_r^2}\left(4n+3-\frac{\lambda}{\beta}\omega_r^2\right)}$$

$$\times\left[\sin\left(\omega_r t+\frac{\beta}{2\omega_r}(x-a)+\frac{\pi}{4}\right)-\sin\left(\omega_r t+\frac{\beta}{2\omega_r}(x+a)+\frac{\pi}{4}\right)\right]. \quad (8.6.17)$$

This describes Rossby waves with frequencies of zero group velocities and decaying with time as $(\omega_r t)^{-1/2}$. Both the inertial waves and the Rossby waves of vanishing group velocities have westward phase velocity.

Next we estimate the contribution of the integral along Γ_3. This amounts to the estimation of the residue of the integrand at $\bar{\omega} = 0$. For small $\bar{\omega}$ we find from (8.6.12)

$$k_1 - k_2 \approx -\frac{\beta}{\bar{\omega}}\left[1 - 2\bar{\omega}^2\frac{\lambda}{\beta}(4n+3)\right]$$

and, hence,

$$v^{(3)} \sim \frac{\hat{I}(\bar{\omega},x)}{\omega} \approx \frac{-1}{\omega\beta(4n+3)}\Big[-\theta(a-|x|)+\theta(a-x)\,e^{i\bar{\omega}\lambda(4n+3)(a-x)}$$
$$-\theta(-x-a)\,e^{-i\lambda(4n+3)(a+x)\bar{\omega}}\Big]$$
$$-\frac{\bar{\omega}}{\omega\beta}\left[\theta(x-a)\,e^{\frac{-i\beta}{\bar{\omega}}(x-a)}-\theta(x+a)\,e^{-i\frac{\beta}{\bar{\omega}}(x+a)}\right] \quad (8.6.18)$$

for small ω.

The first term is readily evaluated. The second term containing an essential singularity can be calculated in the same way as outlined in Section 7.3., see in particular (7.3.9). The result is

$$v^{(3)}(x,\eta,t) = -u_*^2\cdot 2\left(\frac{\lambda}{\beta}\right)^{1/2}\sum_{n=0}^{\infty}\left(\frac{2n+1}{2}\right)^{1/2}\frac{I_{2n}\psi_{2n+1}(\eta)}{4n+3}$$
$$\times\Big\{-\theta(t)\,\theta(a-|x|)+\theta(a-x)\,\theta(t+\lambda(4n+3)(x-a))\,e^{r\lambda(4n+3)(x-a)}$$
$$-\theta(-x-a)\,\theta(t+\lambda(4n+3)(x+a))\,e^{r\lambda(4n+3)(x+a)}$$
$$-\frac{\lambda}{\beta}(4n+3)\frac{d^2}{dt^2}\left[\theta(x-a)J_0(2\sqrt{\beta t(x-a)})-\theta(x+a)J_0(2\sqrt{\beta t(x+a)})\right]\Big\}$$
$$(8.6.19)$$

This part of the response consist of (1) a steady contribution associated with a Yoshida jet within the forcing area, (2) westward propagating long Rossby waves, and (3) eastward propagating short Rossby waves. The complete response is the sum of $v^{(1,5)}$, $v^{(2,4)}$ and $v^{(3)}$.

Next we note also the corresponding expressions of u and p. The first terms of u and p, see (8.6.13) and (8.6.14), which correspond to the contributions of the Kelvin poles in (8.6.5a, b) will be discussed in the next subsection. The contributions of the branch points $\pm\omega_i$ follows from the integrals along Γ_1 and Γ_5. For large times the asymptotic solutions are

$$u^{(1,5)}(t, x, \eta) = -\frac{u_*^2}{2h}\sum_{n=0}^{\infty}\frac{\sin\left(\omega_i t + \dfrac{\beta}{2\omega_i}(x-a) + \dfrac{\pi}{4}\right) - \sin\left(\omega_i + \dfrac{\beta}{2\omega_i}(x+a) + \dfrac{\pi}{4}\right)}{\left(4n+3-\dfrac{\lambda}{\beta}\omega_i^2\right)(\omega_i^2-\omega_r^2)^{1/2}} \times \frac{\sqrt{2n+1}\,\psi_{2(n+1)}(\eta) + \sqrt{2(n+1)}\,\psi_{2n}(\eta)}{\sqrt{2(n+1)}}\,\frac{I_{2n}}{\sqrt{\pi\omega_i t}} \tag{8.6.20a}$$

and

$$p^{(1,5)}(t, x, \eta) = -\frac{u_*^2}{2\lambda h}\sum_{n=0}^{\infty}\frac{\sin\left(\omega_i t + \dfrac{\beta}{2\omega_i}(x-a) + \dfrac{\pi}{4}\right) - \sin\left(\omega_i t + \dfrac{\beta}{2\omega_i}(x+a) + \dfrac{\pi}{4}\right)}{\left(4n+3-\dfrac{\lambda}{\beta}\omega_i^2\right)(\omega_i^2-\omega_r^2)^{1/2}} \times \frac{\sqrt{2n+1}\,\psi_{2(n+1)}(\eta) - \sqrt{2(n+1)}\,\psi_{2n}(\eta)}{\sqrt{2(n+1)}}\,\frac{I_{2n}}{\sqrt{\pi\omega_i t}}\,. \tag{8.6.20b}$$

The contributions from the branch points $\pm\omega_r$ are

$$u^{(2,4)}(t, x, \eta) = -\frac{\beta u_*^2}{4\lambda h}\sum_{n=0}^{\infty}\frac{\cos\left(\omega_r t + \dfrac{\beta}{2\omega_r}(x-a) + \dfrac{\pi}{4}\right) - \cos\left(\omega_r t + \dfrac{\beta}{2\omega_r}(x+a) + \dfrac{\pi}{4}\right)}{\omega_r^2\left(4n+3-\dfrac{\lambda}{\beta}\omega_r^2\right)(\omega_i^2-\omega_r^2)^{1/2}} \times \frac{\sqrt{2n+1}\,\psi_{2(n+1)}(\eta) + \sqrt{2(n+1)}\,\psi_{2n}(\eta)}{\sqrt{2(n+1)}}\,\frac{I_{2n}}{\sqrt{\pi\omega_r t}} \tag{8.6.21a}$$

and

$$p^{(2,4)}(t, x, \eta) = -\frac{\beta u_*^2}{4\lambda^2 h}\sum_{n=0}^{\infty}\frac{\cos\left(\omega_r t + \dfrac{\beta}{2\omega_r}(x-a) + \dfrac{\pi}{4}\right) - \cos\left(\omega_r t + \dfrac{\beta}{2\omega_r}(x+a) + \dfrac{\pi}{4}\right)}{\omega_r^2\left(4n+3-\dfrac{\lambda}{\beta}\omega_r^2\right)(\omega_i^2-\omega_r^2)^{1/2}} \times \frac{\sqrt{2n+1}\,\psi_{2(n+1)}(\eta) - \sqrt{2(n+1)}\,\psi_{2n}(\eta)}{\sqrt{2(n+1)}}\,\frac{I_{2n}}{(\pi\omega_r t)^{1/2}}\,. \tag{8.6.21b}$$

Finally we note the expressions arising from the singularities at $\bar{\omega} = 0$ and $\omega = 0$, i.e., the residues of the integral along Γ_3

$$u^{(3)}(x,t,\eta) = -u_*^2 \sum_{n=0}^{\infty} \frac{I_{2n}}{h}\left(\left(\frac{2n+1}{2n+2}\right)^{1/2} \psi_{2n+2}(\eta) - \psi_{2n}(\eta)\right)$$
$$\times\left\{-\frac{\theta(t)\,\theta(a-|x|)}{4n+3}\,\frac{(1-e^{-rt})}{r} + \frac{\theta(a-x)}{4n+3}\,\theta(t+\lambda(4n+3)(x-a))\,\frac{e^{r\lambda(4n+3)(x-a)}-e^{-rt}}{r}\right.$$
$$-\frac{\theta(-a-x)}{4n+3}\,\theta(t+\lambda(4n+3)(x+a))\,\frac{(e^{r\lambda(4n+3)(x+a)}-e^{-rt})}{r}$$
$$\left.+\left(\frac{\lambda}{\beta}\right)^2 \frac{d^3}{dt^3}\left[\theta(x-a)\,J_0(2\sqrt{\beta t(x-a)}) - \theta(x+a)\,J_0(2\sqrt{\beta t(x+a)})\right]\right\} \tag{8.6.22a}$$

$$p^{(3)}(t,x,\eta) = -\frac{u_*^2}{\lambda}\sum_{n=0}^{\infty}\frac{I_{2n}}{h}\left[\left(\frac{2n+1}{2n+2}\right)^{1/2}\psi_{2n+2}(\eta) + \psi_{2n}(\eta)\right]$$
$$\times\left\{-\frac{\theta(t)\,\theta(a-|x|)}{4n+3}\,\frac{1-e^{-rt}}{r} + \frac{\theta(a-x)}{4n+3}\,\theta(t+\lambda(4n+3)(x-a))\,\frac{e^{r\lambda t(4n+3)(x-a)}-e^{-rt}}{r}\right.$$
$$-\frac{\theta(-a-x)}{4n+3}\,\theta(t+\lambda(4n+3)(x+a))\,\frac{e^{r\lambda(4n+3)(x+a)}-e^{-rt}}{r}$$
$$\left.+\frac{\lambda^2}{\beta^2}\frac{d^3}{dt^3}\left[\theta(x-a)\,J_0(2\sqrt{\beta t(x-a)}) - \theta(x+a)\,J_0(2\sqrt{\beta t(x+a)})\right]\right\}. \tag{8.6.22b}$$

The sums of $u^{(1,5)}$, $u^{(2,4)}$, and $u^{(3)}$, and of $p^{(1,5)}$, $p^{(2,4)}$, and $p^{(3)}$ describe apart from the Kelvin wave contribution the complete response for large time. In particular the response consists of inertial waves and Rossby waves, see (8.6.21) and (8.6.22), of zero group velocities. In the present case these waves decay algebraically. However, for an oscillating forcing with frequencies close to ω_i or ω_r even resonantly excited waves can be expected. The occurrence of such waves can be expected to be a common feature in equatorial oceans.

Moreover, the response contains a Yoshida jet within the forcing area and westward propagating long Rossby waves and eastward propagating short Rossby waves, see (8.6.22).

8.6.3. Kelvin waves and the equatorial undercurrent

In this subsection we discuss the remaining contributions to u and p which are associated with the Kelvin pole. We use the letters u and p without a special subscript although they refer only to the Kelvin pole contributions, i.e., the first terms in (8.6.5a, b), within this subsection.

Fourier transforming of the first terms in (8.6.13) and (8.6.14) yields

$$u(x,\eta,t) = -\frac{u_*^2}{2h} I_0\psi_0(\eta)\left[\theta(t)\,\theta(a-|x|)\,\frac{(1-e^{-rt})}{r}\right.$$
$$-\theta(x+a)\,\theta(t-\lambda(x+a))\,\frac{e^{-r\lambda(x+a)}-e^{-rt}}{r}$$
$$\left.+\theta(x-a)\,\theta(t-\lambda(x-a))\,\frac{e^{-r\lambda(x-a)}-e^{-rt}}{r}\right] \tag{8.6.23a}$$

$$p(x, \eta, t) = \frac{1}{\lambda} u(x, \eta, t) \tag{8.6.23 b}$$

where according to (8.6.3) $I_0 = \pi^{1/4} \sqrt{2}$. If we ignore the friction, i.e., $rt \ll 1$, we get

$$u(x, \eta, t) = -\frac{u_*^2}{2h} I_0\psi_0(\eta) \{\theta(t)\, \theta(a - |x|)\, t - \theta(x + a)\, \theta(t - \lambda(x + a)) [t - \lambda(x + a)] + \theta(x - a)\, \theta(t - \lambda(x - a)) [t - \lambda(x - a)]\} \tag{8.6.24 a}$$

$$p(x, \eta, t) = \frac{1}{\lambda} u(x, \eta, t) . \tag{8.6.24 b}$$

Immediatly after the switch-on of the wind an accelerating jet develops within the forcing area. Simultaneously Kelvin waves radiate with the phase speed λ^{-1} from the wind edges. The Kelvin wave originating from the western edge arrests after its arrival the jet, while the waves starting from the eastern edge propagate outside the forcing region. This behaviour is sketched in Fig. 8.6.3. The modal vertical velocity, i.e., with regard to the vertical mode m, is given by $w = p_t$ and so we get

$$w(x, \eta, t) = \frac{1}{\lambda} \frac{u_*^2}{2h} I_0\psi_0(\eta) [\theta(t)\, \theta(a - |x|) - \theta(x + a)\, \theta(t - \lambda(x + a)) + \theta(x - a)\, \theta(t - \lambda(x - a))]\, e^{-rt} . \tag{8.6.25}$$

Thus, the upwelling associated with the gravest meridional mode, ψ_0, of the equatorial jet within the forcing region vanishes after the passage of the Kelvin front. Moreover, the Kelvin front which propagates from the eastern edge of the wind exports upwelling outside the forcing area.

Let us consider a fixed point within the forcing region, say $x = 0$. Then after $t > \lambda a$ the Kelvin front has passed through this point and from (8.6.24) it follows

$$u(0, \eta, t) = -\frac{u_*^2}{2h} I_0\psi_0(\eta)\, \lambda a .$$

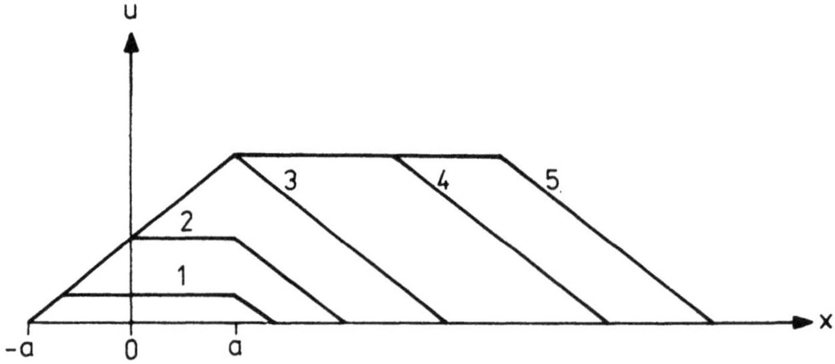

Fig. 8.6.3
Sketch of the zonal distribution of the response of u and p according to (8.6.24a, b)
The labels 1 to 5 indicate different times. Shown is the arresting of the Yoshida jet within the wind band and the eastward propagation of the Kelvin waves.

That is, the equatorial jet was arrested. The occurrence of the factor λ in (8.6.25) indicates that an undercurrent develops in the wake of the Kelvin front, analogous to the coastal undercurrent discussed in Section 5.10.

For very large times practically all Kelvin fronts have passed through the forcing region and from (8.6.23) we find

$$u(x,\eta,t) = -\frac{u_*^2}{2h} I_0\psi_0(\eta)\,\theta(t)\left[\theta(x+a)\frac{1-e^{-r\lambda(x+a)}}{r} - \theta(x-a)\frac{1-e^{-r\lambda(x-a)}}{r}\right]. \qquad (8.6.26)$$

Since the λ's increase with the number of the vertical modes, m, it appears that the damping effect is the stronger the higher the mode number.

In the inviscid case (8.6.26) reduces to

$$u(x,\eta,t) = -\frac{u_*^2}{2h} I_0\psi_0(\eta)\,\theta(t)\,[\theta(x+a)\,\lambda(x+a) - \theta(x-a)\,\lambda(x-a)]\,. \qquad (8.6.27)$$

The complete result follows if the sum over all baroclinic vertical eigenfunctions has been carried out, i.e.,

$$u(x,y,z,t) = -\frac{u_*^2}{2^{1/2}}\frac{\theta(t)}{r}\sum_{m=1}^{\infty}\frac{F_m(z)}{h_m}e^{-\frac{y^2}{2}\lambda_m\beta}$$
$$\times\,[\theta(x+a)(1-e^{-r\lambda_m(x+a)}) - \theta(x-a)(1-e^{-r\lambda_m(x-a)})]\,.$$

For a slowly varying BVF we can perform the summation. Let us assume, just for simplicity, a constant BVF. The we get

$$u(x,y,z,t) = -\frac{u_*^2}{H_{\text{mix}}r}\frac{\theta(t)}{\sqrt{2}\,\pi}\Bigg\{\theta(a-|x|)\left[\arctan\left(\frac{\sin\frac{\pi}{H}(H_{\text{mix}}+z)}{e^{\frac{y^2}{2}\lambda_1\beta}-\cos\frac{\pi}{H}(H_{\text{mix}}+z)}\right)\right.$$
$$\left.+\arctan\left(\frac{\sin\frac{\pi}{H}(H_{\text{mix}}-z)}{e^{\frac{y^2}{2}\lambda_1\beta}-\cos\frac{\pi}{H}(H_{\text{mix}}-z)}\right)\right]$$
$$-\,\theta(x+a)\left[\arctan\left(\frac{\sin\frac{\pi}{H}(H_{\text{mix}}+z)}{e^{\frac{y^2}{2}\lambda_1\beta+\lambda_1 r(x+a)}-\cos\frac{\pi}{H}(H_{\text{mix}}+z)}\right)\right.$$
$$\left.+\arctan\left(\frac{\sin\frac{\pi}{H}(H_{\text{mix}}-z)}{e^{\frac{y^2}{2}\lambda_1\beta+\lambda_1 r(x+a)}-\cos\frac{\pi}{H}(H_{\text{mix}}-z)}\right)\right]$$

$$+\,\theta(x-a)\left[\arctan\left(\frac{\sin\frac{\pi}{H}(H_{mix}+z)}{e^{\frac{y^2}{2}\lambda_1\beta+\lambda_1 r(x-a)}-\cos\frac{\pi}{H}(H_{mix}+z)}\right)\right.$$

$$\left.\left.+\arctan\left(\frac{\sin\frac{\pi}{H}(H_{mix}-z)}{e^{\frac{y^2}{2}\lambda_1\beta+\lambda_1 r(x-a)}-\cos\frac{\pi}{H}(H_{mix}-z)}\right)\right]\right\}. \tag{8.6.28}$$

For $r \to 0$ this reduces to

$$u(x,y,t) = -\frac{u_*^2\lambda_1}{H_{mix}\sqrt{2}}\,\theta(t)\,[\theta(a-|x|)(x+a)+\theta(x-a)\cdot 2a]$$

$$\times\left[\frac{\sin\frac{\pi}{H}(H_{mix}+z)}{\cosh\frac{y^2}{2}\lambda_1\beta-\cos\frac{\pi}{H}(H_{mix}+z)}+\frac{\sin\frac{\pi}{H}(H_{mix}-z)}{\cosh\frac{y^2}{2}\lambda_1\beta-\cos\frac{\pi}{H}(H_{mix}-z)}\right]; \tag{8.6.29}$$

Using (8.6.28) we discuss two meridional sections of the zonal current. The involved parameters are chosen as $a = 2000$ km, $u_*^2 = 0.5\ \mathrm{cm^2/s^2}$, $H = 4000$ m, $H_{mix} = 75$ m, $r = 2\pi/(1\ \text{year})$, and $\lambda_1 = 0.36$ s/m according to Table 1.9.3. The zonal flow near the western edge of the wind pattern is shown in Fig. 8.6.4. We see a westward surface jet in the upper mixed layer with a typical meridional scale of 150 km and a maximum velocity of 100 cm $\mathrm{s^{-1}}$ at the equator. Below an eastward equatorial undercurrent is present. The core of the undercurrent is located at the equator, just beneath the mixed layer with a velocity of about 75 cm/s. The deepest vertical extent of the 10 cm/s isotach is to a depth of about 200 m. The meridional e-folding distance of the undercurrent is roughly half that of the surface jet, i.e., 50 km. Thus some of the listed features compare well with observations. However, the meridional section shown in Fig. 8.6.4 exhibits also features which are not observed in equatorial currents, in particular the occurrence of a core of the surface jet just above the bottom of the mixed layer. This is due to a singularity of the formula (8.6.28) at $y = 0$ and $z = -H_{mix}$, which can be attributed to the idealized step like vertical distribution of our forcing function.

The zonal flow field 500 km east of the wind patch is shown in Fig. 8.6.5. The equatorial undercurrent exists only marginally at this location. The surface jet has the same intensity as within the forcing region, however, the meridional as well the vertical scales of the surface jet are larger outside the wind patch. The surface jet east of the wind forcing area balances the poleward Ekman current inside the wind patch. The equatorial upwelling associated with the gravest Hermite mode has been terminated by Kelvin waves, compare (8.6.25).

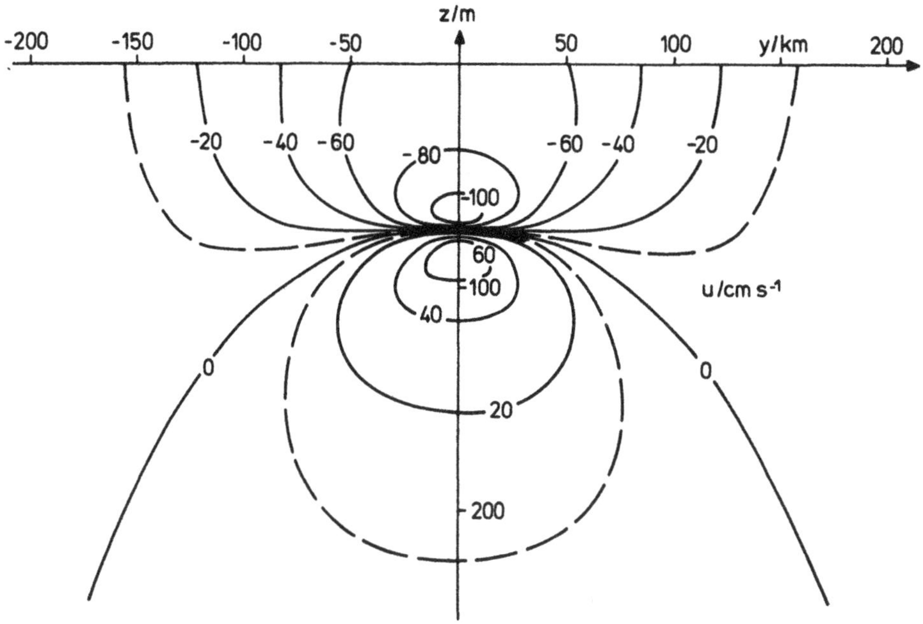

Fig. 8.6.4
Contours of zonal velocity in cm/s within the forcing area (located 500 km east of the western edge of the wind patch) after the passage of all Kelvin waves

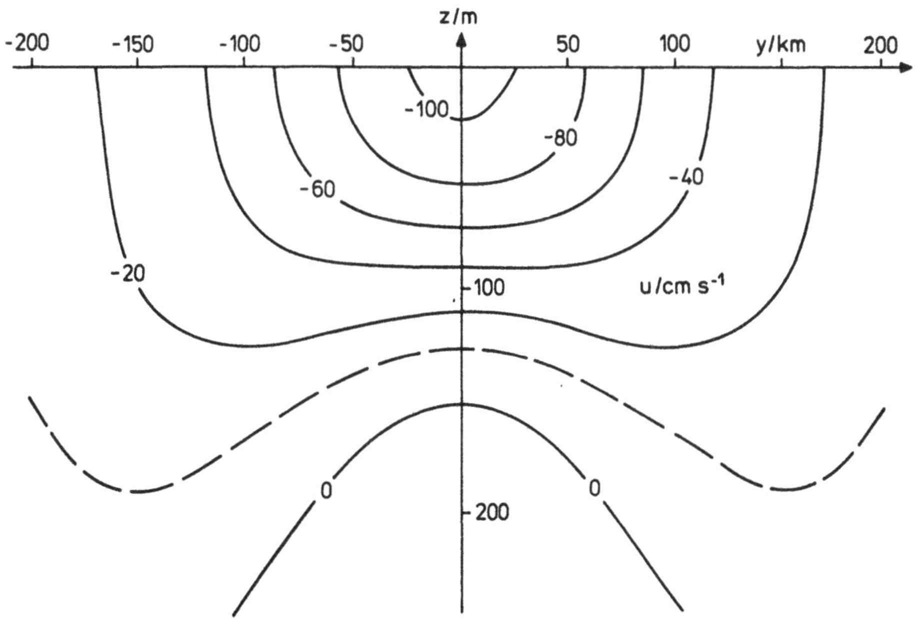

Fig. 8.6.5
Contours of zonal velocity in cm/s outside the forcing area (located 500 km east of the eastern edge of the wind patch) after the passage of all Kelvin waves

Thus, as pointed out by McCreary (1981), an undercurrent exists if the wind stress is spatially bounded. A more realistic picture follows if a variable BVF profile is used. For example, the WKB treatment yields expressions proportional to $N^{1/2}(z)$, compare (1.9.47), and then the surface jet has a smaller amplitude than the undercurrent. Moreover, it was shown by McCreary (1981) that the existence of ocean boundaries weakens the surface flow and intensifies the undercurrent.

8.6.4. Summary of the response scenario

In this subsection we shall give a summary of the scenario of the baroclinic response of the equatorial ocean to a constant, zonally bounded east wind switched on at $t = 0$ as estimated in the preceding sections.

In the first phase there develops an accelerating surface jet, trapped with the baroclinic Rossby radius at the equator which is referred to as Yoshida jet. Upwelling (or downwelling) will occur in the equatorial wave guide if the wind blows westward (or eastward). Simultaneously, inertial wave fronts propagate equally as well eastward as westward with the speed of the inverse vertical eigenvalues, λ_m^{-1}, from the edges of the wind patch, leaving behind an area with inertial waves of vanishing zonal group velocity. These waves decay algebraically as $t^{-1/2}$.

Moreover, Kelvin wave fronts propagate eastward from the westward edge of the forcing area with the same velocities, i.e., λ_m^{-1}, arresting the jet and stopping the upwelling associated with the gravest meridional mode ψ_0 behind the fronts. After the passage of the Kelvin wave fronts an equatorial undercurrent develops. Furthermore, from the eastward edge of the wind patch a Kelvin wave travels into the eastern, i.e., the unforced region, and exports upwelling (or downwelling) into this area. The first vertical mode of this wave moves rather fast and crosses basin wide distances within a few months. Friction becomes particularly important for the Kelvin waves associated with higher vertical modes.

The Rossby wave fronts excited by the switched on wind, travel eight times faster westward than eastward. However, even the speed of the fastest Rossby wave front is three times smaller than that of the Kelvin wave front. Within this area stretching between the eastward and westward propagating Rossby wave fronts establish three kinds of Rossby waves, namely Rossby waves of zero group velocity decaying as $t^{-1/2}$, eastward propagating short Rossby waves decaying as $t^{-3/2}$, and westward propagating non-dispersive long Rossby waves. The long Rossby waves excited at the western edge of the wind patch propagate signals into the unforced region westward of the wind patch. Far from the equator, i.e., at distances exceeding the Rossby radius $(\lambda_1 \beta)^{-1/2}$ the response consists within the forcing area mainly of the Ekman transport. The results reveal significant differences between the responses of the equatorial ocean to a zonally bounded forcing and a unbounded forcing as discussed in Section 8.5.

Some important modifications occur if meridional boundaries of the ocean are taken into account.

Responses of a zonally bounded equatorial ocean were studied by Moore and Philander (1977), Cane and Sarachik (1977), and McCreary (1980). Thus, at the western boundary an incoming long Rossby wave produces reflected Kelvin waves and, moreover, a western boundary layer which is formed by short Rossby waves. At the eastern boundary incoming Kelvin waves transform into reflected long Rossby waves and an eastern boundary layer built up by Rossby waves. Within this boundary layer a part of the incoming Kelvin waves is fed into poleward propagating Kelvin waves.

Observational evidence for the dynamic features of the linear equatorial ocean has been reported in several articles. Significant peaks at frequencies of vanishing group velocities of inertial-gravity wave modes have been observed by WUNSCH and GILL (1976), LUTHER (1980), and ERIKSEN (1982) in sea level records in the Pacific. It was shown by WEISBERG (1979), WEISBERG et al. (1979), and ERIKSEN (1982) that the gravest inertial-gravity modes possess westward phase velocities and wavenumbers which are consistent with those given in (8.6.16) and (8.6.20). Moreover, LUKAS et al. (1984) analyzed cross correlations of sea level along the equatorial Pacific and found a westward travelling inertial-gravity wave front.

Impressive examples of Kelvin waves excited in the western Pacific by wind pulses of a few weeks duration, which propagate eastward with a speed close to λ_1^{-1} over the entire equatorial Pacific, were observed by KNOX and HALPERN (1982), ERIKSEN et al. (1983), and LUKAS et al. (1984). Furthermore, a signal which propagates westward along the equator with a speed of approximately the gravest mode of long Rossby waves was also observed by LUKAS et al. (1984).

Thus it appears that many of the essential dynamic features of the equatorial ocean can well be understood in the frame of the linear theory.

8.7. FREQUENCY SPECTRA ON THE EQUATOR

8.7.1. General remarks

In the Section 8.6., particularly in 8.6.2., we discussed the response of the equatorial ocean to a special wind pattern asymptotically in physical space. It appeared to be impossible to estimate analytically the complete response in the physical space due to the difficulties of inverting completely the Fourier transforms with respect to ω. It was, however, possible to perform the Fourier transform of the general solution (8.3.9) with respect to the zonal wavenumber k. The resulting expressions represent frequency spectra of the dynamical variables at fixed locations.

Theoretical estimations of spectra can be of great value because it is common in oceanography to measure time series of observables such as sea level, horizontal current components, temperature etc. at fixed locations and to transform these time series into frequency spectra. Regarding the Fourier transformation of discrete stationary time series of finite length we refer to standard textbooks, e.g. JENKINS and WATTS (1968).

Thus, we may compare measured frequency spectra with those calculated from an analytical theory. This yields a comparison of observations and theory on the level of frequency spectra. For example frequency spectra of horizontal current components on and close to the equator have been estimated and discussed in terms of ETW's for all three oceans, see e.g. WEISSBERG et al. (1979), ERIKSEN (1980), FENNEL and LASS (1983), ERIKSEN (1985), and FENNEL et al. (1987).

Typical spectra of the current components measured in the surface mixed layer and in the thermocline in the eastern Pacific are depicted in Figs. 8.7.1 to 8.7.4. These current measurements were conducted from March 1980 to April 1982, see HALPERN (1983) for technical details.

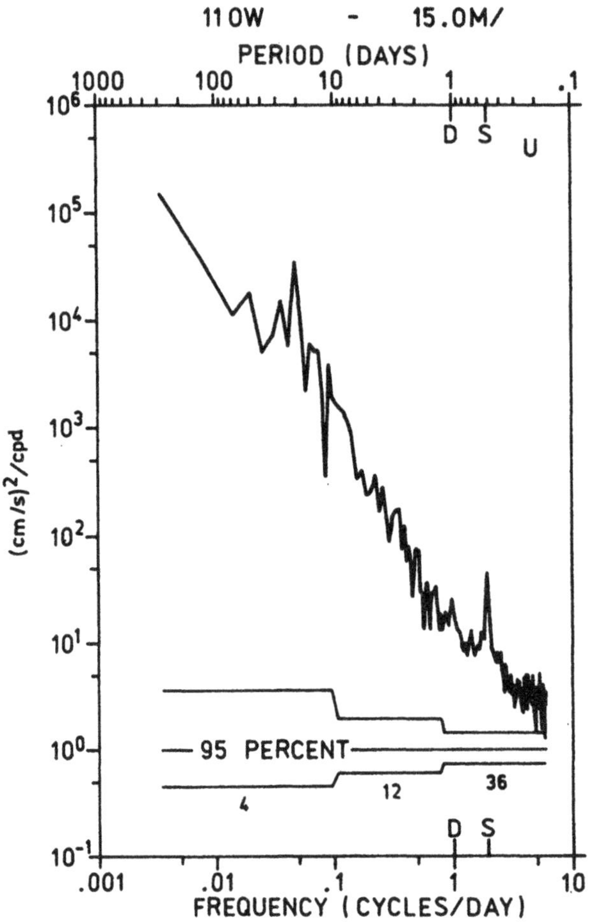

Fig. 8.7.1
Spectrum of the zonal current at the equator in the Eastern Pacific at 111° W in the upper mixed layer, $z = -15$ m (redrawn from FENNEL et al. (1987))

Near the surface, i.e. at 15 m depth, the zonal current spectrum, see Fig. 8.7.1, increases continuously with decreasing frequency. At 100 m depth, however, the spectrum shows some saturation for periods larger than about 100 days, see Fig. 8.7.2.

The meridional spectra appear to be saturated in the low frequency domain in both 15 m and 100 m depth, as can be seen from Figs. 8.7.3 and 8.7.4. All spectra exhibit various peaks in the range of periods less than, say, 50 days. A predominant peak can be observed at about 20 days, particularly, in the near surface spectrum of the meridional current.

The overall structure of the near surface zonal spectrum resembles sea level spectra found at near equatorial islands in the Pacific, see WUNSCH and GILL (1976), and LUTHER (1980).

In the following we wish to estimate theoretically these frequency spectra. To this end we shall consider a model of the wind fields which has the gross features of the observed equatorial wind field. This section is organized as follows: In Subsection 8.7.2. we specify the forcing, then, in 8.7.3, we estimate the theoretical spectra, and in Subsection 8.7.4a comparison of theory and observations is given. In the these sections we follow FENNEL et al. (1987).

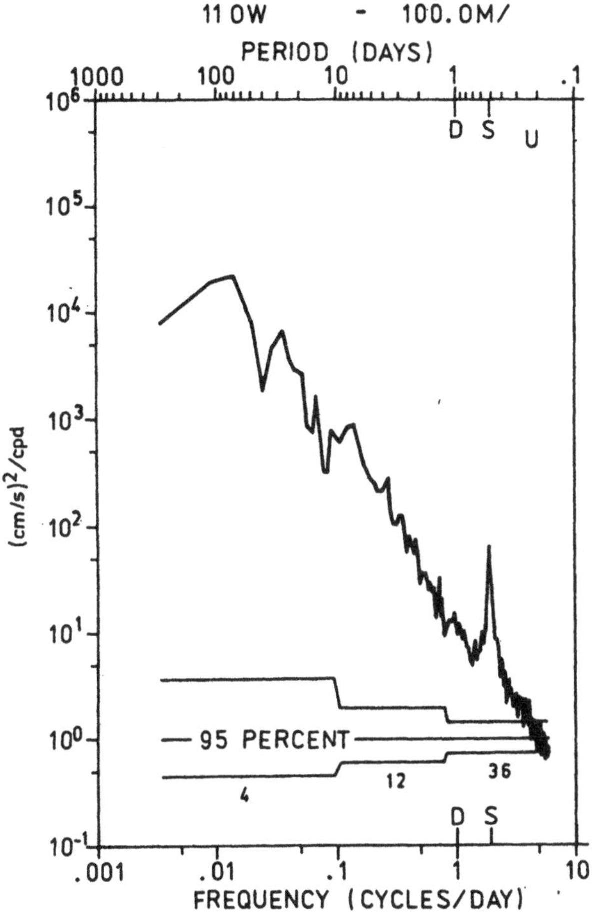

Fig. 8.7.2
Spectrum of the zonal current at the equator in the Eastern Pacific at 111° W within the thermocline, $z = -100$ m (redrawn from FENNEL et al. (1987))

8.7.2. The forcing functions

Frequency spectra of the wind components recorded at several near equatorial sites appear to be approximately "red", i.e., they vary as ω^{-1} to $\omega^{-1/2}$, as reported by LUTHER (1980), and GARZOLI and KATZ (1984). Moreover, some peaks were detected at periods of 4 to 5 days.

This gross behaviour of the wind spectra corresponds in the time domain to functions such as $\theta(t)$ or $\theta(t)\, t^{-1/2}$, whose Fourier transforms vary as ω^{-1} or $\omega^{-1/2}$, respectively. We adopt the step function $\theta(t)$ but being aware that this choice may exaggerate the overall slope of the response spectra.

According to LUTHER (1980) the zonal wavenumber spectra of the wind components are concentrated in a small band in the vicinity of $k = 0$. The energy maxima of the zonal wind spectra are distributed nearly symmetrically around $k = 0$, the most energetic part of the meridional wind spectra appears to be shifted into the domain of small negative wavenumbers. The zonal scales of the wind range from about 3000 km to one circum-

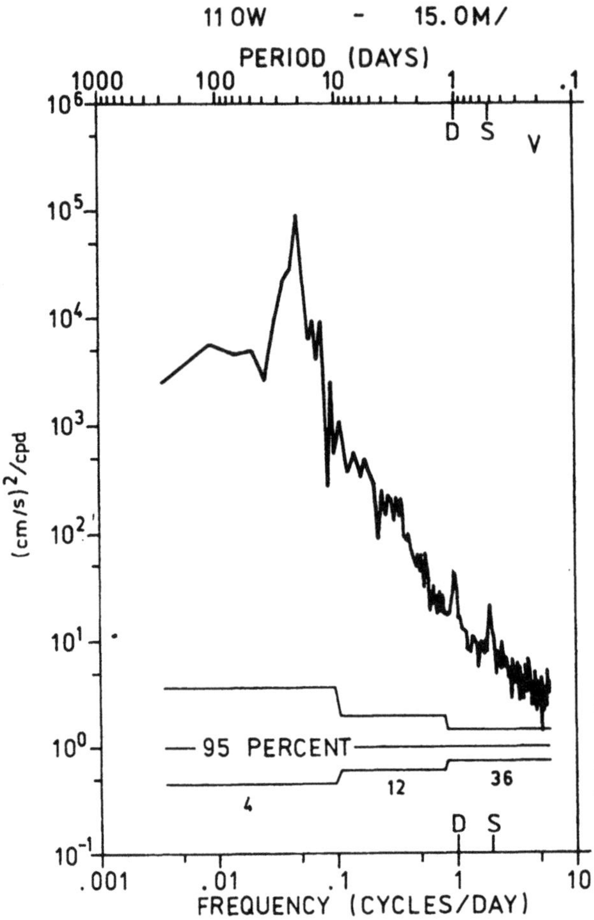

Fig. 8.7.3
Spectrum of the meridional current at the equator in the Eastern Pacific at 111° W in the upper mixed layer, $z = -15$ m (redrawn from FENNEL et al. (1987))

ference of the earth. Thus, we may model the wavenumber spectra of the forcing by Dirac's delta function peaked at a small wavenumber k_0.

Moreover, we assume the wind field to be trapped near the equator with characteristic meridional scale $L_a = 1000$ km.

This is in accordance with the fact that at least during boreal summer the equatorial areas of both the Pacific and the Atlantic are exposed to the South-East trades, being nearly constant around the equator on a scale which is larger than the equatorial Rossby radius.

Thus, we assume

$$(X(x, y, z, t), Y(x, y, z, t)) = \frac{(u_*^2, v_*^2)}{H_{mix}} \theta(z + H_{mix})\, \theta(t)\, e^{ik_0 x}\, \Psi_0(y) \tag{8.7.1}$$

where

$$\Psi_0(y) = e^{-\frac{1}{2}\left(\frac{y}{L_a}\right)^2}$$

is the gravest "atmospheric" Hermite function.

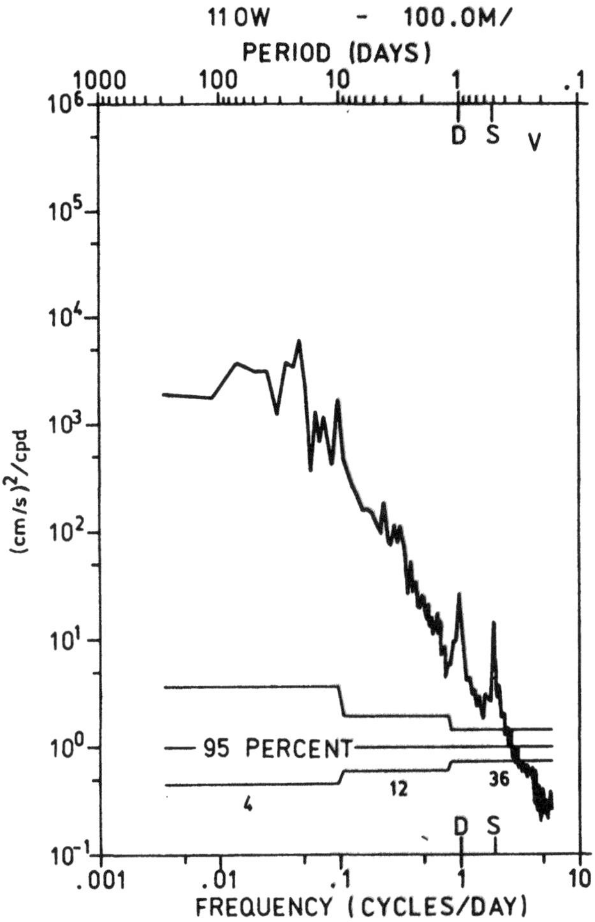

Fig. 8.7.4
Spectrum of the meridional current at the equator in the Eastern Pacific at 111° W within the thermocline, $z = -100$ m (redrawn from FENNEL et al. (1987))

Expanding (8.7.1) in vertical eigenfunctions, and performing the Fourier transforms with respect to t and x we get

$$(X_m(k, s, \omega), Y_m(k, s, \omega)) = \frac{(u_*^2, v_*^2)}{h_m} \frac{2\pi i}{\omega + i\varepsilon} \delta(k - k_0) \Psi_0(s) \tag{8.7.2}$$

where the normalized meridional distance $s = \sqrt{2\beta\lambda}\, y$ has been introduced. Again the vertical mode index m will be dropped below. Obviously $\Psi_0(s)$ is given by

$$\Psi_0(s) = e^{-\zeta\frac{s^2}{4}} \tag{8.7.3}$$

with ζ being the squared ratio of the meridional scale L_a of the atmospheric forcing and the oceanic equatorial Rossby radius, i.e., $\zeta = L_a^2\lambda\beta$.

8.7.3. The general solution

In the following we wish to calculate the spectra of the dynamical variables right on the equator and so we start our calculations with putting $s = 0$ in the general solution (8.3.9). Then we obtain

$$v(k, 0, \omega) = \mathrm{i}\, \frac{\bar{\omega}^2 \lambda^2 - k^2}{2\lambda\beta\bar{\omega}}\, G * Y + \sqrt{\frac{\lambda}{2\beta}}\, G * \frac{s'}{2}\, X + \frac{k}{\sqrt{2\lambda\beta}\,\bar{\omega}}\, G * X_{s'} \tag{8.7.4 a}$$

$$u(k, 0, \omega) = \frac{k}{\bar{\omega}\sqrt{2\lambda\beta}}\, \frac{\partial}{\partial s}\, G * Y + \frac{\mathrm{i}\lambda^2 \bar{\omega}}{\bar{\omega}^2 \lambda^2 - k^2} \left[X - \frac{k}{\bar{\omega}\lambda}\, \frac{\partial}{\partial s} \left(G * \frac{s'}{2}\, X + \frac{k}{\bar{\omega}\lambda}\, G * X_{s'} \right) \right] \tag{8.7.4 b}$$

$$p(k, 0, \omega) = \frac{1}{\sqrt{2\lambda\beta}}\, \frac{\partial}{\partial s}\, G * Y + \frac{\mathrm{i}k}{\bar{\omega}^2 \lambda^2 - k^2} \left[X - \frac{\bar{\omega}\lambda}{k}\, \frac{\partial}{\partial s} \left(G * \frac{s'}{2}\, X + \frac{k}{\bar{\omega}\lambda}\, G * X_{s'} \right) \right]. \tag{8.7.4 c}$$

First we estimate the convolutions occurring in this set. Taking the Green's function (8.3.4), (8.3.5) into account we have to calculate two types of integrals.

The first one is

$$\lim_{s \to 0} \int_{-\infty}^{+\infty} G_q(s, s')\, f(s')\, \mathrm{d}s' = -\frac{\Gamma(-q)}{\sqrt{2\pi}}\, D_q(0) \int_0^{\infty} D_q(s') \left[f(s') + f(-s') \right] \mathrm{d}s' \tag{8.7.5 a}$$

which vanishes for odd $f(s)$, and the second one is

$$\lim_{s \to 0} \frac{\partial}{\partial s} \int_{-\infty}^{+\infty} G_q(s, s')\, f(s')\, \mathrm{d}s' = -\frac{\Gamma(-q)}{\sqrt{2\pi}}\, D_{q+1}(0) \int_0^{\infty} D_q(s') \left[f(s') - f(-s') \right] \mathrm{d}s' \tag{8.7.5 b}$$

which vanishes if $f(s)$ is an even function. In (8.7.5a, b) the relation

$$\lim_{s \to 0} \frac{\partial}{\partial s}\, D_q(\pm s) = \mp D_{q+1}(0)$$

has been used, which follows from the recurrence relation of the parabolic cylinder functions

$$\frac{\mathrm{d}}{\mathrm{d}s}\, D_q(\pm s) = \pm \frac{s}{2}\, D_q(\pm s) \mp D_{q+1}(\pm s). \tag{8.7.6}$$

Substituting the forcing (8.7.2) in the set (8.7.4) and using the symmetry properties of the integrals (8.7.5a, b) yield

$$v(k, 0, \omega) = \frac{v_*^2}{h}\, \frac{2\pi \mathrm{i}}{\omega}\, \delta(k - k_0)\, \frac{\mathrm{i}}{2}\, \frac{\bar{\omega}^2 \lambda^2 - k^2}{\lambda\beta\bar{\omega}}\, G * \Psi_0(s') \tag{8.7.7 a}$$

$$u(k, 0, \omega) = \frac{u_*^2}{h}\, \frac{2\pi \mathrm{i}}{\omega}\, \delta(k - k_0) \left[\frac{\mathrm{i}\lambda^2 \bar{\omega}}{\bar{\omega}^2 \lambda^2 - k^2}\, \Psi_0(0) + \frac{\mathrm{i}k(\zeta k - \bar{\omega}\lambda)}{2(\bar{\omega}^2 \lambda^2 - k^2)\, \bar{\omega}}\, \frac{\partial}{\partial s}\, G * s' \Psi_0(s') \right] \tag{8.7.7 b}$$

$$p(k, 0, \omega) = \frac{u_*^2}{h} \frac{2\pi i}{\omega} \delta(k - k_0) \left[\frac{ik}{\bar{\omega}^2 \lambda^2 - k^2} \Psi_0(0) - \frac{i(\bar{\omega}\lambda - k\zeta)}{2(\bar{\omega}^2\lambda^2 - k^2)} \frac{\partial}{\partial s} G * s'\Psi_0(s') \right]. \tag{8.7.7 c}$$

Using the properties of the integrals (8.7.5a, b) we see that several of the convolutions in (8.7.7) vanish and it remains to calculate the integrals

$$S_1(q) = -2 \frac{\Gamma(-q)}{\sqrt{2\pi}} D_q(0) \int_0^\infty D_q(s')\, e^{-\frac{\zeta s'^2}{4}}\, ds' \tag{8.7.8 a}$$

$$S_2(q) = -2 \frac{\Gamma(-q)}{\sqrt{2\pi}} D_{q+1}(0) \int_0^\infty D_q(s')\, s'\, e^{-\frac{\zeta s'^2}{4}}\, ds' . \tag{8.7.8 b}$$

The integral (8.7.8a) can be found in standard tables, e.g., GRADSTEIN and RYSHIK (1971),

$$\int_0^\infty e^{-\frac{\zeta s'^2}{4}} D_q(s')\, ds' = \frac{\sqrt{\pi}\; {}_2F_1\left(-\frac{q}{2}, \frac{1}{2}; \frac{2-q}{2}; \frac{\zeta - 1}{\zeta + 1}\right)}{2^{\frac{q}{2}} (\zeta + 1)^{1/2}\, \Gamma\left(\frac{1-q}{2}\right)} \tag{8.7.9}$$

where ${}_2F_1$ is the hypergeometric function

$${}_2F_1(a, b; c; x) = \frac{\Gamma(c)}{\Gamma(a)\,\Gamma(b)} \sum_{n=0}^{\infty} \frac{\Gamma(a+n)\,\Gamma(b+n)}{\Gamma(c+n)\, n!} x^n .$$

Thus, with

$$D_q(0) = 2^{\frac{q}{2}} \frac{\Gamma\left(\frac{1}{2}\right)}{\Gamma\left(\frac{1-q}{2}\right)}$$

we have

$$S_1(q) = \left(\frac{2}{\zeta + 1}\right)^{1/2} \frac{1}{q}\; {}_2F_1\left(-\frac{q}{2}, \frac{1}{2}; 1 - \frac{q}{2}; \frac{\zeta - 1}{\zeta + 1}\right). \tag{8.7.10}$$

The integral S_2 can also be expressed by means of (8.7.9). To this end we use the recurrence formula (8.7.6) to put S_2 in the form

$$S_2(q) = -2 \frac{\Gamma(-q)}{\sqrt{2\pi}} D_{q+1}(0) \int_0^\infty ds'\, e^{-\frac{\zeta s'^2}{4}} (D_{q+1}(s') + qD_{q-1}(s'))$$

and with (8.7.9) we get

$$S_2(q) = \frac{2^{\frac{(q+1)}{2}}\sqrt{\pi}}{\sqrt{\zeta+1}\,\Gamma\left(-\frac{q+1}{2}\right)}\left[-\frac{2}{(q+1)}\,{}_2F_1\left(-\frac{q+1}{2}, 1-\frac{q+1}{2}; \frac{\zeta-1}{\zeta+1}\right)\right.$$

$$\left. + \frac{q}{(q+1)\left(\frac{q+1}{2}-1\right)^2}\,{}_2F_1\left(1-\frac{q+1}{2}, \frac{1}{2}; 2-\frac{q+1}{2}; \frac{\zeta-1}{\zeta+1}\right)\right].$$

Taking the corresponding Gauss' relations of contiguous ${}_2F_1$-functions into account, e.g., GRADSTEIN and RYSHIK (1971), this can be written as

$$S_2(q) = \left(\frac{2}{\zeta+1}\right)^{3/2}\frac{1}{q-1}\,{}_2F_1\left(-\frac{q-1}{2}, \frac{3}{2}; 1-\frac{q-1}{2}; \frac{\zeta-1}{\zeta+1}\right). \tag{8.7.11}$$

With the integrals (8.7.10) and (8.7.11) we obtain from (8.7.7)

$$v(k,0,\omega) = \frac{v_*^2}{h}\frac{2\pi\,\mathrm{i}}{\omega}\,\delta(k-k_0)\frac{\mathrm{i}}{2}\frac{\bar\omega^2\lambda^2-k^2}{\lambda\beta\bar\omega}\sqrt{\frac{2}{\zeta+1}}\,{}_2F_1\left(-\frac{q}{2},\frac{1}{2};1-\frac{q}{2};\frac{\zeta-1}{\zeta+1}\right) \tag{8.7.12 a}$$

$$u(k,0,\omega) = \frac{u_*^2}{h}\frac{2\pi\,\mathrm{i}}{\omega}\,\delta(k-k_0)\left[\frac{\mathrm{i}\lambda^2\bar\omega}{\bar\omega^2\lambda^2-k^2} + \frac{\mathrm{i}k(\zeta k-\bar\omega\lambda)}{2\bar\omega(\bar\omega^2\lambda^2-k^2)}\left(\frac{2}{\zeta+1}\right)^{2/3}\right.$$

$$\left.\times\frac{1}{(q-1)}\,{}_2F_1\left(\frac{1-q}{2},\frac{3}{2};\frac{3-q}{2};\frac{\zeta-1}{\zeta+1}\right)\right] \tag{8.7.12b}$$

$$p(k,0,\omega) = \frac{u_*^2}{h}\frac{2\pi\,\mathrm{i}}{\omega}\,\delta(k-k_0)\left[\frac{\mathrm{i}k}{\bar\omega^2\lambda^2-k^2} - \frac{\mathrm{i}(\bar\omega\lambda-k\zeta)}{2(\bar\omega^2\lambda^2-k^2)}\left(\frac{2}{\zeta+1}\right)^{3/2}\right.$$

$$\left.\times\frac{1}{q-1}\,{}_2F_1\left(\frac{1-q}{2},\frac{3}{2};\frac{3-q}{2};\frac{\zeta-1}{\zeta+1}\right)\right]. \tag{8.7.12c}$$

Obviously, right on the equator u and p are forced by the zonal component of the wind, while v is forced by the meridional forcing. We note that u, v, p, λ, q, and ζ are implicitly indexed with respect to the vertical mode number m.

In particular we wish to estimate the spectra of the horizontal current components u and v. These spectra follow from (8.7.12a) and (8.7.12b) after inverting the Fourier transforms with respect to k, and after summation over the vertical eigenfunctions, i.e.,

$$u(x,0,z,\omega) = \frac{u_*^2}{\beta\omega}\,\mathrm{e}^{-\mathrm{i}k_0x}\sum_{m=1}^{\infty}\frac{F_m(z)}{h_m}\,\beta\left[\frac{\bar\omega\lambda_m^2}{\bar\omega^2\lambda_m^2-k_0^2} + \frac{k_0}{2\bar\omega}\frac{(k_0\zeta_m-\bar\omega\lambda_m)}{\bar\omega^2\lambda_m^2-k_0^2}\left(\frac{2}{\zeta_m+1}\right)^{2/3}\right.$$

$$\left.\times\frac{1}{q_m-1}\,{}_2F_1\left(\frac{1-q_m}{2},\frac{3}{2};\frac{3-q_m}{2},\frac{\zeta_m-1}{\zeta_m+1}\right)\right] \tag{8.7.13 a}$$

$$v(x,0,z,\omega) = \frac{u_*^2}{\beta\omega}\,\mathrm{e}^{-\mathrm{i}k_0x}\sum_{m=1}^{\infty}\frac{F_m(z)}{h_m\lambda_m}\frac{\lambda_m^2\bar\omega^2-k_0^2}{2\bar\omega}\left(\frac{2}{\zeta_m+1}\right)^{1/2}$$

$$\times\frac{1}{q_m}\,{}_2F_1\left(-\frac{q_m}{2},\frac{1}{2},\frac{1-q_m}{2},\frac{\zeta_m-1}{\zeta_m+1}\right). \tag{8.7.13 b}$$

In order to perform the mode summation we use the eigenfunctions, $F_m(z)$, as given by (1.9.67b) and (1.9.77), which correspond to the BVF (1.9.53). The numerical values of the occurring parameters are $N_1 = 1.6 \cdot 10^{-3}\,s^{-1}$, $N_2 = 5.8 \cdot 10^{-3}\,s^{-1}$, $a^{-1} = 1400$ m, $H_{mix} = 75$ m, and $H = 4000$ m. The corresponding eigenvalues, projection coefficients, and the values of the vertical eigenfunctions in the surface mixed layer ($z = -10$ m) and in the thermocline ($z = 300$ m) are listed in Table 1.9.3 for the first ten modes. The first few eigenfunctions are shown in Fig. 1.9.8. For the practical calculations we have to terminate the infinite sums (8.7.13a, b) at a finite m. From Fig. 1.9.7 we see that the first ten modes represent nearly 70% of the forcing function. The current, however, is not such a strongly localized function as the forcing and so it should be sufficient for a first order description to take only the first ten vertical modes into account. The results will be discussed in the following section.

8.7.4. Discussion of the spectra on the equator

The spectral functions of the horizontal currents on the equator, which are in response to the forcing (8.7.1), are given by (8.7.13a, b). Let us define the amplitude spectra S_u and S_v by the modulus of (8.7.13a, b), normalized by the strength of the forces, u_*^2/H_{mix} and v_*^2/H_{mix}, and β

$$S_u(x, 0, z, \omega) = \frac{\beta H_{mix}}{u_*^2} |u(x, 0, z, \omega)| \qquad 8.7.14a)$$

$$S_v(x, 0, z, \omega) = \frac{\beta H_{mix}}{u_*^2} |v(x, 0, z, \omega)| \,. \qquad (8.7.14b)$$

Generally, the expressions (8.7.13a, b), and so (8.7.14a, b), have poles in the lower half plane of the complex ω-plane which correspond to peaks of finite height on the real ω-axis. The peak frequencies follow from the intersections of the line $k = k_0$ and the dispersion curves in the dispersion diagram shown in Fig. 8.2.1. For the meridional and zonal components only the even and odd equatorial trapped modes, respectively, contribute to the spectra. For positive k_0 the inertial waves, and the Yanai wave are excited, while for negative k_0 also the Rossby waves contribute. Moreover, in the zonal current spectra a jet, which for $k_0 = 0$ is confined to the upper mixed layer, and in case of positive k_0 the equatorial Kelvin wave may occur. Thus the zonal spectrum should be expected to be more energetic in the low frequency range than the meridional one.

The calculated frequency spectra of both the meridional and zonal current component are shown in Fig. 8.7.5 and Fig. 8.7.6, respectively, for frequencies ranging between 1/300 and 1/2 cycles per day. The chosen parameters are $L_a = 1000$ km, $k_0 = -2\pi/(3000$ km), and $r = 2\pi/(1$ year). An analysis of the sensitivity of the spectra with respect to the parameters, and a more detailed comparison between calculated and measured spectra is given by FENNEL et al. (1987).

It appears that the best resemblence between observed and calculated spectra could be achieved with the numerical values of the parameters L_a, r, and k_0 adopted above. However, the resulting shape of the spectra turns out to be not very sensitive to variations of L_a and r. Generally, the model frequency spectra of the meridional component (Fig. 8.7.5) are flat for periods exceeding 30 days, and have a peak in the period range of 10 to 30 days, which is, however, not as pronounced as in the observed spectra.

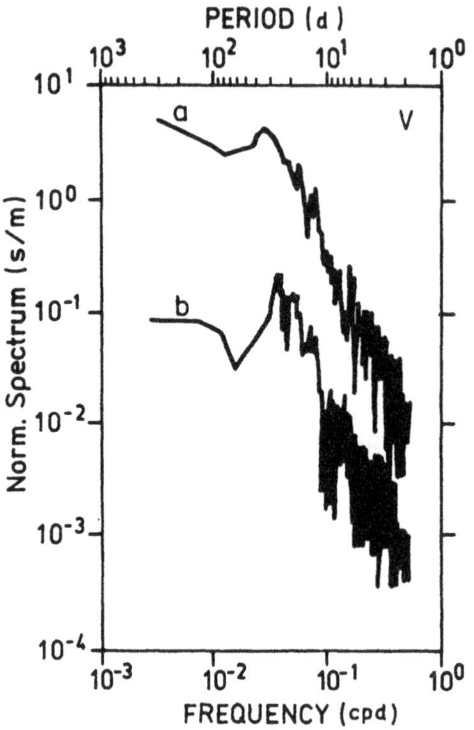

Fig. 8.7.5
Theoretical spectra of the meridional current at the equator for $z = -10$ m (a) and $z = -300$ m (b) (redrawn from FENNEL et al. (1987))
The lower curve is displaced downward by one decade.

For the periods smaller than 10 days the spectra decay approximately like ω^{-1} and there is an ensemble of superimposed peaks. Obviously the prominent peak at some 10 days and the peaks in the high frequency part are due to the excitation of the Yanai wave and the even modes of the inertial gravity waves, respectively.

Friction seems to be less important at these time scales. However, the saturation of the spectrum at periods longer than 30 days indicates the stronger impact of friction on the Rossby modes.

There is no essential difference between the v-spectra within the surface mixed layer and in the thermocline, Fig. 8.7.5 label a and label b, respectively.

Contrary to the v-spectra both the observed and the modelled u-spectra are, at least in the surface layer, red in the low frequency range, i.e., for periods larger than 30 days. The zonal current component in the low frequency range can be understood in terms of accelerating surface jets. These jets drain much more energy from the wind field into the ocean as the Rossby waves do.

The decay of the low frequency part of the u-spectrum with depth, significantly in the observed, but only marginally in the modelled spectra, as depicted in Fig. 8.7.6, suggests that these jets are confined to the surface area but may radiate partly into the deep ocean. Thus we find good agreement of the general shape of observed and modelled near surface current component spectra at the equator. This suggests that at first order the equatorial ocean responds to wind forcing by a dynamic pattern which can be described by a linear superposition of inertia-gravity waves, Yanai and Rossby waves, Kelvin waves, and surface jets.

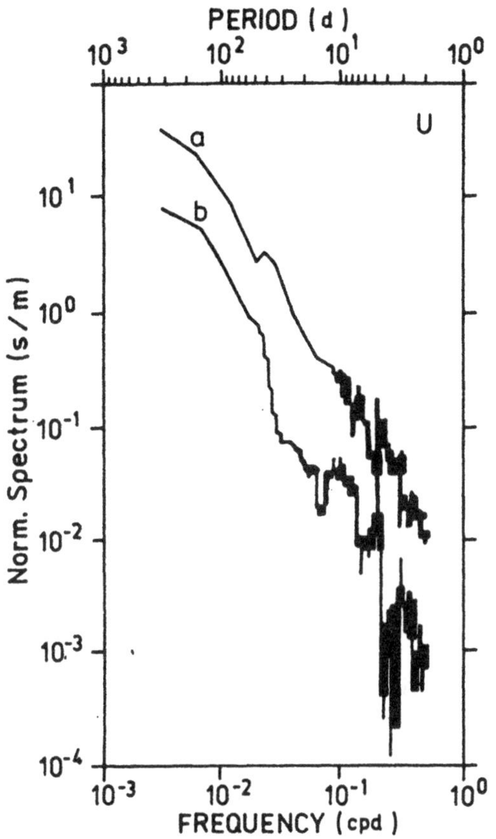

Fig. 8.7.6
Theoretical spectra of the zonal current at the equator for $z = -10$ m (a) and $z = -300$ m (b) (redrawn from FENNEL et al. (1987))
The lower curve is displaced by one decade.

Finally we note two differences between the observed and the theoretical spectra. In the high frequency range the decay of the model spectra appears to be exaggerated. As mentioned above this is due to the chosen switch on of the wind, i.e., a forcing being proportional to $\theta(t)$. In accordance with the observed wind spectra, e.g. LUTHER (1980), the wind would be better modelled by $\theta(t)\, t^{-1/2}$ with the Fourier transformed proportional to $\omega^{-1/2}$. In other words a wind which decays after its 'switch on' would feed more energy into the high frequency part of the spectra. Moreover, the peak at 20 days is more pronounced in the observed spectra within the surface mixed layer. This may be due to instabilities caused by the large horizontal and vertical current shear of the equatorial mean currents in the surface layer which are not considered in the linear model.

It was hypothesized by PHILANDER (1978) that 2—3 week oscillations arise via barotropic instability as the equatorial surface currents intensify seasonally. WEISBERG (1985) reported observational evidence which implies that the 20 day oscillations found in the equatorial Atlantic are consistent with this hypothesis.

Appendix

In the following we append some mathematical relations which were utilized in this book.

A. PROPERTIES OF THE DELTA FUNCTION

We start with the step function which is defined by

$$\theta(x) = \begin{cases} 1 & x > 0 \\ \frac{1}{2} & x = 0 \\ 0 & x < 0 . \end{cases} \tag{A 1}$$

The delta function may be introduced by the derivative of the step function

$$\delta(x) = \frac{d}{dx}\theta(x) . \tag{A2}$$

Obviously $\delta(x)$ is zero everywhere except at $x = 0$ where $\delta(x)$ is singular in such a way that

$$\int_{-\infty}^{+\infty} dx\, \delta(x) = \int_{-\varepsilon}^{+\varepsilon} dx\, \delta(x) = 1 . \tag{A 3}$$

This implies

$$\int_{-\infty}^{+\infty} dx\, f(x)\, \delta(x) = f(0) \int_{-\infty}^{+\infty} dx\, \delta(x) = f(0) . \tag{A 4}$$

A simple change of variable shows

$$\int_{-\infty}^{+\infty} dx\, f(x)\, \delta(x-a) = f(a) . \tag{A 5}$$

As a special case it follows

$$x\,\delta(x) = 0 . \tag{A6}$$

Further evident properties of the delta function are

$$\delta(x) = \delta(-x) \tag{A7}$$

$$\delta(ax) = \frac{\delta(x)}{|a|} . \tag{A8}$$

The delta function has the inverse dimension of its argument. Let $x = \tilde{x}[x]$, where $\tilde{x}$ is a number and $[x]$ is the dimension of x, e.g. length in meter. Then from (A8) it follows

$$\delta(x) = \frac{\delta(\tilde{x})}{[x]}.$$

Let $f(x)$ be a function with N first order zeros, i.e., $f(x_n) = 0$ with $f'(x_n) \neq 0$, $n = 1, 2, \dots, N$. Then the generalization of (A8) yields

$$\delta(f(x)) = \sum_{n=1}^{N} \frac{\delta(x - x_n)}{|f'(x_n)|}. \tag{A 9}$$

In particular it follows

$$\delta(x^2 - a^2) = \frac{1}{2a}[\delta(x - a) + \delta(x + a)], \qquad a > 0. \tag{A10}$$

Next we consider a derivative of the delta function

$$\int_{-\infty}^{+\infty} \mathrm{d}x\, \delta'(x)\, f(x) = [f(x)\, \delta(x)]_{-\infty}^{+\infty} - \int_{-\infty}^{\infty} \mathrm{d}x\, f'(x)\, \delta(x) = -f'(0). \tag{A 11}$$

Here the rule of integration by parts was used. Repeating this process we find

$$\int_{-\infty}^{+\infty} \mathrm{d}x\, f(x) \frac{\mathrm{d}^n}{\mathrm{d}x^n} \delta(x) = (-1)^n f^{(n)}(0)$$

or

$$\frac{\mathrm{d}^n}{\mathrm{d}x^n} \delta(x) = (-1)^n \delta(x) \frac{\mathrm{d}^n}{\mathrm{d}x^n}. \tag{A12}$$

We add some further relations

$$\frac{\mathrm{d}}{\mathrm{d}x} x\theta(x) = x\,\delta(x) + \theta(x) = \theta(x) \tag{A13}$$

$$\frac{\mathrm{d}}{\mathrm{d}x} |x| = \operatorname{sgn}(x) = 2\theta(x) - 1 \tag{A14}$$

$$\frac{\mathrm{d}}{\mathrm{d}x} \operatorname{sgn}(x) = 2\delta(x). \tag{A15}$$

B. FOURIER TRANSFORMS

In this part of the appendix we summarize some special relations of Fourier transforms which are utilized in the book. For a thorough discussion we refer to standard texts, e.g. SNEDDON (1955).

The Fourier transforms are defined by

$$\int_{-\infty}^{+\infty} \mathrm{d}t\, \mathrm{d}x\, \mathrm{d}y\, \mathrm{e}^{i\omega t - ikx - i\varkappa y} F(x, y, t) = F(\omega, k, \varkappa) \tag{B 1}$$

and the inverse relation

$$\int_{-\infty}^{+\infty} \frac{d\omega}{2\pi} \frac{dk}{2\pi} \frac{d\varkappa}{2\pi} e^{-i\omega t + ikx + i\varkappa y} F(\omega, k, \varkappa) = F(x, y, t). \tag{B 2}$$

Let $F(\omega)$ be the Fourier transforms of $F(t)$ and $G(\omega)$ that of $G(t)$. Then the convolution theorem states

$$\int_{-\infty}^{+\infty} \frac{d\omega}{2\pi} e^{-i\omega t} F(\omega)\, G(\omega) = \int_{-\infty}^{+\infty} dt'\; F(t - t')\, G(t') \tag{B 3}$$

$$\int_{-\infty}^{+\infty} dt\, e^{i\omega t} F(t)\, G(t) = \int_{-\infty}^{+\infty} \frac{d\omega'}{2\pi} F(\omega - \omega')\, G(\omega'). \tag{B 4}$$

From (A4) it follows

$$\int dt\, e^{i\omega t}\, \delta(t) = 1, \text{ implying} \int \frac{d\omega}{2\pi} e^{-i\omega t} = \delta(t)$$

and, hence,

$$\int_{-\infty}^{+\infty} dt\, e^{i\omega t} = 2\pi\, \delta(\omega). \tag{B 5}$$

Analogously we find

$$\int_{-\infty}^{+\infty} dt\, e^{ift}\, e^{i\omega t} = 2\pi\, \delta(\omega + f). \tag{B6}$$

Taking the real and the imaginary part of e^{ift} in (B6) we get

$$\int dt \cos ft\, e^{i\omega t} = \pi[\delta(\omega + f) + \delta(\omega - f)] \tag{B7}$$

$$\int dt \sin ft\, e^{i\omega t} = \frac{\pi}{i} [\delta(\omega + f) - \delta(\omega - f)]. \tag{B 8}$$

With (A9) we may write

$$\begin{aligned} &\delta(\omega + f) + \delta(\omega - f) = 2f\, \delta(\omega^2 - f^2), \\ &\delta(\omega - f) - \delta(\omega + f) = 2f \operatorname{sgn}(\omega)\, \delta(\omega^2 - f^2). \end{aligned} \tag{B9}$$

Next we consider

$$\int dt\, t\, e^{i\omega t} = \frac{1}{i} \frac{d}{d\omega} \int dt\, e^{i\omega t} = \frac{2\pi}{i} \frac{d}{d\omega} \delta(\omega) \tag{B 10}$$

and repeating this process n-times

$$\int dt\, t^n\, e^{i\omega t} = \frac{2\pi}{i^n} \frac{d^n}{d\omega^n} \delta(\omega). \tag{B 11}$$

As already mentioned in Section 2.3., compare (2.3.6), the Fourier transform of the step function is given as

$$\int_{-\infty}^{+\infty} dt\, \theta(t)\, e^{i\omega t} = \frac{i}{\omega + i\varepsilon}, \qquad \varepsilon \to +0 \tag{B 12}$$

where ε is a positive infinitesimal which gives the prescription how to go around the pole at $\omega = 0$. Note that (B12) is the one-sided Fourier transform of 1, because

$\int_{-\infty}^{+\infty} dt\, \theta(t)\, e^{i\omega t} = \int_{0}^{\infty} dt\, e^{i\omega t}$. By means of (B 12)

and the convolution theorem we may readily calculate further one-sided Fourier transforms, e.g.,

$$\int_{-\infty}^{+\infty} dt\, \theta(t) \sin ft\, e^{i\omega t} = \int \frac{d\omega'}{2\pi} \frac{\pi}{i} [\delta(\omega' + f) - \delta(\omega' - f)] \frac{i}{(\omega - \omega') + i\varepsilon}$$

$$= \frac{f}{f^2 - (\omega + i\varepsilon)^2} \tag{B 13}$$

where (B8) has been used. Analogously we find

$$\int dt\, \theta(t) \cos ft\, e^{i\omega t} = \int \frac{d\omega'}{2\pi} [\delta(\omega' + f) + \delta(\omega' - f)] \frac{i\pi}{(\omega - \omega') + i\varepsilon}$$

$$= \frac{i\omega}{(\omega + i\varepsilon)^2 - f^2}. \tag{B 14}$$

With (B10) and (A11) we get

$$\int dt\, \theta(t)\, t\, e^{i\omega t}\, dt = \int \frac{d\omega'}{2\pi} \frac{i}{\omega - \omega' + i\varepsilon} \frac{2\pi}{i} \frac{d}{d\omega'} \delta(\omega')$$

$$= -\int d\omega'\, \delta(\omega') \frac{d}{d\omega'} \frac{1}{(\omega + i\varepsilon - \omega')} = \frac{-1}{(\omega + i\varepsilon)^2}. \tag{B 15}$$

In the same manner the relations

$$\int dt\, \theta(t)\, (1 - \cos ft)\, e^{i\omega t} = \frac{if^2}{(\omega + i\varepsilon)[f^2 - (\omega + i\varepsilon)^2]} \tag{B 16}$$

and

$$\int dt\, \theta(t) \left(t - \frac{\sin ft}{f}\right) e^{i\omega t} = \frac{f^2}{(\omega + i\varepsilon)^2 [(\omega + i\varepsilon)^2 - f^2]} \tag{B 17}$$

can be obtained.

Analogously to (B12) we have for a spatial step function

$$\int dt\, \theta(\pm x)\, e^{-ikx} = \frac{(\mp i)}{k \mp i\varepsilon}. \tag{B 18}$$

We may readily show that

$$\int \frac{dk}{2\pi} \frac{-e^{ikx}}{(k \mp i\varepsilon)^2} = \pm x\theta(\pm x). \tag{B 19}$$

We list some further important Fourier integrals

$$\int \frac{dk}{2\pi} \frac{e^{ikx}}{1 + k^2R^2} = \frac{1}{2R} e^{-\frac{|x|}{R}}. \tag{B 20}$$

Differentiating this equation yields

$$\int \frac{dk}{2\pi} \frac{ik\, e^{ikx}}{1 + k^2R^2} = -\operatorname{sgn}(x) \frac{1}{2R} e^{-\frac{|x|}{R}} \tag{B 21}$$

From (B18) and (B20) we get

$$\int \frac{dk}{2\pi} \frac{(\mp i)\, e^{ikx}}{(k \mp i\varepsilon)(1 + k^2R^2)} = \theta(\pm x) \mp \operatorname{sgn}(x) \frac{1}{2} e^{-\frac{|x|}{R}} \tag{B 22}$$

Furthermore we consider

$$\int \frac{dk}{2\pi} \frac{2 \sin ka}{k} e^{ikx} = \theta(x + a) - \theta(x - a) = \theta(a - |x|) \tag{B 23}$$

and

$$\begin{aligned} \int \frac{dk}{2\pi} \frac{2 \sin ka}{k} \frac{e^{ikx}}{1 + k^2R^2} &= \int_{-\infty}^{+\infty} dx' \frac{1}{2R} e^{-\frac{|x-x'|}{R}} \theta(a - |x'|) \\ &= \theta(a - |x|) + \operatorname{sgn}(x - a) \frac{1}{2} e^{-\frac{|x-a|}{R}} - \operatorname{sgn}(x + a) \frac{1}{2} e^{-\frac{|x+a|}{R}} \end{aligned} \tag{B24}$$

where the convolution theorem was used.

Next we estimate

$$\int_{-\infty} \frac{dk}{2\pi} \frac{2 i \sin ka}{k^2} e^{ikx}.$$

It can readily be seen that the integrand is not singular. This integral can be reduced to a form given in standard tables. However, we may put the integral in the form

$$\begin{aligned} \int \frac{dk}{2\pi} \frac{2 i \sin ka}{k^2} e^{ikx} &= \lim_{\varepsilon \to 0} \int \frac{dk}{2\pi} \frac{2 i \sin ka\, e^{ikx}}{\varepsilon^2 + k^2} \\ &= \lim_{\varepsilon \to 0} \frac{1}{2\varepsilon} (e^{-\varepsilon|x+a|} - e^{-\varepsilon|x+a|}) = \frac{1}{2} (|x - a| - |x + a|). \end{aligned} \tag{B 25}$$

Moreover, we add the integrals

$$\int \frac{\mathrm{d}k}{2\pi}\,\frac{2\,\mathrm{i}\sin ka\,\mathrm{e}^{\mathrm{i}kx}}{1+k^2R^2}=\frac{1}{2R}\left(\mathrm{e}^{-\frac{|x+a|}{R}}-\mathrm{e}^{-\frac{|x-a|}{R}}\right) \tag{B 26}$$

$$\int \frac{\mathrm{d}k}{2\pi}\,\frac{-2k\sin ka\,\mathrm{e}^{\mathrm{i}kx}}{1+k^2R^2}=\frac{1}{2R^2}\left(\operatorname{sgn}(x-a)\,\mathrm{e}^{-\frac{|x-a|}{R}}-\operatorname{sgn}(x+a)\,\mathrm{e}^{-\frac{|x+a|}{R}}\right) \tag{B 27}$$

which follow by differentiating (B24).

Finally we discuss some Fourier integrals containing $\sqrt{1+k^2R^2}$ which have been used, particularly, in Chapter 5 and which in several cases can only be expressed in terms of convolutions.

First we note the integral

$$\int_{-\infty}^{+\infty} \frac{\mathrm{d}k}{2\pi}\,\frac{\mathrm{e}^{-\sqrt{1+k^2R^2}\,\frac{y}{R}}}{\sqrt{1+k^2R^2}}\,\mathrm{e}^{\mathrm{i}kx}=\frac{1}{\pi R}\,K_0\left(\frac{1}{R}\sqrt{x^2+y^2}\right) \tag{B 28}$$

which can be found in standard tables, e.g. OBERHETTINGER (1957), p. 13. With (B28) and with the convolution theorem we obtain the following relationships

$$\begin{aligned}\int_{-\infty}^{+\infty} \frac{\mathrm{d}k}{2\pi}\,\frac{(\mp\mathrm{i})\,\mathrm{e}^{\mathrm{i}kx}}{(k\mp\mathrm{i}\varepsilon)}\,\frac{\mathrm{e}^{-\sqrt{1+k^2R^2}\,\frac{y}{R}}}{\sqrt{1+k^2R^2}} &= \int_{-\infty}^{+\infty}\mathrm{d}x'\,\theta(\pm x')\,K_0\left(\frac{\sqrt{(x-x')^2+y^2}}{R}\right)\\ &= \int_0^{\infty}\frac{\mathrm{d}x'}{\pi R}\,K_0\left(\frac{\sqrt{(x\mp x')^2+y^2}}{R}\right)\end{aligned} \tag{B29}$$

$$\int_{-\infty}^{+\infty} \frac{\mathrm{d}k}{2\pi}\,\frac{-\mathrm{e}^{\mathrm{i}kx}}{(k\mp\mathrm{i}\varepsilon)^2}\,\frac{\mathrm{e}^{-\sqrt{1+k^2R^2}\,\frac{y}{R}}}{\sqrt{1+k^2R^2}}=\int_0^{\infty}\frac{\mathrm{d}x'}{\pi R}\,x'K_0\left(\frac{1}{R}\sqrt{(x\mp x')^2+y^2}\right) \tag{B 30}$$

$$\int_{-\infty}^{+\infty} \frac{\mathrm{d}k}{2\pi}\,\frac{(\mp\mathrm{i})\,\mathrm{e}^{\mathrm{i}kx}}{(k\mp\mathrm{i}\varepsilon)}\,\frac{\mathrm{e}^{-\sqrt{1+k^2R^2}\,\frac{y}{R}}}{1+k^2R^2}=\int_0^{\infty}\mathrm{d}x'\int_y^{\infty}\frac{\mathrm{d}y'}{\pi R^2}\,K_0\left(\frac{\sqrt{(x\mp x')^2+y'^2}}{R}\right) \tag{B31}$$

$$\int_{-\infty}^{+\infty} \frac{\mathrm{d}k}{2\pi}\,\frac{-\mathrm{e}^{\mathrm{i}kx}}{(k\mp\mathrm{i}\varepsilon)^2}\,\frac{\mathrm{e}^{-\sqrt{1+k^2R^2}\,\frac{y}{R}}}{1+k^2R^2}=\int_y^{\infty}\mathrm{d}y'\int_0^{\infty}\frac{\mathrm{d}x'}{\pi}\,\frac{x'}{R^2}\,K_0\left(\frac{1}{R}\sqrt{(x\mp x')^2+y'^2}\right) \tag{B 32}$$

$$\begin{aligned}&\int_{-\infty}^{+\infty} \frac{\mathrm{d}k}{2\pi}\,2\sin ka\,\mathrm{e}^{\mathrm{i}kx}\,\frac{\mathrm{e}^{-\sqrt{1+k^2R^2}\,\frac{y}{R}}}{\sqrt{1+k^2R^2}}\\ &=\frac{\mathrm{i}}{\pi R}\left[K_0\left(\frac{1}{R}\sqrt{(x-a)^2+y^2}\right)-K_0\left(\frac{1}{R}\sqrt{(x+a)^2+y^2}\right)\right]\end{aligned} \tag{B33}$$

$$\int_{-\infty}^{+\infty} \frac{dk}{2\pi} \frac{2\sin ka}{k^2} e^{ikx} \frac{e^{-\sqrt{1+k^2R^2}\frac{y}{R}}}{\sqrt{1+k^2R^2}}$$

$$= -\frac{i}{2\pi R} \int_{-\infty}^{+\infty} dx'(|x'-a| - |x'+a|)\, K_0\left(\frac{1}{R}\sqrt{(x-x')^2+y^2}\right) \tag{B 34}$$

$$\int_{-\infty}^{+\infty} \frac{dk}{2\pi}\, 2\sin ka\, e^{ikx} \frac{e^{-\sqrt{1+k^2R^2}\frac{y}{R}}}{1+k^2R^2}$$

$$= \int_{y}^{\infty} \frac{dy'}{\pi R^2}\left[K_0\left(\frac{1}{R}\sqrt{(a-x)^2+y'^2}\right) - K_0\left(\frac{1}{R}\sqrt{(a+x)^2+y'^2}\right)\right] \tag{B 35}$$

$$\int_{-\infty}^{\infty} \frac{dk}{2\pi} \frac{2\sin ka}{k} e^{-\sqrt{1+k^2R^2}\frac{y}{R}} = \frac{y}{R}\int_{-a}^{a} \frac{dx'}{\pi} \frac{K_1\left(\frac{1}{R}\sqrt{(x-x')^2+y^2}\right)}{\sqrt{(x-x')^2+y^2}} \tag{B 36}$$

References

ABRAMOWITZ, M., and I. A. STEGUN (1984): Pocketbook of mathematical functions. Verlag Harri Deutsch, Thun, Frankfurt/Main, 468 pp.

ALENIUS, P. (1982): Inertial wave parameters in the Bothnian sea. Proceedings of the XIII. Conference of Baltic Oceanographers, Helsinki pp. 87—94

ANDERSON, D. L. T., and A. F. GILL (1975): Spin-up of a stratified ocean with applications to upwelling. Deep Sea Research **22**, pp. 583—596

ANDERSON, D. L. T., and P. B. ROWLANDS (1976): The Somali current response to southwest Monsoon: the relative importance of local and remote forcing. Journal of Marine Research **34**, pp. 395—417

BENDER, C. M., and S. A. ORSZAG (1978): Advanced mathematical methods for scientists and engineers. McGraw-Hill, New York, 593 pp.

CANE, M. A., and F. S. SARACHIK (1976): Forced baroclinic motions I, the linear equatorial unbounded case. Journal of Marine Research **34**, pp. 629—665

CANE, M. A., and E. S. SARACHIK (1977): Forced baroclinic motions II — the linear bounded case. Journal of Marine Research **35**, pp. 395—432

CHARNEY, J. G. (1955): The generation of ocean currents by winds Journal of Marine Research **14**, pp. 477—498

COURANT, R., and D. HILBERT (1968): Methoden der mathematischen Physik. Vol. I and II, Springer, Berlin (West), 1018 pp.

CREPON, M. (1967): Hydrodynamique marine en regime impulsionel. Cahiers Oceanographique **19**, pp. 627—655, 847—880

CREPON, M. (1969): Hydrodynamique marine en regime impulsionel. Cahiers Oceanographique **21**, pp. 333—353, 863—877

CREPON, M., and C. R. RICHEZ (1982): Transient upwelling generated by the two-dimensional atmospheric forcing and variability in the coastline. Journal of Physical Oceanography **12**, pp. 1437—1457

CSANADY, G. T. (1977): The coastal jet conceptual model in the dynamics of shallow seas. In: the Sea, Vol. 6, Wiley, New York, pp. 117—144

CSANADY, G. T. (1982): Circulation in the coastal ocean. Reidel Publishing Company, Dordrecht, 279 pp.

ECKART, C. (1960): Hydrodynamics of oceans and atmospheres. Pergamon Press, Oxford, 290 pp.

EKMAN, V. W. (1904): On dead water. In: Scientific results of the Norwegian North Polar expedition 1893—1896. **5**, no. 15

EKMAN, V. W. (1905): On the influence of the earth's rotation on ocean currents. Arkiv för Matematik, Astronomi och Fysik **2**, 52 pp.

ERIKSEN, C. C. (1980): Evidence for a continuous spectrum of equatorial waves in the Indian Ocean. Journal of Geophysical Research **85**, pp. 3285—3303

ERIKSEN, C. C. (1982): Equatorial wave vertical modes observed in a western Pacific island array. Journal of Physical Oceanography **12**, pp. 1206—1227

ERIKSEN, C. C. (1985): Moored observations of deep low frequency motions in the central Pacific ocean. Journal of Physical Oceanography **15**, pp. 1085—1113

FENNEL, W. (1986): Theory of inertial response of channel models. Beiträge zur Meereskunde, No. 54, pp. 3—18

FENNEL, W., and H. U. LASS (1979): On the vertical eigenvalue problem of equatorial trapped waves. Gerlands Beiträge zur Geophysik **88**, pp. 279—293

FENNEL, W., and H. U. LASS (1983): Theory of frequency spectra of equatorial trapped waves. In: Hydrodynamics of the Equatorial Ocean. Elsevier Oceanography Series **36**, Amsterdam pp. 319—338

FENNEL, W., D. HALPERN, and H. U. LASS (1987): Current spectra at the equator. Beiträge zur Meereskunde, No. 56, pp.

FJELDSTAD, J. E. (1933): Interne Wellen. Geophysike Publikasjoner **X**, pp. 313—356

FOFONOFF, N. P. (1969): Spectral characteristics of internal waves in the ocean. Deep Sea Research **16**, Supplement, pp. 58—71

FRANKIGNOUL, C., and P. MÜLLER (1979): Quasi-geostrophic response of an infinite beta-plane ocean to stochastic forcing by the atmosphere. Journal of Physical Oceanography **9**, pp. 104—127

FU, L. L. (1981): Observations and models of inertial waves in the deep ocean. Review of Geophysics and Space Physics **19**, pp. 141—170

GARRETT, C., and W. MUNK (1972): Space-time scales of internal waves. Geophysical Fluid Dynamics **2**, pp. 225—264

GARRETT, C., and W. MUNK (1975): Space-time scales of internal waves — a progress report. Journal of Geophysical Research **80**, pp. 291—297

GARRETT, C., and W. MUNK (1979): Internal waves in the ocean. Annual Review of Fluid Dynamics 11, pp. 339—369

GARZOLI, S. L., and E. J. KATZ (1984): Winds at St. Peter and St. Paul rocks during the first sequal year. Geophysical Research Letters **11**, pp. 715—718

GASCARD, J. C. (1973): Vertical motions in a region of deep water formation. Deep Sea Research **20**, pp. 1011—1027

GEISLER, J. E., and R. E. DICKINSON (1972): The role of variable Coriolis parameter in the propagation of inertiagravity waves during the process of geostrophic adjustment. Journal of Physical Oceanography **2**, pp. 263—272

GILL, A. E. (1982): Atmosphere-ocean dynamics. Academic Press, New York, 662 pp.

GOULD, W. J. (1971): Spectral characteristics of some deep current records fron the eastern north Atlantic. Philosophical Transactions of the Royal Society of London A **270**, pp. 437—450

GRADSTEIN, I. S., and I. M. RYSHIK (1971): Tables of series, products, and integrals. Moscow (in Russian)

GUSTAFSON, T., and B. OTTERSTEDT (1932): Observations de courants dans la baltique 1931. Svenska Hydrografisk-Biologiska Kommissionens Skrieter, by Serie Hydrografi XIII, pp. 3—14

HALPERN, D. (1983): Variability of the Cromwell current at 110 W before and during the 1983—84 warm event. Tropical Ocean-Atmosphere Newsletter, No. 21, pp. 31—34

HOLLAND, W. R. (1983): Simulations of midlatitude variability. In: Proceedings Hawaiian Winter Workshop, University of Hawaii at Manoa, pp. 97—112

JENKINS, G. M., and D. G. WATTS (1968): Spectral analysis and its applications. Holden-Day, San Francisco

KANG, Y. Q., and L. MAGAARD (1980): Annual Rossby waves in the central north Pacific. Journal of Physical Oceanography **10**, pp. 1159—1167

KÄSE, R. H., W. ZENK, T. B. SANDFORD, and W. WILLER (1985): Current, fronts and eddy fluxes in the Canary basin. Progress in Oceanography **14**, pp. 231—257

KIELMANN, J., W. KRAUSS, and K. H. KEUNECKE (1973): Currents and stratification in the Belt sea and in the Arkona basin dusing 1962—1968. Kieler Meeresforschungen **29**, pp. 90—111

KNOX, R. A., and D. HALPERN (1982): Long range Kelvin wave propagation in the Pacific ocean equatorial currents. Journal of Marine Research **40**, pp. 329—339

KRAUSS, W. (1966): Methoden und Ergebnisse der theoretischen Ozeanographie. Vol. 2, Gebrüder Bornträger, Berlin (West)

KRAUSS, W. (1979): Inertial waves in an infinite channel of rectangular cross section. Deutsche Hydrographische Zeitschrift **32**, pp. 248—266

KRAUSS, (1981): The erosion of a thermocline. Journal of Physical Oceanography **11**, pp. 415—433

KRAUSS, W., and C. WÜBBER (1982): Response of the north Atlantic to annual wind variations along the eastern coast. Deep Sea Research **29**, pp. 851—868

KUNDU, P. K., S. Y. CHAO, and J. P. MCCREARY (1983): Transient coastal currents and inertia-gravity waves. Deep Sea Research **30**, pp. 1054—1082

KUNDU, P. K. (1984): Generation of coastal inertial oscillations by time-varying wind. Journal of Physical Oceanography **14**, pp. 1901—1913

LANDAU, L. D., and E. M. LIFSCHITZ (1967): Lehrbuch der theoretischen Physik, Band III, Quantenmechanik, Akademie-Verlag, Berlin

LASS, H. U., and W. FENNEL (1980): On the filtering of stochastic processes and the applicability of the Reynolds rules. Beiträge zur Meereskunde, No. 44/45, pp. 145—153

LEBLOND, P. H., and L. A. MYSAK (1978): Waves in the ocean. Elsevier Oceanography Series **20**, Amsterdam, 602 pp.

LIGHTHILL, M. J. (1967): On waves generated in dispersive systems by travelling forcing effect, with application to the dynamics of rotating fluids. Journal of Fluid Mechanics **27**, pp. 725—752

LIGHTHILL, M. J. (1969): Dynamic response of the Indian ocean to onset of the south west monsoon. Phisolophical Transactions of the Royal Society of London A **265**, pp. 45—92

LONGUET-HIGGINS, M. S. (1965): The response of a stratified ocean to stationary or moving wind-systems. Deep Sea Research **12**, pp. 923—973

LUKAS, R., S. P. HAYES, and K. WYRTKI (1984): Equatorial sea level response during the 1982—1983 El Nino. Journal of Geophysical Research **89**, pp. 10425—10430

LUTHER, O. S. (1980): Observations of long period waves in the tropical oceans and atmospheres. Technical Report W.H.O.I.-80-17, Woods Hole, 210 pp.

MÄLKKI, P. (1975): On the variability of currents in a coastal region of the Baltic sea. Merentutk. Julk. Havsforskninginst. Skr. no. 240, pp. 3—56

MCCOMAS, C. H., and P. MÜLLER (1981): The dynamical balance of internal waves. Journal of Physical Oceanography **11**, pp. 970—986

MCCREARY, J. P. (1980): Modelling wind-driven ocean circulation. Hawaii Institute of Geophysics, Technical Report HIG-80-3

MCCREARY, J. P. (1981): A linear stratified ocean model of the equatorial undercurrent. Philosophical Transactions of the Royal Society of London **298**, pp. 603—635

MCCREARY, J. P. (1985): Modeling equatorial ocean circulation. Annual Review of Fluid Mechanics **17**, pp. 559—409

MEINCKE, J. (1975): Evidence for atmospheric forcing of arctic water overflow events. ICES C.M. 1975/ C:22, 11 pp. (Unpublished Document)

MILLOT, C., and M. CREPON (1981): Inertial oscillations on the continental shelf of the gulf of Lion-observations and theory. Journal of Physical Oceanography **11**, pp. 639—657

MONIN, A. S., and A. M. YAGLOM (1965): Statistical hydrodynamics. Nauka, Moscow (in Russian)

MOORE, D. W., and S. G. H. PHILANDER (1977): Modelling the tropical ocean circulation. In: The Sea, Vol. VI, Wiley, New York, pp. 319—361

MORSE, P. M., and H. F. FESHBACH (1953): Methods of theoretical physics, Vol. I and II. McGraw-Hill, New York, 1978 pp.

MORTIMER, C. H. (1968): Internal waves and associated currents observed in Lake Michigan during the summer of 1963. University of Wisconsin-Milwaukee, Center for Great Lakes Studies, special report No. 1

MUNK, W. (1981): Internal waves and small scale processes. In: Evolution of physical oceanography, MIT-Press, Cambridge, pp. 264—290

MUNK, W., and N. PHILLIPS (1968): Coherence and band structure of inertial motion in the sea. Review of Geophysics **6**, pp. 447—472

NANSEN, F. (1902): The oceanography of the north pole basin. In: Scientific Results of the Norwegian North Pole Expedition 1893—1896, 3—9, 427 pp.

OBERHETTINGER, F. (1957): Tabellen zur Fouriertransformation. Springer-Verlag, Berlin (West), 213 pp.

OLBERS, D. J. (1976): Nonlinear energy transfer and the energy balance of the internal wave field in the deep ocean. Journal of Fluid Mechanics **74**, pp. 375—399

OLBERS, D. J. (1983): Models of the oceanic internal wave field. Review of Geophysics and Space Physics **21**, pp. 1567—1606

OLBERS, D. J., M. MENZEL, and J. WILLEBRAND (1985): The inference of north atlantic circulation patterns from climatological hydrographic data. Review of Geophysics **23**, pp. 313—356

PEDLOSKY, J. (1979): Geophysical fluid dynamics. Springer-Verlag, Berlin (West), 624 pp.

PHILANDER, S. G. H. (1977): The effects of coastal geometry on equatorial waves. Journal of Marine Research **35**, pp. 509—523

PHILANDER, S. G. H. (1978): Instabilities of equatorial currents: part II. Journal of Geophysical Research **83**, pp. 3679—3682

PHILANDER, S. G. H. (1983): El Nino Southern Oscillation phenomena. Nature **302**, pp. 295—301

PHILANDER, S. G. H., and J. H. YOON (1982): Eastern boundary currents and coastal upwelling. Journal of Physical Oceanography **12**, pp. 862—879

PHILLIPS, O. M. (1977): The dynamics of the upper ocean (2-nd Edition). Cambridge University Press, London, 336 pp.

RAYLEIGH, Lord (1883): Investigation of the character of the equilibrium of an incompressible heavy fluid of variable density. Proceedings of the London Mathematical Society **14**, pp. 170—178

RICHEZ, C., S. G. H. PHILANDER, and M. CREPON (1984): Oceanic response to coastal winds with shear. Oceanologica Acta **7**, pp. 409—416

ROBERTS, J. (1975): Internal gravity waves in the ocean. Dekker Inc., New York, 274 pp.

RODEN, G. I. (1975): On the north Pacific temperature, salinity, soundvelocity, and density fronts and their relation to the wind and energy flux fields. Journal of Physical Oceanography **5**, pp. 557—571

ROSSBY, C. G. (1939): Relation between variations in the intensity of the zonal circulation of the atmosphere and the displacement of the semi-permanent centers of action. Journal of Marine Research **2**, pp. 38—55

SIEDLER, G., W. ZENK, and W. J. EMERGY (1985): Current events related to a subtropical front in the north-east Atlantic. Journal of Physical Oceanography **15**, pp. 886—897

SJØBERG, B., and M. MORK (1985): Wind-induced stratified ocean response in the ice edge region: an analytical approach. Journal of Geophysical Research **90**, pp. 7273—7285

SNEDDON, I. N. (1955): Functional analysis. In: Handbuch der Physik, Vol. II, Flügge (ed.), Springer Verlag, Berlin (West), pp. 198—348

STOKES, G. G. (1847): On the theory of oscillatory waves. Transactions of the Cambridge Philosophical Society **8**, pp. 441—455

SUTTON, W. G. (1934): The asymptotic expansion of a function whose operational equivalent is known. Journal of the London Mathematical Society **9**, pp. 23—33

THORPE, S. A. (1975): The excitation, dissipation, and interaction of internal waves in the deep ocean. Journal of Geophysical Research **80**, pp. 328—338

TREWARTHA, G. T., and L. H. HORN (1980): An introduction to climate. McGraw-Hill, New York, 416 pp.

TURNER, J. S. (1981): Small scale mixing processes. In: Evolution of physical oceanography, MIT-Press, Cambridge, pp. 236—262

UNESCO (1981): Tenth report of the joint panel on oceanographic standards. UNESCO Technical Papers in Marine Science, No. 36, UNESCO, Paris

VERONIS, G. (1956): Partition of energy between geostrophic and non-geostrophic oceanic motion. Deep Sea Research **3**, pp. 157—177

VERONIS, G. (1958): On the transient response of a β-plane ocean. Journal of the Oceanographic Society of Japan **14**, pp. 1—5

VERONIS, G., and H. STOMMEL (1956): The action of variable wind stresses on a stratified ocean. Journal of Marine Research **15**, pp. 43—75

VOORHIS, A. D. (1968): Measurement of vertical motions and the partition of energy in the New England slope water. Deep Sea Research **15**, pp. 599—608

WEARN, R. B., and D. J. BAKER (1980): Bottom pressure measurements across the Antarctic circumpolar current and their relation to the wind. Deep Sea Research **27**, pp. 875—878

WEISBERG, R. H. (1979): Equatorial waves during GATE and their relation to the mean zonal circulation. Deep Sea Research **26**, Supplement II, pp. 179—198

WEISBERG, R. H. (1985): Equatorial Atlantic velocity and temperature observations: February—November 1981. Journal of Physical Oceanography **15**, pp.533—543

WEISBERG, R. H., L. MILLER, A. HORIGAN, and J. A. KNAUSS (1979): Velocity observations in the equatorial thermocline during GATE. Deep Sea Research **26**, Supplement II, pp. 217—248

WHITE, W. B. (1977): Annual forcing of baroclinic long waves in the tropical north Pacific ocean. Journal of Physical Oceanography **7**, pp. 50—61

WILLEBRAND, J., S. G. H. PHILANDER, and R. C. PACANOWSKI (1980): The oceanic response to large scale atmospheric disturbances. Journal of Physical Oceanography **10**, pp. 411—429

WLADIMIROW, W. S. (1972): Gleichungen der mathematischen Physik. Verlag der Wissenschaften, Berlin, 379 pp.

WOODS, J. D. (1968): Wave induced shear instability in the summer thermocline. Journal of Fluid Mechanics **32**, pp. 791—800

WOODS, J. D., and R. L. WILEY (1972): Billow turbulence and ocean microstructure. Deep Sea Research **19**, pp. 87—121

WUNSCH, C., and A. E. GILL (1976): Observations of equatorial trapped waves in the Pacific sea level variations. Deep Sea Research **23**, pp. 371—390

WYLD, H. W. (1976): Mathematical methods for physics. Benjamin Inc., London 628 pp.

YAMAGATA, T., and S. G. H. PHILANDER (1985): The role of damped equatorial waves in the oceanic response to winds. Journal of the Oceanographic Society of Japan **41**, No. 5, pp. 345—357

YOON, J. H., and S. G. H. PHILANDER (1982): The generation of coastal undercurrents. Journal of the Oceanographic Society of Japan **38**, pp. 215—224

YOSHIDA, K. (1959): A Theory of the Cromwell current and of the equatorial upwelling. Journal of the Oceanographic Society of Japan **15**, pp. 154—170

Index

Bei Fragen zur Produktsicherheit wenden Sie sich bitte an:
If you have any questions regarding product safety,
please contact:

Walter de Gruyter GmbH
Genthiner Straße 13
10785 Berlin
productsafety@degruyterbrill.com